HURRICANE KATRINA'S EFFECT ON GASOLINE SUPPLY AND PRICES

WEDNESDAY, SEPTEMBER 7, 2005

HOUSE OF REPRESENTATIVES,
COMMITTEE ON ENERGY AND COMMERCE,
Washington, DC.

The committee met, pursuant to notice, at 11:05 a.m., in room 2123, Rayburn House Office Building, Hon. Joe Barton (chairman) presiding.

Members present: Representatives Barton, Hall, Upton, Stearns, Gillmor, Deal, Whitfield, Norwood, Shimkus, Wilson, Shadegg, Fossella, Radanovich, Bass, Pitts, Bono, Walden, Terry, Ferguson, Rogers, Otter, Myrick, Sullivan, Murphy, Burgess, Blackburn, Dingell, Waxman, Markey, Pallone, Brown, Rush, Eshoo, Stupak, Engel, Wynn, Green, Strickland, DeGette, Capps, Doyle, Allen, Davis, Schakowsky, Solis, Gonzalez, Inslee, Baldwin, and Ross.

Staff present: Bud Albright, staff director; Andy Black, deputy staff director/policy coordinator; Mark Menezes, chief counsel for energy and the environment; Margaret Caravelli, majority counsel; Maryam Sabbaghian, majority counsel; Tom Hassenboehler, majority counsel; Kelly Cole, majority counsel; Peter Kielty, legislative clerk; David Schooler, minority general counsel; Sue Sheridan, minority senior counsel; Michael Goo, minority counsel; Bruce Harris, minority professional staff; and Reed Stuntz, minority staff director.

Chairman BARTON. The committee will come to order. We are going to have a very important hearing this morning and this afternoon but also a very long hearing.

The Chair should not have to announce this, but the Chair is going to announce it: There will be regular order.

All members that wish to will be allowed to give their opening statements. Those members that wish to defer the opening statements will be given extra time in the Q and A period.

At noon today, approximately, we are going to have a video presentation by the Governor of Mississippi, Mr. Barbour.

We asked the Governor of Louisiana if she would also like to participate by video conference. She is not able to do so, but she is going to ask one of her assistants, who I believe is in the room, to read her statement into the record.

So whenever Governor Barbour is able to teleconference with us, we will suspend what we are doing at that moment for that to happen. Then we will give the Governor of Louisiana's representative an opportunity to read a statement into the record, and then we will resume.

I also want to make a point of personal privilege before beginning our opening statements to announce that Baby Barton has not yet joined us in the world. He is due any day now. And I know the airline schedule back to Texas. I have it memorized. So if you see me hopping out and running out of here, it means that I have received a phone call that I need to get home. But we expect Baby Barton to be here any time between today and next Friday.

Mr. HALL. Will the chairman yield?

Chairman BARTON. Briefly.

Mr. HALL. Did you ever know where the term "son of a gun" came from?

Chairman BARTON. I hesitate to ask.

Mr. HALL. Sailors used to take their wives to sea with them. And when they were enceintes and they could not deliver, they would walk them past the big guns, shoot the big guns off. That is a son of a gun.

Chairman BARTON. Okay. That is one theory.

Now we are going to resume regular order, and the Chair recognizes himself for an opening statement.

I want to begin by expressing the deep sorrow that everybody on this committee, on both sides of the aisle, has for the families and friends who have lost loved ones and who are experiencing, as we speak, the tragedy and loss as a result of Hurricane Katrina.

This is one of the worst natural catastrophes to ever hit our country, and I would remind us that we are the United States of America, so our hearts reach out to those citizens in Louisiana, Alabama, and Mississippi. Our thoughts and prayers go out to them.

Many of the constituents hurt by Katrina are represented directly on this committee. Vice-Chairman Chip Pickering of Mississippi has had the benefits of representing his constituents in Mississippi for a number of years.

Our former chairman, Billy Tauzin of Louisiana, represented his constituents on this committee for years and years.

This storm is not a burden on any one State, it is a burden for the entire Nation, and we will deal with it as a united Nation.

Some States have come forward already to give aid and comfort. To name a few: Texas, Arizona, Alabama, Tennessee, Arkansas, Georgia, Florida, Kansas, Utah, and Ohio have all opened their doors to Katrina refugees; we thank them for that.

In my congressional district in Texas, I know of at least 2,000 refugees in shelters as of the day before yesterday. The Energy and Commerce Committee is going to do the very best it can to help and alleviate pain and suffering and hopefully prevent future events of this type from having the kind of impact that it has had.

This hearing is the first of several hearings that we hope to hold on the impact that Katrina has had on our energy policy, our health care policy, and our telecommunications policy.

Unlike hurricanes of the recent past, Katrina has been destructive and disruptive. The disruptions have had an impact on energy, telecommunications, health, interstate commerce, and all sectors of our Nation's economy. These are all areas that are within the purview of the Energy and Commerce Committee.

HURRICANE KATRINA'S EFFECT ON GASOLINE SUPPLY AND PRICES

HEARING

BEFORE THE

COMMITTEE ON ENERGY AND COMMERCE
HOUSE OF REPRESENTATIVES

ONE HUNDRED NINTH CONGRESS

FIRST SESSION

SEPTEMBER 7, 2005

Serial No. 109–32

Printed for the use of the Committee on Energy and Commerce

Available via the World Wide Web: http://www.access.gpo.gov/congress/house

U.S. GOVERNMENT PRINTING OFFICE

24–246PDF WASHINGTON : 2005

For sale by the Superintendent of Documents, U.S. Government Printing Office
Internet: bookstore.gpo.gov Phone: toll free (866) 512–1800; DC area (202) 512–1800
Fax: (202) 512–2250 Mail: Stop SSOP, Washington, DC 20402–0001

COMMITTEE ON ENERGY AND COMMERCE

JOE BARTON, Texas, *Chairman*

RALPH M. HALL, Texas
MICHAEL BILIRAKIS, Florida
Vice Chairman
FRED UPTON, Michigan
CLIFF STEARNS, Florida
PAUL E. GILLMOR, Ohio
NATHAN DEAL, Georgia
ED WHITFIELD, Kentucky
CHARLIE NORWOOD, Georgia
BARBARA CUBIN, Wyoming
JOHN SHIMKUS, Illinois
HEATHER WILSON, New Mexico
JOHN B. SHADEGG, Arizona
CHARLES W. "CHIP" PICKERING, Mississippi, *Vice Chairman*
VITO FOSSELLA, New York
ROY BLUNT, Missouri
STEVE BUYER, Indiana
GEORGE RADANOVICH, California
CHARLES F. BASS, New Hampshire
JOSEPH R. PITTS, Pennsylvania
MARY BONO, California
GREG WALDEN, Oregon
LEE TERRY, Nebraska
MIKE FERGUSON, New Jersey
MIKE ROGERS, Michigan
C.L. "BUTCH" OTTER, Idaho
SUE MYRICK, North Carolina
JOHN SULLIVAN, Oklahoma
TIM MURPHY, Pennsylvania
MICHAEL C. BURGESS, Texas
MARSHA BLACKBURN, Tennessee

JOHN D. DINGELL, Michigan
Ranking Member
HENRY A. WAXMAN, California
EDWARD J. MARKEY, Massachusetts
RICK BOUCHER, Virginia
EDOLPHUS TOWNS, New York
FRANK PALLONE, Jr., New Jersey
SHERROD BROWN, Ohio
BART GORDON, Tennessee
BOBBY L. RUSH, Illinois
ANNA G. ESHOO, California
BART STUPAK, Michigan
ELIOT L. ENGEL, New York
ALBERT R. WYNN, Maryland
GENE GREEN, Texas
TED STRICKLAND, Ohio
DIANA DEGETTE, Colorado
LOIS CAPPS, California
MIKE DOYLE, Pennsylvania
TOM ALLEN, Maine
JIM DAVIS, Florida
JAN SCHAKOWSKY, Illinois
HILDA L. SOLIS, California
CHARLES A. GONZALEZ, Texas
JAY INSLEE, Washington
TAMMY BALDWIN, Wisconsin
MIKE ROSS, Arkansas

BUD ALBRIGHT, *Staff Director*
DAVID CAVICKE, *Deputy Staff Director and General Counsel*
REID P.F. STUNTZ, *Minority Staff Director and Chief Counsel*

CONTENTS

This is not a hearing today to engage in a blame game, or to pose recriminations against anybody at any level. This is a hearing to begin to understand the effect Katrina has had on our committee's area of responsibility. There will be numerous opportunities to determine where the blame should be placed. I hope that we can spend time learning from our mistakes and taking positive actions, if possible, to correct those mistakes.

I want to thank the witnesses that are here today for their time and their preparation to appear before the committee. Many of you are here to discuss energy security. As we confront the human tragedy from Katrina, the consequences force us to think more expansively about energy security and to focus harder on matters that the recently passed energy bill have already emphasized.

If there is a silver lining, and I am not saying there is, but if there is, it may be that our country is beginning to realize how fragile our energy sector is and how easy it is to disrupt it. It is my opinion that we have an energy infrastructure based on a 1970's population and a 1970's demand, and obviously we are in the 21st century and we have not kept pace.

The U.S. oil infrastructure is operating at maximum capacity. It has done so for the past 2 to 3 years. It was stressed before Katrina. We have just signed the most comprehensive energy bill I think that the Congress has ever passed. We did that on a bipartisan basis. A majority of the Democrats on this committee voted for that bill, and I wish to thank them. I wish to thank the ranking member, Mr. Dingell, for his support in that effort. So this committee should not have to have a wake-up call, but Hurricane Katrina is definitely a reminder that there is more to be done. It is always easier to say after the fact what we should have done before the fact.

Katrina reminds us of the need to protect and expand resources and infrastructure not just in the Gulf-producing States but in all areas of our Nation. I am pleased that our recently passed energy bill did include a $500 million provision directed at coastal restoration. So we have already made a start in helping that region.

My time has expired, so I am going to put the rest of my statement in the record.

[The prepared statement of Hon. Joe Barton follows:]

PREPARED STATEMENT OF HON. JOE BARTON, CHAIRMAN, COMMITTEE ON ENERGY AND COMMERCE

I want to begin by expressing my deep sorrow to the families and friends who are experiencing such unimaginable loss from Hurricane Katrina. I honestly can't imagine the terrible feeling of loss and displacement that so many fellow Americans are being forced to face right now. People have lost their loved ones, their friends, their homes and their livelihoods. My thoughts and prayers and the thoughts and prayers of America go out to the many.

Many of the constituents hurt by Katrina's wrath are represented directly on this Committee by Vice Chairman Pickering, and the Committee for years has had the benefit of representation from Louisiana, including its former Chairman, Chairman Tauzin. But this storm is not the burden of any one state, it has damaged a nation and if the nation is going to survive, we must turn to one another for support. To name a few, the States of Texas, Arizona, Alabama, Tennessee, Arkansas, Georgia, Florida, Kansas, Utah and Ohio have all opened their doors to Katrina refugees and we thank them for that.

The Committee too will do its part to help. This hearing will be the first of several hearings that the Committee plans to hold on the impact that Katrina had on energy, health and telecommunications. Unlike hurricanes of the recent past, Katrina has been both destructive and disruptive. The disruptions have had an impact on the energy, telecommunications, health and commerce sectors of the nation's economy—all areas within the purview of this Committee's broad jurisdiction. Let me say at the very beginning today that this is not a hearing to engage in a blame game or to pose recriminations against one another. This is a hearing to begin to understand the effect Katrina had on our committee's areas of responsibility. There will be plenty of time to determine where blame should be placed. I hope we spend more time learning from our mistakes, and taking positive action to correct our mistakes, than we do in finger pointing. The American people deserve no less.

I want to thank our witnesses for their time and preparation to appear before the Committee today. Many of you are here today to discuss energy security. As we confront the human tragedy from Katrina, the consequences force us to think more expansively about energy security, and to focus harder on matters that the recently passed Energy Bill already emphasized.

If there is a silver lining in this tragic situation, it is that it may finally bring home to the American people how fragile our energy sector is and our energy infrastructure is.

The U.S. oil infrastructure is operating at maximum capacity and has done so for the past 2-3 years. It's stressed. We just signed the most comprehensive energy bill in the last 15 years. There are lots of things in that bill to help and we're fortunate to have it. This hurricane is a wake up call that we need to do things across-the-board on infrastructure and to also expand the base.

It's always easier to say after the fact what should have been done before the fact. Katrina reminds us of the need to protect and expand resources and infrastructure in the Gulf producing states to encourage continued operations. For example, the recently passed Energy Bill included a $500 million provision directed at Coastal Restoration, but we should and will need to do more.

Katrina also reminds of how centralized our nation's energy infrastructure is and the need to encourage investment and diversification. For a sense of the numbers, 29% of our oil production and 20% of the natural gas is in the Gulf of Mexico. It doesn't have to be that way. We could be drilling in Alaska right now; we could be drilling off the coasts of several other states. It would make a difference today if we were not as restrictive as we've been the last 20 years in where we drill. We can't just get our oil and gas from Texas, Louisiana, Mississippi and the Gulf of Mexico. We need to diversify our domestic oil resources.

We have not built a new refinery in the U.S.A. in over 30 years and Katrina has shown us that our refinery capacity is inadequate. Last week Katrina forced a shutdown of approximately 25% of our refining capacity. Relief efforts have brought much of this capacity back on line, but my understanding from recent Department of Energy reports is that 10% of our gasoline refining capacity will nevertheless be out of commission for some time. To encourage new refineries the Energy Bill has a provision that allows a governor of a state to petition the Environmental Protection Agency for technical and financial assistance in the refinery permitting process.

We need to encourage states outside of the Gulf to take advantage of Energy Bill provisions like this.

Also today I expect to hear more about gasoline pricing. I think a good case can be made today that some retailers may have taken advantage of the Katrina emergency. If that's true, that is something that needs to be investigated and, in all probability, prosecuted. Among other issues, we're going to investigate the price increase at retail today. I believe in a market economy and there is no need for price controls and price freezes at any level, but I think there may be a need at the retail level to make sure we have adequate enforcement tools to prevent pure price gouging

Also today we welcome witnesses that will help us begin developing an understanding of Hurricane Katrina's effect on the communications systems in the region and begin understanding the road to rebuilding critical infrastructure that has been damaged or lost.

Again, this hearing will be the first of several hearings that the Committee plans to hold on Katrina. We will have further hearings in other areas of Committee jurisdiction. I thank you all for your time in appearing today and look forward to hearing what you have to say. And without objection, the Committee will proceed pursuant to Committee Rule 4(e), allowing Members the opportunity to defer opening statements for extra questioning time.

Chairman BARTON. As I said at the beginning, we are proceeding pursuant to Committee Rule 4(e), which will allow members that wish to defer opening statements additional time on their question-and-answer period.

I would now like to recognize the distinguished ranking member of the committee, Mr. Dingell of Michigan.

Mr. RUSH. Mr. Chairman, I have a unanimous consent request.

Chairman BARTON. If it is in order. I am going to assume it is. What is the gentleman's unanimous consent request?

Mr. RUSH. Mr. Chairman, I would just ask for unanimous consent that all members of this committee please join with the President in refraining from calling American citizens who are distressed refugees. They are not refugees. They are American citizens. They pay taxes. They have been involved in helping to build this country, and they are not refugees, and I think it is a disservice to them.

Chairman BARTON. I am not sure that is a unanimous consent. The Chair would encourage members to use the appropriate terminology.

Mr. RUSH. I just ask, Mr. Chairman, if they would just join the President and others in refraining from using the word refugee.

Chairman BARTON. I support the gentleman of Illinois' intention.

Mr. RUSH. Thank you.

Chairman BARTON. I am going to recognize the distinguished ranking member from Michigan for an opening statement. Mr. Dingell.

Mr. DINGELL. Mr. Chairman, thank you; and I commend you for holding this hearing. It is very important that it should be held at this time. The committee has many matters of interest here related to the events that have followed Katrina; and under your leadership, as under your leadership on the recent energy bill, I am satisfied that we will address them well.

The hearing today takes place while vital rescue relief efforts are still under way in New Orleans and our Gulf State communities devastated by Hurricane Katrina. As we continue to consider how this Nation will recover, we must also be mindful of the scale and severity of the destruction in the Gulf and the challenge of caring for those whose homes have been destroyed and whose lives will not soon return to anything resembling normal.

Our first efforts must be to take care of those who are suffering, their families, and the families of those who have been killed or seriously injured in these events. In the coming months, the effects of this disaster will continue to ripple through the economy. Several critical sectors are affected by Katrina: health, energy and telecommunications. All of these fall within this committee's responsibilities; and, again, Mr. Chairman, I commend you for holding this hearing to help our members focus on the work that lies ahead.

We know that the Federal response to Hurricane Katrina, particularly that of the Federal Emergency Management Agency, FEMA, has been just plainly disgraceful. But we must now focus our attention on the tasks ahead. As a preliminary but very important matter, I have introduced legislation to restore FEMA to an independent agency with Cabinet-level status reporting directly to the President.

With respect to gasoline, which was part of the original focus of this hearing, it is important that the committee provide strong oversight to ensure that consumers are not subject to price gouging for gasoline and other energy supplies and that recovering energy markets are not manipulated.

While local gas stations are usually the easiest target, we learned in the 1970's—and again this is a little bit like deja vu—that the major violators were elsewhere. They were the oil price controls that were in effect then, and they were also the persons who were compelled to disgorge billions of dollars in overcharges. Those were people who—largely who were traders in the industry, and people at the different major oil companies and in major institutions inside that industry. Any examination of price gouging must begin with a review of practices by persons like this.

Already, a number of States have acted to stem gasoline price increases, from suspension of State gas taxes to invoking State emergency authorities limiting price increases; and it appears that the announcements of release from the Strategic Petroleum Reserve, the "SPR," and from the International Energy Agency Stockpile will temper escalating prices to some degree.

But we cannot focus solely on gasoline. Natural gas and heating oil prices could very well pose an even greater challenge for our constituents as winter approaches. I applaud Saturday's release of $27.25 million in LIHEAP funds to the affected States, but we should recognize that we will need a significant increase in LIHEAP funding in the coming months.

While the Nation's energy needs are critically important, we cannot forget the real human need that exists in the Gulf States right now.

First, how has our public health infrastructure met the challenge? I hope that we can have hearings focusing on this vital question.

Second, what do we do to provide for ongoing care of those who were suffering in this area, for the industries and for the institutions and for the States in the area and for hundreds of thousands of displaced families? I note Medicaid is going to be a lifeline in the coming months.

Earlier this year, Democrats strongly opposed the budget plan that included $106 billion in new tax cuts benefiting mostly wealthy people while requiring our committee to cut a likely $10 billion in Medicaid. That budget must be scraped and instead immediately replaced with a package of assistance to assure that the health care needs of families and children do not go unmet.

Hurricane Katrina has created an environmental catastrophe for the Gulf region that will require significant Federal assistance. This committee should monitor the environmental issues that are arising, from Clean Air Act waivers to the rebuilding of the safe water drinking water infrastructure, and we must pay careful attention to the environmental consequences as we consider best how to make the needed improvements in our refining capacity.

Finally, the committee must also look closely at how the communications and media sectors responded to Katrina and what steps should be taken to better prepare for and to warn people about and how to respond to emergencies. Functioning communications networks are critical for first responders to do their jobs efficiently and safely as possible and for victims to call for help or communicate with loved ones.

I look forward to hearing from today's witnesses about the response of the Federal Government and industry to the current disruptions in these and all of the other vital sectors which are affected and their plans for the coming months as we try to help the Gulf region and its people to recover.

I thank you, Mr. Chairman.

Chairman BARTON. I thank the gentleman for the opening statement.

We want to recognize the distinguished chairman of the Energy and Air Quality Subcommittee, Mr. Hall, for an opening statement.

Mr. HALL. I thank you, Mr. Chairman, for holding this hearing.

Just as a foreigner attacked New York on 9/11 and devastated a great city and a great State, another foreigner called Katrina hit New Orleans, Louisiana, and sister cities in various States. I think we have a lot of work to do to meet the challenges that are posed by the devastation that Katrina inflicted on thousands of families and communities along the Gulf; and we look for answers, not accusations. I think we must also address the disruption to our Nation's infrastructure in the wake of Katrina, particularly the impact on our energy supply and delivery system.

Gasoline prices were already too high in August as a result of increased worldwide demand and limited spare capacity. The disruption of our energy infrastructure from Katrina compounded the program. Actually, Americans are alarmed at the raising cost of gasoline and the projected higher cost of natural gas, and they are looking to Congress to address their concerns. The Energy Policy Act of 2005 is certainly a step in the right direction, and Katrina lends a sense of urgency to provisions in that Act that need to be expedited.

A diversification of energy supplies is an important component. Diversification could help ensure energy security and thereby national security from disruption due to natural disasters or terrorist acts.

Too much of our national gas supply comes from one region, the Gulf of Mexico. By ultra deep provisions—the amendment that we passed, we passed it two sessions ago, it got by, the conference committee had accepted it, we passed it this time—drilling deeper in the Gulf is going to make drilling operations less susceptible to hurricane damage, for one thing.

Another thrust for diversification would be to streamline the permit process for new refineries in each of the 50 States as outlined in Section 392 of the Energy Act, and Governors have been alerted and are alerted and are looking at that at this time because that will allow them to seek at least a refinery per State with a lot of encouragement from the Act itself.

Drilling off the other coast is another option that would us far less susceptible to disruption.

So these and other policies will help us achieve energy security in the long term, but we also need to consider what actions will give us immediate relief. Our citizens are paying the price for our dependance on foreign sources for too many years, and we need to stop that. The margins are just too thin in our energy market to absorb the fluctuation in supplies and prices due to catastrophic occurrences such as Hurricane Katrina.

The hearing gives us opportunity to hear from experts in the Government and industry about the magnitude of the problems we face and suggestions for corrective action.

Mr. Chairman, thank you again for scheduling this hearing; and I thank the panelists for giving us their time, their time for preparation, their time for attending.

I thank you very much. I yield back my time.

Chairman BARTON. We thank the gentleman.

The Chair recognizes the gentleman from California, Mr. Waxman, for an opening statement.

Mr. WAXMAN. Thank you, Mr. Chairman.

The Federal response to Hurricane Katrina has been woefully inadequate. Hurricane Katrina was an unstoppable force of nature, but it is plain that the Federal Government could have done much more far sooner to respond to the immediate survival needs of the residents of Louisiana and Mississippi.

Congress has a responsibility to understand what went wrong and why, and unless Congress conducts thorough oversight investigations to examine the preparation for and response to Hurricane Katrina, few lessons will be learned and the Nation will remain vulnerable to future natural disasters.

The administration has told us that they were prepared for this kind of disaster. Two years ago, FEMA Director Michael Brown testified before Congress that FEMA would be able to respond to disasters within 12 hours. Well, FEMA failed miserably. Relief and supplies took days, not hours, to arrive; and the toll on those affected was terrible.

Today's hearing focuses on the energy implications of Katrina, but the pattern is the same. The administration policies that we were told would protect Americans from skyrocketing fuel prices and price gouging have failed. The administration's energy policy is based on a trickle-down theory: If we give the big energy companies enough subsidies, tax cuts, and regulatory relief, then they will keep gasoline prices low. This policy is great for the oil companies, but it simply does not work. For the past few years, long before Katrina, gasoline prices have been on a steady march upwards; and the oil company profits tripled between 2002 and 2004 to $87 billion.

Since last month, gasoline prices have shot up another 30 percent. Oil companies appear to have taken advantage of this crisis to earn even higher profits, and now some Republicans are saying that the answer is to give the industry even more subsidies and breaks.

Our energy policy is fundamentally broken. As the hurricane proved, we are exactly on the wrong track. To keep gas prices down and to protect our energy security, we need conservation, increased fuel efficiency, new technologies and not another round of industry handouts.

Hurricane Katrina showed the bankruptcy of our policies. It is not enough to look after the interests of the special interests. We need to be focused on providing good government and life-saving services to all Americans.

Thank you, Mr. Chairman.

Chairman BARTON. We thank the gentleman.

The gentleman from Michigan, Mr. Upton.

Mr. UPTON. I am going to take the extra time and defer.

Chairman BARTON. The gentleman defers.

The gentleman from Florida, Mr. Stearns.

Mr. STEARNS. Thank you, Mr. Chairman; and I want to thank you for holding this hearing.

As chairman of Commerce, Consumer Protection and Trade, I have jurisdiction over the Federal Trade Commission; and I want to welcome their counsel this morning for coming here.

Maybe in the near future we can also have a hearing out of my subcommittee where we talk about the study that you did recently. I hope that Cecil will mention a little bit about the study, and I compliment the Federal Trade Commission, because that study was done well before Katrina.

I have to tell my colleagues that the Federal Trade Commission looks at the price of gasoline in a continuous operation mode with a modeling, a simulator; and I think a lot of Americans do not realize that they have this jurisdiction in which they can stop price gouging and can stop collusion between oil companies. Some of the actions that they have done, the American people should realize, have been beneficial in stopping some of this monopoly practices. So I went to commend them this morning.

But I think Katrina, the hurricane, has highlighted a very serious problem that we have in this Nation with crude oil and gasoline supply and demand that is out of balance. Before Katrina, this balance was already very tight and prices were already at record highs. Thus, by removing nearly a third of the United States' crude oil production and 10 percent of the Nation's refining capacity at a time of very high demand, we caused gas prices to spike even further. This confluence of events is precisely the situation the United States faced during the Labor Day holiday.

The future is not bleak, though. We have new technologies that are being developed in this country. As Chairman Barton has mentioned, we tried to give incentives, we tried to give alternative ways for Americans to view the problem.

For example, in Alberta, Canada, for example, a method of producing oil out of deposits of Bitzium buried in the ground—this is called oil shale—oil sands—is now finally becoming very profitable and a viable alternative for crude oil production. So I think the United States should realize right there, close by in Canada, with oil sands we have a possibility of a viable alternative for crude oil. Alberta's oil sands deposits are second only to Saudi Arabia's reserves, and estimates have shown it could satisfy the world's demand for petroleum for the next hundred years. So there is some light at the end of the tunnel.

In closing, Mr. Chairman, even under the best circumstances, a storm like Hurricane Katrina would have had a noticeable impact on gas prices no matter what we did. However, at a time of extremely high demand and tight supplies practically shutting down the United States largest oil refining region, obviously gas prices are going to spike even higher.

So I look forward to our witnesses today, and again I commend the Federal Trade Commission for the study that they did much before the hurricane.

Thank you, Mr. Chairman.

Chairman BARTON. We thank the gentleman.

The gentleman from New Jersey.

Mr. PALLONE. Thank you, Mr. Chairman.

Mr. Chairman, I have to say that I am kind of torn today. Because, on the one hand, I admire the fact that this committee, under your chairmanship, is holding this hearing today and is trying to take quick action to address the crisis from Hurricane Katrina. On the other hand, I feel that the Bush administration has been totally incompetent in handling this situation and the emergency response in particular; and I do—I, amongst others, have called for the FEMA director to resign, because I think that he has acted in a totally unconscionable way.

I also agree with Mr. Dingell's comments about how we need to change FEMA because of it. But, in addition to that, I must say that my constituents are outraged and actually shameful about the way our government reacted in terms of the emergency response but also feel that the oil companies are taking advantage of the situation to gouge and to increase prices in a way that is also unconscionable.

So I appreciate the fact that you are having the hearing today. I think it shows leadership on your part. But as far as the Bush administration, they have acted in a shameful way, and my constituents are absolutely outraged by what this administration has done in response to the hurricane and by what they think the oil companies are doing to gouge prices.

Now, in our committee, of course, we deal with the energy issues; and I think that the devastation in the Gulf region and the spike in prices is a wake-up call for our Nation, which is accustomed to cheap oil, and raises several important questions.

First, why were gas prices rising even before the hurricane while oil companies were seeing record profits? Second, how can a country that consumes 25 percent of the world's oil but produces only 3 percent continue to use as much oil as we do without being left vulnerable to severe price volatility? And, third, how much price gouging occurred in the wake of the hurricane? Do we need to consider implementing Federal anti-gouging authority?

I introduced a bill on Friday which tries to deal with some of these things because of the gouging and because of the high prices. The bill would specifically limit the profits of big oil companies that sell on the wholesale market to their average over the past 5 years so profits do not continue to skyrocket as consumers struggle in the wake of the hurricane. The bill would also reduce gas price volatility by limiting companies that sell on the wholesale gasoline market to only one price increase per day. It also directs the FTC to investigate whether there has been gas price gouging in the wake of the hurricane. But, most important, the bill requires the President to find ways to reduce our national oil consumption.

The truth of the matter is that, over the long term, the only way we will be able to keep gas prices down will be to reduce our consumption of oil. That means increased fuel efficiency of our cars and trucks. It does not mean the administration's recently announced new rules concerning light truck fuel efficiency, which will do little to solve the problem. Indeed, it may encourage manufactures to make existing models even bigger.

Again, I want to thank you, Mr. Chairman, for having this hearing, but your response is so different from that of the Bush administration.

Chairman BARTON. Does Mr. Deal wish to make an opening statement?

Mr. DEAL. I would reserve my time for questions, Mr. Chairman.

Chairman BARTON. Does Mr. Whitfield wish to make an opening statement?

Mr. WHITFIELD. I do, Mr. Chairman.

Chairman BARTON. The gentleman from Kentucky is recognized for 3 minutes.

Mr. WHITFIELD. Thank you, Mr. Chairman.

Today's hearing is vitally important. Despite the widespread destruction and personal tragedy inflicted by Katrina, it does raise an issue of utmost importance not only for our country but I think for the world, and that relates to this whole question of energy.

Gasoline prices are skyrocketing. There is no question about it. There are examples of price gouging. We know that. But I think that Katrina has demonstrated that we have a more systemic problem relating to energy.

First, worldwide consumption of oil is presently at a staggering 83 million barrels a day; and worldwide production is around 84 to 85 million barrels a day. Worldwide demand has been increasing at a faster rate than at any time in history. As a matter of fact, in China alone last year demand increased by 16 percent.

A new refinery has not been built in the United States since 1976, but half of the refineries in the U.S. since that time have been closed. In the U.S. alone, consumers are using right around 21 million barrels of oil a day.

We use six times as much fuel per day as people in Europe. Their gasoline taxes are much higher in Europe than they are in the U.S., so we have become accustomed to low prices compared to the rest of the world, and all of a sudden we find ourselves in a situation that we do not particularly like.

I might also add that contributing to the situation today we have speculators in oil futures more than at any other time. That is putting a burden on higher prices.

We see other countries nationalizing oil reserves more than at any other time in our Nation's history. Reserves available to U.S. companies are not being produced the way and located the way that they have in the past. We are, for the first time in a long time, being forced to use reserves from our Strategic Petroleum Reserve. Even the European reserve is going to be providing the U.S. 2 million barrels a day for the next 30 days.

So we have some significant issues affecting this country in the area of energy. It is going to require us as a Nation to reexamine the way that we need to go. I think the energy bill that we passed is going to help answer some of those questions.

But I want to commend the chairman for having this hearing and allowing us to focus on an issue of utmost importance not only for us but for the world. I think that is the only bright spot that

I have seen from Katrina, is it is going to require us to focus on this issue.

Chairman BARTON. We thank the gentleman.

Does the gentleman from Ohio wish to make an opening statement?

Mr. BROWN. Yes, I do.

Chairman BARTON. The gentleman is recognized.

Mr. BROWN. Thank you, Mr. Chairman.

As I wrote to you yesterday, I believe the committee staff should be begin a jurisdiction-wide review to identify policy areas where our committee can act; and particularly it is important to examine the public health consequences of this disaster, again with an eye toward identifying unmet needs. I hope these questions will be the subject of another hearing in the near future, especially Medicaid, hospital funding, long-term health consequences to those victims of the hurricane, and especially in New Orleans.

This hearing was originally focused on gas price consequences of the disaster. I want to talk for a moment about that.

One of our witnesses today, Dr. Mark Cooper of the Consumer Federation of America,reminds us that Congress has missed opportunities to provide a cushion to protect consumers when supply disruptions cause price spikes. His testimony attaches a 2001 report calling for a regional reserve of gasoline similar in concept to the Strategic Petroleum Reserve. Twice in our committee, once on the House floor, we failed to take that commonsense step.

It is also indefensible that, as gas prices break record after record after record, that we continue to pump oil into the ground without regard to price. Before 2002, the Energy Department took price into account before deciding whether to take oil off the market, but, since then, price has literally been no object.

Congresswoman Baldwin and I offered an amendment during this year's energy bill debate to correct that. Our amendment would have required the Department to consider price before making SPR acquisitions. It would have allowed the agency to weigh the further energy security merits of acquiring oil at times of high price against the cost to consumers. That proposal was also rejected by this committee.

This committee and this Congress have not taken the lead from States that have already acted to protect their consumers. My State of Ohio and other States have enacted quote, unquote, unconscionable sales practices laws that have been used to enforce gasoline price gouging. But many States have no such protections; and even for those States like mine that do, the absence of a Federal standard contributes to a confusing and chaotic environment that, frankly, provides ample cover for those who would take advantage of national tragedy to abuse consumers and pad their profit margins.

Our first priority should be to find unmet needs and act to meet them, but we also must look at the lessons learned from this tragedy. As we do so, we ought to begin by looking at the opportunities that we ourselves have missed.

Thank you again, Mr. Chairman, for scheduling today's hearing.

Chairman BARTON. I thank the gentleman.

Does the gentleman from Georgia wish to make an opening statement?

Mr. NORWOOD. I do, Mr. Chairman.

Thank you, Mr. Chairman, for having this hearing; and I do want to thank all of our witnesses for their time and their willingness to help analyze this dire situation. We know that all of you are working overtime and your staffs have been working overtime and probably will have to continue to do so for days and weeks to come.

I would like to join all of my colleagues in expressing my deepest and most heartfelt sorrow for those struggling along the gulf coast. The devastation and the scope of the tragedy there is beyond anything we have recently or ever seen in this country. Our thoughts and our prayers have been with our fellow Americans. Our assistance in just about anything that Congress can do is coming and will continue. Hopefully, this panel can help us identify exactly how to provide that assistance and the best way to deliver it.

At some point in time, I hope that we have an opportunity to examine how to better deal with situations like this regarding energy and telecommunications, if and when there is another time.

First, the short term. What is needed now and in the near future to help deal with this tragedy, is dealing with the human suffering. Of course, all of us recognize that dealing with the human side of this will take longer than a few days. Lives, not just homes, need to be rebuilt in so many cases.

The effects of this tragedy also reach beyond the Gulf. As many of you know, the original scope of this hearing was high gas prices, but, smartly, the chairman changed it and expanded it to be much more than that.

I am very interested in this important issue because so much of our Nation's infrastructure, energy infrastructure, is in the Gulf.

Second, long term. We have had the reports of what went wrong already. But I think many want to know what we can do to prevent those same problems in the future. By the very nature of a disaster like this, unexpected things happen. We need to expect the unexpected and be prepared with a comprehensive plan B, C and even D. A future tragedy maybe averted or at least our response improved by learning lessons. I value your insight, gentlemen, on this point.

Thank you, Mr. Chairman. I yield back.

Chairman BARTON. Does Mr. Rush wish to make an opening statement?

Mr. RUSH. I would, Mr. Chairman.

Thank you, Mr. Chairman. I also want to join with my colleagues in thanking you for holding this on-time hearing.

Mr. Chairman, when I think of the devastating effects of Hurricane Katrina, I cannot help but wonder at the value of some human life in this country, along with the rest of the world, outraged by the slow response to take action and provide aid to the thousands of individuals who were left to die and fend for themselves in the aftermath of the worst national disaster in American history.

Those who did not die were subject to the most dehumanizing conditions imaginable. The demoralizing squalor in the Superdome and other relief centers in New Orleans has been compared to the conditions in the hulls of slave ships, and this is not an exaggeration. This is an example of how government failed, a complete breakdown when responding to the needs of those who needed help in critical times. In times of national crisis, the cries from Louisiana, Mississippi, and Alabama went unheard.

Mr. Chairman, I am concerned about price gouging at America's pumps, but at the same time I am more concerned about the price of human suffering being paid by the most vulnerable segment of our society. I fully realize this committee does not have jurisdiction over FEMA or the National Guard, but this committee has jurisdiction over multiple areas of immediate and emergency concerns including water, the purification of drinking water, the abatement of dreaded diseases, including e-coli, Hepatitis A, cholera, salmonella, West Nile and other mosquito and other waterborne diseases.

Clearly, the public health concerns of this Nation and particularly the devastated Gulf region are of paramount importance. That said, I want the record to reflect that we will be quite intentional regarding conducting hearings to determine what is the appropriate Federal response to this acute and critical crisis.

Additionally, Mr. Chairman, I join with the ranking member, Mr. Dingell, and I share the opinion that we as an authorizing committee of jurisdiction has the authority to increase our commitment to the LIHEAP program.

Mr. Chairman, with that, I yield back the balance of my time.

Chairman BARTON. The gentleman yields back.

Does Mr. Shimkus wish to make an opening statement?

Mr. SHIMKUS. Yes, sir, thank you. Just a couple comments.

I want to thank the panelists for being here.

When we passed the energy bill, we set out on a process so that we could have a diversified energy portfolio; and I think the Chairman was correct in that, in saying that what this tragedy highlights is how fragile our infrastructure has been for many, many years. Obviously, we hope that with a new look at energy we can start reclaiming some independence and diversifying our portfolio.

That is not just electricity generation but also in the fuel arena. I, like everybody else, travel around our districts numerous times, and there are parts of our policies on the energy issues that there is some optimism out there.

I drive a Ford Explorer flexible-fuel vehicle. It runs on 85 permanent ethanol. Years ago, I had a flexible-fuel Ford Taurus. Two years ago, I could not fill up this Taurus at a single retail location in my district. Now I can fill it up all throughout my district, probably 30 retail sites. In fact, I have a picture of one.

Now the prices are still pretty scary: unleaded, $3.69; E85, $3.09. That is a 60 cent price deferential for a vehicle that runs on 85 permanent ethanol. Now this is an example of public policy moving in the right direction.

I also have another article from a stop in Nashville, Illinois, at a—this is a State and Federal addressing of our reliance on imported crude oil; and the State has also pushed and helped the rollout of biodiesel. Now most of—a lot of the fleets in Illinois are moving to 11 percent bio, soy diesel, or another formulation; and this article says trucking firm embraces biodiesel. So this—over the long haul a truck company of 150 tractor trailers runs his operations across the Nation with 11 percent. Now you might say 11 percent is not much. Well, just add that 11 percent back into the petroleum-based diesel fuel and see what happens to prices.

So we have great challenges to deal with. The energy bill talks about a hydrogen economy and moving to hydrogen fuels. So we need to diversify our energy fuel.

The infrastructure is weak. The hurricane showed that. Let's go about the job of diversifying our fuel portfolio.

Thank you, Mr. Chairman. I yield back.

Chairman BARTON. The gentleman yields back.

The gentlewoman from California.

Ms. ESHOO. Thank you, Mr. Chairman, for your leadership in calling this hearing. It is an important one and I think underscores the very broad and powerful jurisdictions of the House Energy and Commerce Committee. We are the oldest committee in the Congress, and we are one of the most powerful. Today and I think in subsequent hearings and the action that this committee can take are going to flow from the power that this committee has.

I want to express my sympathy and the sympathy of my constituents very directly to members of this committee whose congressional districts have been hardest hit.

Now, having said that, Mr. Chairman, there are two things that I want to highlight today.

First of all, is there a commitment of the participants, the leaders of the energy industry, most specifically the oil industry, to go on record that they will not tolerate price gouging? The answer has to come from them. If we try to do this and address it legislatively through the various agencies of the Federal Government, we are going to get hung up on the ropes. I would like to hear very directly from the leadership of the oil industry in our country that they will not engage or tolerate price gouging. It is the simplest, most eloquent way for this to be handled. So, No. 1, I think we need to have an answer from them.

No. 2, Mr. Chairman, I think the next answer needs to come from the leadership here, certainly yourself and the leadership in the House, that the cuts to Medicaid will be suspended.

Front and center, we heard from more than one Secretary last night as they came to the floor of the House to address the Congress of what the safety net is in this country, how it will be used and put out there effectively for tens of thousands of our follow citizens, that they need this program, wherever they are, whether they have been moved to different cities in Texas, your home State, to California, to the District of Columbia, to other places in our Nation.

This is not the time, this is not the time to be moving forward with the cuts that the committee took up and that the Congress is considering. This is wrong, it is hurtful, and it is not the message to send to the victims. So when we speak about compassion, when we speak about being effective, when we speak about standing next to our colleagues whose districts have been wiped out, whose constituents are seen floating in contaminated waters, this committee has to respond and respond effectively.

So those are the two things that I would highlight today. We have our work cut out for us.

Now when the words "blame game" are used, I really resent that; and I think that we all should. This is not a game. This is not a game. People are dying, have died, people have been displaced, taken away from their communities. We have long-term and short-term work to do.

One of the great hallmarks of our Nation is that the American Congress, that the Congress of the United States of America, has been able to take up both a critical role of being critical so that we learn from the mistakes that have been made.

So this is not a game, Mr. Chairman. This is sobering work as we try to adjust to the horrific catastrophe that has happened to our country.

So, with that, I do not have time to yield back, but I believe that these two issues need to be taken up front and center. Thank you.

Chairman BARTON. We thank the gentlelady.

We encourage members to try to stay within their 3-minute limit, if possible.

The gentlewoman, Mrs. Wilson, is recognized for 3 minutes. We are trying to get the Governor of Mississippi up on the live video. So I think we can get you in before that happens.

Mrs. WILSON. Thank you, Mr. Chairman.

We have had a devastating storm, and it is not over. We are still in the middle of the process of saving life and sustaining life and recovering and rebuilding, and that will go on for a long time. Nothing should distract us from those priorities.

Sometimes, you know, my husband is kind of—he has got a great sense of humor. And sometimes when he watches people often criticizing with only partial information, he just kind of laughs and says, you know, they should shut up and start bailing. I think that is good advice, and a lot of ordinary Americans have taken it.

I think we have seen across this country people opening their homes and their wallets and their churches, our wonderful National Guard and medical doctors embracing the displaced and doing what they can from where they are with what they have got. One of the great lessons of this disaster is that the real strength of America is in the goodness of ordinary Americans, and we have seen that again and again and again across this country.

We also need here to continue to pursue policies that create jobs and keep our economy on track. A disaster and a tragedy should not be windfall, a windfall for opportunists. All of us have seen prices go up at the gas pump and in some communities exorbitantly. Most gouging laws are State laws, but only 23 States have anti-gouging laws, and the standards and definitions vary widely. I think we need to take a serious look at how we dissuade and deter and punish those who would gouge people in a time of tragedy.

We also have an opportunity here to put politics aside and to look at our energy policy anew, with conservation, exploration, production and refining, things that we look at routinely here, but also to look at our own perhaps failures of imagination. What are we going to do as a Nation to get beyond the gasoline engine? We are here at a historic turning point to make some serious decisions and have a serious debate about the follow-on to the gasoline engine.

Because world oil supplies are not increasing, and we need to make those decisions and investments now so that we change the way in which we get things and people across the country and back and forth to work.

Mr. Chairman, thank you for this hearing. I am sure it is going to be the first of many. And God bless the people of the gulf coast.

Chairman BARTON. Does the gentleman from Michigan wish to make an opening statement?

Mr. STUPAK. No, Mr. Chairman, I will waive.

Chairman BARTON. Does Mr. Green wish to make an opening statement?

Mr. GREEN. Yes, sir, Mr. Chairman.

Chairman BARTON. The gentleman is going to start, understanding that we are trying to get this thing set up.

Mr. GREEN. I would like to have the full statement placed in the record.

One, I want to thank you for having this timely hearing on the second day we are actually back. I am glad we broadened the scope to beyond just energy impacts, which is quite severe; and I respectfully suggest a further hearing on the serious public health impacts and our response in the near future.

Our pressing need in the Houston area, where we are home now to about 140,000 plus residents from Louisiana, is health care. With thousands in tight quarters, infectious disease a real threat, we need to provide the necessary assurances to our States who are the recipients that the health care providers, that they will be reimbursed.

I asked Secretary Levitt last night at our briefing to agree to provide a 100 percent Medicare/Medicaid reimbursement rate when caring for out-of-State Medicaid beneficiaries. I hope the administration will ease the Medicaid eligibility requirements for Hurricane Katrina evacuees.

Again, the State of Texas is the biggest recipient; and our Medicaid budgets are already stretched with our own constituents, much less adding rolls from Louisiana and the neighboring States. We want to be welcome neighbors, and we are. In fact, I am so proud of what Houston has done and the State of Texas.

The neighboring States of this disaster need massive Federal assistance to care for these victims. When a neighbor is in need, our neighboring States have opened—again, Texas has 250,000 out-of-State evacuees. That is unprecedented.

I have been first hand every day we have been home by both the Reliant Astrodome and the George R. Brown to see the massive shelters. Again, we need to be able to eliminate red tape now and get those folks out of those shelters into some reasonable living conditions, both for health reasons but also to try to return them to normalcy.

I am glad that, just today, we were notified that yesterday at our dealing meeting that the Houston area leaders, the mayor and the county judge and the business community, we found out that people are having their cell phone service disconnected from Louisiana. That is the only number most of the time their relatives know how to reach them. The FCC this morning announced, in working with CTIA, that those numbers and in compelling businesses and companies, not to disconnect those cell phones. One, that is a great declaration. The Federal Government should interpret our ability broadly and flexibly to make sure that we can handle the disaster and the relief that we need.

Turning to energy, gasoline prices are already high due to tight global supply and stretched energy infrastructure. Now that has gotten pounded by a hurricane. Gas, oil and natural gas will even be higher after most of the Gulf's production is halted; and, thank goodness, a lot of those platforms are trying to get back in use and even some of the refineries.

All of the pieces are connected when there is huge action on the market like Hurricane Katrina and a huge reaction throughout the system, and can we help without doing more harm than good? I gather from the Senate hearing yesterday that the FTC has no authority to investigate price gouging. We need to know who does, if anyone; and if there is some stations taking advantage, we need to stop them.

Even my Texas constituents want price caps, but if the Government tries caps for any length of time, supply will literally disappear. Let us not repeat the mistakes of President Richard Nixon. Large companies typically don't set the price at the pump, which is up to the individual station owner.

Chairman BARTON. Mr. Green, will you suspend so we can hear from the Governor?

Mr. GREEN. Mr. Chairman, I would be glad to yield to the Governor of Mississippi.

Chairman BARTON. We are going to suspend our opening statements. We do have video contact and, apparently, audio contact with the Honorable Governor of the great State of Mississippi, Governor Haley Barbour.

Governor, if you can hear me, you have got the full panoply of the Energy and Commerce Committee waiting for your statement; and then, once you have spoken, we are going to have a written statement read in the record by a representative of your companion Governor, Governor Blanco of Louisiana.

So, Governor Barbour, our hearts and our prayers are with you; and you have our undivided attention.

STATEMENT OF HON. HALEY BARBOUR, GOVERNOR, STATE OF MISSISSIPPI

Governor BARBOUR. [Via teleconference.] Mr. Chairman, thank you very much; and to all of the members of the committee, I appreciate the chance to try to share with you what has been going on in Mississippi for the last 9 days.

I do not have to tell you that this was the worst hurricane to ever hit the United States, and it struck us a grievous blow in Mississippi. The devastation is genuinely unimaginable and indescribable. Total obliteration of many things, some of which are the things that your committee is interested in.

I want to say to you that we appreciate you and the Federal Government. Nothing is perfect when you have an epic disaster like this. I told my wife as the week went on, every day we made progress. But there was not any day that we made as much progress as I wanted to.

Our Federal partners were great help, but there were days when we wished they would have been faster. There were days when we wished they would have done more. But when you consider the way all of our systems were overwhelmed, we are very grateful, and so thank you all.

Let me just say on the terms of energy, our energy situation the first few days was cataclysmic. This disaster is not just a coastal calamity. It goes 150 miles north. We had 130-mile gusts 90 miles inland. We had 90-mile an hour winds 150 miles inland. There is tremendous damage way, way north of the coast. But the 80 miles across the Mississippi gulf coast is largely destroyed. A town like Waveland, Mississippi has no inhabitable structures. None. The fire, the 26 policemen on Monday of last week went to the second floor, then got on the roof of their headquarters, and then all 26 of them swam off. Some of them hung in trees holding on until the storm was over. The destruction is unbelievable, and it overwhelmed our infrastructure.

Our utility that serves the coast in the southeastern part of the State lost every transmission line, had two power plants put out of commission, and virtually 100 percent of their customers lost power. The company that serves the southwestern part of the State, which is well inland, 75 percent of their customers lost electricity. Our rural electric power associations had similar percentages based on the geography. Even the Tennessee Valley Authority, as far north as it is, had tens of thousands of customers lose power.

When you lose power, the telecommunications systems falls down because of the need for electricity, not to mention the fact that virtually all the towers are blown down. We lost water because the water systems run and the sewer systems run on electricity. So we had a huge need, and one of our first goals was to try to get fuel, particularly diesel fuel, to run the generators that were powering our hospitals, our emergency operations centers, the ones that weren't destroyed, our sheriff departments, police departments, fire departments. So from an energy standpoint, for about 5 days we were hustling to keep people from running out.

Ultimately, the Federal Government started on Friday by the activities of the U.S. Department of Transportation, the U.S. Coast Guard, and FEMA was able to provide us with enough fuel for all of our emergency vehicles, and since Friday we have had an assured source of fuel for all our emergency vehicles, whether it is fire trucks, police cars, National Guard trucks, et cetera, and we are appreciative of that.

Today, we have about 288,000 customers who still don't have power. The peak was about 1 million, on the report Tuesday was about 1 million customers; we are down to 288,000. Mississippi Power Company, which is the southeastern coastal industrial utility, reports that they will have power to every customer who can receive power by Saturday, which is incredibly remarkable that in less than 2 weeks they can have restored power, because every one of their customers had just about lost power and their power plants are out.

They have about 7,000 people on the ground, pole climbers and tree cutters, and Entergy Mississippi is making the same kind of effort. And we are grateful. We have power, we have linemen and tree people from all over the United States and Canada who are down here helping our people getting electricity back on.

As I say, we are about 75 percent recovered, and because of their response. Except for the rural electric power associations who don't have as much equipment, they are further apart, you know, you may have to put back up 10 poles to serve one customer. We are getting over the hump, and by the end of the week should be over the hump on electrical power.

For telecom, the phone companies have really humped it to get service back. The first few days, there was as close to literally no communications, as you can believe. We couldn't in Jackson get people on the phone, even the emergency operations centers on the coast. And where there was particularly bad is that people in the affected areas and near the affected areas, they had no phone service, they didn't know what was going on, they had no television so they didn't have that as a source of information. A few of them had done like we asked people, and that is to have battery powered radios, but most people had no way to communicate and they were utterly isolated after living a life with our information-rich environment. It was a huge problem. It also led to some of the worst rumor mongering that you ever can imagine. But the phone companies have restored at least cellular telephone service to most of the populated areas, and they are getting it better out into the countryside.

But Cellular South, which is our home-owned cellular company, and BellSouth, which is our biggest provider and also is a partner in Cingular, again, their people have worked untold hours just like the electric utility people and made huge sacrifices.

And, Mr. Chairman, we have got a lot of people here who are first responders or utility people whose homes are blown down, and they are out getting the other people's electricity back on, or they are out digging through debris, firemen and search and rescue, while their wives and children are having to stay somewhere inland because their house isn't there anymore. The stories of sacrifice and selflessness that come out of this are pretty remarkable. In fact, they are not pretty remarkable, they are mighty remarkable.

The U.S. Coast Guard helicopter team, starting Monday night when the wind was still howling, have taken 1,700 Mississippians off of roofs or out of isolated places where people couldn't get out because of the debris and wreckage, 1,700 by the Coast Guard alone. Over 5,000 when you include the other first responders like firemen and policemen and National Guardsmen. We appreciate all the States that have let us have National Guard. We have more than 11,000 National Guard here. And they were particularly critical last week when our law enforcement people who had worked 18-hour and 20-hour days, 120 patrolmen, narcotics officers, and investigators from the State law enforcement down on the coast who slept in cars for 5 nights but worked 18-hour days to help people. It has been an incredible effort, and lots of people deserve credit.

I know that health is one of your issues. And I want to report to you that we in the last 24 hours had four deaths in Mississippi from a Vibrio type diarrheal disease, but the CDC and the Health Department report to us that this is not contagious, that this is the kind of disease that we common folks think and get from eating bad oysters; and, that people that have diminished immune capacity because of some other disease like HIV or cirrhosis or something, that all four of these people died of that disease. Because of HIPPA, we can't tell, we are not allowed to know any more about who those people were and what their conditions are. But the CDC tells us it is not contagious, and the disease is normally gotten by somebody that eats bad food, drinks bad water, or perhaps gets an open wound. But, again, we have had that in the last 4 days, which is I know a significant health thing that you would want to know about.

The search and rescue wasn't as fast as we wanted, but if you could come down here and see the devastation. We have areas, tens of square blocks in a row, that have debris waist high, head high, and search and rescue means people walking through there and moving all that stuff out and looking to see what is under it. As late as Friday we were finding people alive buried in the debris, but unfortunately we are finding people buried in the debris that are not alive. The official fatality as reported is about 148—that is not right, 154. The news reports, which we consider credible and relative and reliable, are closer to 200, and the likelihood is that the number will go up.

Let me just close by saying I am old enough to remember Camille. As a college boy I drove a dump truck full of blankets and pillows and baby clothes down to Gulfport in the wake of Camille. Down here, we have always thought Camille was the benchmark for what a hurricane could do. Katrina was worse than Camille. The devastation is wider spread in terms of breadth. Where Senator Lott's home was totally wiped off the beach in Pascagoula that is about 75 miles east of the eye of the storm. This storm's breadth was unbelievable, but its power was, too. You know, I am not a meteorologist or a scientist. For some reason, this storm's storm surge was much, much worse than Camille. Places where people thought it was safe because Camille didn't do any damage got 10 feet of water, and we had some people that died because they thought it can't be worse than Camille.

Again, in all of these things that we have talked about, the Federal agencies have worked very hard to help us, and their people have been down here busting it just like I talked about, the Coast Guard and others, and we appreciate that. We are going to need a lot more help. We are kind of turning the corner to where we are starting recovery, we are starting cleanup in most of our towns, we are going to start rebuilding.

Our attitude is on the future, and we are going to rebuild. We are going to rebuild the gulf coast bigger and better than ever it was, and all of the south part of Mississippi is going to be improved when we get finished, but we are going to need a lot of help and it is going to take a lot of time.

Thank you for letting me have a chance to tell you what is going on, Mr. Chairman.

Chairman BARTON. Well, Governor, we first of all want to commend you for what you have done for the citizens of Mississippi the last week or so. Your leadership has been invaluable. We are not going to take questions because we still have about 20 members that need to give an opening statement and we have five witnesses that have waited patiently for the last hour to give their statements. But we do want to commend you for what you have done. You have got two United States Senators and a United States Congressman who is a member of this committee, plus several other Congressmen in the House. Whatever you need from the Federal Government, if you will work through them or directly with us if it is within this committee's jurisdiction, we are going to do everything we can to make it happen sooner rather than later and bigger rather than smaller. And God bless you and God bless the great State of Mississippi.

Governor BARBOUR. Thank you, Mr. Chairman.

Chairman BARTON. We have as a representative of the Governor of Louisiana, Governor Blanco, we have Mr. Scott Angelle, who is the Secretary for the Louisiana Department of Natural Resources. We would now recognize you, Mr. Angelle, to read the Governor's statement. And Governor Barbour, you are welcome to leave. Mr. Angelle.

Mr. ANGELLE. Thank you, Chairman Barton, and committee members. Governor Blanco sends her greetings and her thanks for all the prayers and support that are flowing into the gulf coast and southeast Louisiana.

I am pleased to be here as a member of the second panel to give you Louisiana's views on energy policies post Katrina, but in her absence Governor Blanco has asked me to share this brief statement with you.

Katrina dealt southeast Louisiana a devastating blow, but I also know that this storm did not and will not destroy the spirits or the hope of our citizens. I wish I could join you today, but we, all of us here, are working hard and working together to finish the rescues and begin the reconstruction.

The people of southeast Louisiana are already making plans to rebuild their lives and their communities, and we will help them do it. Our people, our most valuable asset, have been forced to take shelter all across the country. We know Louisiana will not fully recover until those displaced by this storm rejoin their families and rebuild their communities. Part of rebuilding Louisiana will be rebuilding our oil and gas infrastructure. In the wake of all of this, we still understand that America counts on Louisiana to produce the energy to fuel this great Nation. We will focus on restoring and repairing the offshore and onshore assets that are so vital to this region's economy and so vital to America's economy. At this moment, while we are focusing on the immediate needs of our people, we also are looking forward to the rebuilding.

Thank you again for your prayers and your aid, and thank you for also looking forward to the future of Louisiana and the future of America's energy economy. Thank you, sir.

Chairman BARTON. Mr. Angelle, I know you are going to be on the second panel. But in response to the Governor's statement, if you talk to her later today, you tell her that our prayers are with the great State of Louisiana and with her, and that we make the same offer to your Governor that I just made by teleconference to the Governor of Mississippi: Whatever we can do to help, if it is within our jurisdiction, we are going to try to do it sooner rather than later and larger rather than smaller.

Mr. ANGELLE. Thank you, sir.

Chairman BARTON. We are now going to go back to our opening statements. And I believe Mr. Green had finished his, so it would be Mr. Shadegg's opportunity if he wishes to make his opening statement.

Mr. SHADEGG. Thank you, Mr. Chairman, for holding this important hearing on the devastating impact of Hurricane Katrina. My heart and my prayers go out to those whose lives have been impacted and devastated by this disaster.

In 1969, I was stationed at Keesler Air Force Base in Biloxi, Mississippi, and I arrived there literally days after Hurricane Camille struck. It is tragic to see this kind of devastation again to the gulf coast, and as the Governor pointed out, to see that it is even worse.

I wholeheartedly agree with my colleague Mrs. Wilson regarding the importance of moving beyond the gasoline engine in the long run. But today, whether we like it or not, America runs on refined oil products, and our transportation sector, airlines, trucking industry, and railroads, require a steady supply of fuel to keep our economy moving. In addition, families across our Nation require that fuel to heat their homes, and they will need it this winter and for winters to come.

The damage that Hurricane Katrina has done to this energy infrastructure, which has rippled from coast to coast, raises many important policy questions for this Congress and this committee to address, not the least of which are: Do we have the facilities that we need to meet America's demands? And, is our energy infrastructure too heavily concentrated along the gulf coast?

Hurricane Katrina's impact on an already strained refining industry has had a dramatic impact, most notably on the recent stunning price spikes seen by Americans at their local gas stations.

While I am encouraged that some refineries closed by Katrina have already opened or are close to reopening, reports indicate that several large refineries have experienced significant flood damage and will not reopen for some time to come. This is especially troubling because U.S. refineries were already operating at over 97 percent capacity before Hurricane Katrina hit.

As has already been noted here this morning, we have not built a new refinery in the United States since 1976, a span of 29 years. Currently, we import roughly 12 percent of the gasoline and diesel fuel we consume in this country from foreign refineries. Yet, not long ago we refined all of the gasoline and diesel fuel used in the U.S. from refineries here in the U.S.

We should not be outsourcing the refining of the fuels we need to run this country's economy. We must do more to bring our refining capacity in line with all of our domestic demands.

Currently, this critical portion of our industry is operating with no margin for error. Whenever a U.S. refinery needs to interrupt production for any reason, including just routine maintenance, Americans pay an unnecessary price because we have insufficient domestic refining capacity. When a disaster like Katrina strikes, we are in much worse shape. This problem is exacerbated by the fact that there is a worldwide shortage of refining capacity.

It is preposterous to argue that we do not need to fix this system or that we can continue down a path of reliance on foreign refining capacity. As America grows, total miles driven each year go up. Demand for refined petroleum products also goes up. The price of a barrel of oil is ever increasing, yet just last year this committee heard testimony that investors would frown on any decision by an energy firm to meet the rising demand here in the U.S. for refined product by building new refineries.

Let me illustrate how this point impacts us directly and why it is more than a crude oil problem. Crude oil futures have gone up over 60 percent over the last year, but refined gasoline futures have more than doubled. We must address this critical problem. Mr. Chairman, I thank you for holding this hearing.

Mr. WHITFIELD [presiding]. Thank you. At this time I recognize the gentleman from New York, Mr. Engel, for his opening statement.

Mr. ENGEL. Thank you, Mr. Chairman. Let me first of all say that my heart and prayers go to the brave people of Louisiana, Mississippi, and Alabama, and anything that we can do to help them, we should and we will.

Mr. Chairman, in the 1970's there was a movie where the lead character gets up and he says: I am mad as hell, and I am not going to take it anymore.

Well, I think the American people are rightfully mad as hell, and we are not going to take it anymore. We are mad as hell about rising gas prices, price gouging, and all things that we have seen disgracefully over the past week. We have seen on TV many pictures of people looting stores. Well, I would say that the biggest looters have been the big oil companies. They are looting the American public. There is no way that increased gas prices at the pump could have been reflected in 2 days after the hurricane with spikes of 30 to 50 cents per gallon. It is absolutely shameful and unconscionable that big oil companies are making profits off people's misery with this hurricane. There is no other way to say it. Because when the cost of oil drops a barrel—a gallon drops, it takes several weeks for it to be reflected at the pumps. So how could this be reflected in a matter of 2 days? These increases in gasoline prices are unconscionable and should not stand. The oil companies own the means of cost and production. They have long-term contracts on the oil fields. They own their drilling equipment, they own their tankers. These haven't changed. Their costs haven't changed. That is why their profits are soaring to record levels. Why make profit off people's misery and cause the entire American public to suffer? Gasoline over $3 a gallon? Unconscionable. Now, they are saying that prices will drop, and it will only be $2 and change a gallon what was before. We are supposed to be grateful that it is going to drop to $2 and some odd cents a gallon. There is no way that this should continue.

Now, it is not a matter of the blame game. I ask unanimous consent for an editorial of the New York Times today called "It is not a blame game." I ask unanimous consent for it to be inserted into the record.

Mr. WHITFIELD. Without objection, so ordered.

[The article follows:]

[Wednesday, September 7, 2005—The New York Times]

IT'S NOT A "BLAME GAME"

With the size and difficulty of the task of rescuing and rebuilding New Orleans and other Gulf Coast areas still unfolding, it seemed early to talk about investigating how this predicted cataclysm had been allowed to occur and why the government's response was so slow and inept. Until yesterday, that is, when President Bush blithely announced at a photo-op cabinet meeting that he, personally, was going to "find out what went right and what went wrong." We can't imagine a worse idea.

No administration could credibly investigate such an immense failure on its own watch. And we have learned through bitter experience—the Abu Ghraib nightmare is just one example—that when this administration begins an internal investigation, it means a whitewash in which no one important is held accountable and no real change occurs.

Mr. Bush signaled yesterday that we are in for more of the same when he sneered and said, "One of the things that people want us to do here is to play a blame game." This is not a game. It is critical to know what "things went wrong," as Mr. Bush put it. But we also need to know which officials failed—not to humiliate them, but to replace them with competent people.

It's obvious, for instance, that Michael Brown has met the expectations of those who warned that he would be a terrible director of the Federal Emergency Management Agency. This is no time to be engaging in a wholesale change of leadership, but in Mr. Brown's case there seems to be precious little leadership to lose. He should be replaced with someone who can do the huge job that remains to be done.

But the questions go way beyond Mr. Brown—starting with why federal officials ignored predictions of a disastrous flood in New Orleans—and the answers can come only from an independent commission. We agree with the Senate minority leader, Harry Reid, Senator Hillary Clinton and others who say that such a panel should follow the successful formula of the 9/11 commission: bipartisan leadership and members chosen by the White House and both parties in Congress on the basis of real expertise. It should have subpoena power and a staff expert enough to find answers and offer remedies.

Mrs. Clinton has also proposed pulling FEMA out of the Homeland Security Department and restoring its cabinet-level status. That is premature. The current setup makes sense, at least in theory. The nation should not have to support two different bureaucracies for dealing with sudden disasters.

Before throwing the system into chaos again, an investigation should determine whether the problem lies in the structure or in execution. Yesterday, The Wall Street Journal showed how the Bush administration had systematically stripped power and money from FEMA, which had been painfully rebuilt under President Bill Clinton but had long been a target of Republican "small government" ideologues. The Journal said state officials had been warning Washington—as recently as July 27—that the homeland secretary, Michael Chertoff, was planning further disastrous cuts.

This page supported the creation of Mr. Chertoff's department. But it was poorly run by the first secretary, Tom Ridge, with his maddening color-wheel alerts.

It is clearly in need of a hard look and perhaps serious reorganization. Senators Susan Collins, Republican of Maine, and Joseph Lieberman, Democrat of Connecticut, have plans for hearings, which is fine. But they created the department in the first place and may have more of a stake in the outcome than a panel of impartial experts.

The panel should also look at the shortcomings of local officials and governments. It was chilling, to put it mildly, to read Mayor Ray Nagin's comment in The Journal that New Orleans's hurricane plan was "get people to higher ground and have the feds and the state airlift supplies to them."

But disasters like this are not a city or a state issue. They concern the entire nation and demand a national response—certainly a better one than the White House comments that "tremendous progress" had been made in Louisiana. We're used to that dismissive formula when questions are raised about Iraq. Americans deserve better about a disaster of this magnitude in their own country.

Mr. ENGEL. Thank you. One of the things that we ought to have is we ought to have an independent commission to investigate what happened. The panel should follow the successful formula, as the New York Times says today, of a September 11 Commission, bipartisan leadership, members chosen by the White House and both parties in Congress on the basis of real expertise. It should have subpoena power and a staff expert enough to find answers and offer remedies. We cannot allow the administration to investigate itself to have a whitewash and a coverup.

Now, as soon as the enormity of the approaching storm became clear, obviously preparations should have been immediately ramped up. It wasn't. FEMA failed miserably. Provisions and assistance should have been ready so that, hours after the storm moved on, food, water, medical supplies would be on their way. We must not ignore the mistakes that have been made. We must fix them immediately and learn from them for the future. And I want to add my voice to the other members who have said that it is now again unconscionable to have these huge Medicaid cuts. As hundreds of thousands of people have lost their jobs and net worth, it is more clear than ever how much our citizens need Medicare and to be flexible and responsive in times of crisis.

Now, we need to look to the future. For years I have been talking about the need to wean ourselves off of oil because we have to rely on sheikdoms that are either unstable, unfriendly to the U.S., or even supporters of terrorism. We need to improve the fuel economy of passenger cars and SUVs to a level of our advanced technology that makes it possible, not issue CAFE standards as the administration did last month, which do nothing to improve fuel efficiency.

I hope that this committee will continue to hold hearings, and I hope that we will get to the bottom, again, not because of the blame game, but the people of the United States particularly in those three States affected deserve nothing less, and I thank you.

Mr. WHITFIELD. Thank you. And I would remind the members that these opening statements are 3 minutes. And we do have a number of witnesses today and we have a lot of other people. So I would urge you to try to confine yourself to 3 minutes.

At this time I would recognize the gentleman from Pennsylvania, Mr. Pitts.

Mr. PITTS. Thank you, Mr. Chairman. Thank you for holding this hearing. I would like to thank the panelists for coming. And first, again, our thoughts and prayers are with those undergoing this disaster.

Mr. Chairman, we need to look at price gouging today, and we need to encourage fuel efficiency and new technology and conservation. But we also need to look at refining capacity in our deliberations. And I would like to make a few comments on that issue.

There are 149 oil refineries in the United States. And before the disruption of Hurricane Katrina, the tragedy that occurred in the gulf coast, they were all running at full capacity. But we have yet to build any new refineries, despite the fact that our aging system—none has been built in 30 years—cannot handle the increasing demand that we are placing on them. ABC News reported last month that, "analysts say just a few new big refineries could produce enough extra gasoline to make a dent in prices."

The problem, according to ABC's report, is that building even the smallest refinery is an uphill task. Faced with a complicated morass of local and State and Federal regulations, as well as residents who do not want a refinery in my back yard, companies simply are not willing to shoulder the cost of complying with regulations or fighting protracted legal battles over land use, and so the problem remains.

Rising gas prices are the result of supply problems. And supply problems are the result of refining capacity that cannot keep pace with demand. And this is most apparent during times of crisis, such as we face now. Hurricane Katrina knocked out a significant portion of our refining capacity. Because we have been unable to build refineries in other areas of the country, our economy must wait until these refineries come back on line.

We need only to look as far as Arizona to see the obstacles that the government has placed in front of those trying to build new refineries. The Maricopa Refining Company received a permit to build a 50,000 BPD refinery on January 16, 1992. MRC, operating under the name of Arizona Clean Fuels, continued to develop its refinery project through the 1990's, and because of delays presented by the government, lost a significant investor; in 1999 the project scope was changed, and ACF applied for a new permit. That permit, however, was lost in red tape as the EPA and other agencies squabbled about whether a refinery could be built on the originally proposed site. The permit application is still under review as ACF attempts to hit the moving target presented by bureaucrats, EPA, and Federal regulations.

This story is not unusual. It is not an anomaly. It is common. And it is one reason we are facing these shortages. No one is suggesting that we sacrifice environmental stewardship to power SUVs. However, we must face the reality that our economy, whether we have SUVs or not, needs oil to run. And while there might come a day, and I hope this day comes, when we find a suitable alternative to oil and gas, we are still far away from discovering or developing a source of energy as potent or reliable as oil. So we must find an environmentally responsible way to increase our refining capacity. We simply cannot go any longer without expanding our capacity to refine oil.

Since my time is up, I will submit the rest of my statement for the record. I look forward to the hearing today, and thank the witnesses for sharing their expertise.

[The prepared statement of Hon. Joseph R. Pitts follows:]

PREPARED STATEMENT OF HON. JOE PITTS, A REPRESENTATIVE IN CONGRESS FROM THE STATE OF PENNSYLVANIA

Mr. Chairman, thank you for holding this hearing and thank you to the panelists for coming.

Our thoughts and prayers are will those enduring this disaster.

We need to look at price gouging today, and we need to encourage fuel efficiency, new technology, and conservation.

But we also need to look at refining capacity in our deliberations.

I'd like to make a few comments on that issue.

Today, there are 149 oil refineries in the United States.

Before the disruption of Hurricane Katrina and the tragedy that occurred on the Gulf Coast, they were running at full capacity.

But we have yet to build any new refineries despite the fact that our aging fleet—none have been built in thirty years—cannot handle the increasing demand we are placing on them.

ABC News reported that last month that "analysts say just a few new big refineries could produce enough extra gasoline to make a dent in prices."

The problem according to ABC's report is that building even the smallest refinery is an uphill task.

Faced with a complicated morass of local, state, and federal regulations as well as residents who do not want a refinery "in my back yard," companies simply are not willing to shoulder the costs of complying with regulations or fighting protracted legal battles over land use.

So, the problem remains.

Rising gas prices are the result of supply problems.

Supply problems are the result of refining capacity that cannot keep pace with demand.

This is most apparent during times of crisis such as we face now.

Hurricane Katrina knocked out a significant portion of our refining capacity.

Because we have been unable to build refineries in other areas of the country, our economy must wait until these refineries come back on line.

We need only look as far as Arizona to see the obstacles the government has placed in front of those trying to build new refineries.

The Maricopa Refining Company received a permit to build a 50,000 BPD refinery on January 16, 1992.

MRC, operating under the name Arizona Clean Fuels continued to develop its refinery project through the nineties.

However, because of delays presented by the government, it lost nificant investor.

In 1999, the project's scope was changed and ACF applied for a new permit.

That permit however was lost in red tape as the EPA and other agencies squabbled about whether the refinery could be built on the originally proposed site.

The permit application is still under review as ACF attempts to hit the moving target presented by bureaucrats at the EPA and federal regulations.

This story is not unusual.

It's not an anomaly.

It's quite common.

And it's one reason why we're facing these shortages.

No one is suggesting that we sacrifice environmental stewardship to power SUVs.

However, we must face the reality that our economy, whether we have SUVs or not, needs oil to run.

And while there might come a day—and I hope this day comes—when we find a suitable alternative to oil and gas, we are still far away from discovering or developing a source of energy as potent or reliable as oil.

So, we must find an environmentally-responsible way to increase our refining capacity.

We simply cannot go any longer without expanding our capacity to refine oil.

Even if we wanted to import more oil or produce more, it wouldn't matter.

This harms our ability to respond to increased demand or deal with crises that disrupt oil refining.

Our economy depends on a reliable and affordable source of energy.

Frivolous and costly regulations make it impossible to build new refineries.

Whatever their intent, these regulations harm the economy and drive up the price of gas more than they protect the environment.

There must be a middle-ground between no regulation and so many regulations that consumers suffer.

We can find that middle ground and build new refineries while still protecting the environment.

I look to hearing today.

Thank you again to the witnesses for sharing their expertise.

Mr. WHITFIELD. I thank you very much, and at this time recognize the gentlelady from Colorado, Ms. DeGette, for her opening statement.

Ms. DEGETTE. I believe Mr. Strickland——

Mr. WHITFIELD. Well, I was told at the time the gavel went down that Mr. Strickland was not here at that time, and that we are going down the order of appearance.

Ms. DEGETTE. All right. Thank you, Mr. Chairman. Thank you, Mr. Strickland.

I have been sitting here listening to everybody, and I agree with a lot of what everyone has said. We are angry and sickened by what happened on the gulf coast, and we all hope that we can get as much help as we can. It looks like maybe a million people have either lost their homes or their loved ones or both, and their lives will be changed forever. But as we assess the damage and we bury the dead and we begin to rebuild, we also really do have to have a full accounting of the actions. And it would frankly be political malfeasance of us not to do that, which is why it is good we are having the hearing today.

I think what we are seeing in Louisiana, Mississippi, and Alabama is an echo of the Federal Government's failings on September 11. We have seen an appalling lack of imagination, planning, or preparation for a mass casualty disaster, and an inept response to the disaster once it occurred that cost people their lives. Now, all of this was supposed to be solved when we created the Department of Homeland Security. And instead, it seems to me like things just got worse. My constituents are flooding my office with calls saying that the Federal Government failed Americans in their time of need. And I know that this is common to all of us in this room.

So what we need to talk about in this hearing is within this committee's jurisdiction: What can we do to fix the problems and make sure we can minimize disasters in the future. And I don't mean the disaster of the hurricane. I mean the disaster of the response.

Just talking about energy for a minute, because that is what this hearing is about, the Nation faced a surge in gas prices in the hours and days after Hurricane Katrina. In my district of Denver, Colorado, far from the eye of the hurricane, we saw gas prices going up almost hourly at some of these pumps. And I know that there were some disruptions in service in the Southeast and mid-Atlantic, expectedly so, prices expected to jump everywhere in the country without reason.

I went to the briefing, as many of us did last night. The members of the cabinet briefed the Members of Congress about what happened. And it was all very Pollyannish and everything was going well. What really struck me about energy was when Secretary Bodman said there were no real long-term disruptions in supply. So what I want to know is why were prices of gas skyrocketing in Colorado even though there was ample supply at that time and frankly no connection to the distribution network in the gulf? To us, this looks like price gouging, not disaster impact, and it is frankly immoral and it is illegal in a lot of places, too.

Now, we have been struggling for months with rising costs, and it has been fueled by surging worldwide demand for oil, infrastructure operating at near capacity, and also the increasing profits of oil companies. So why did we have to add to this price gouging as a result of a naturally occurring disaster? I think it is wrong. I am glad we are having this hearing. And I am very interested in hearing the testimony.

Mr. WHITFIELD. The gentleman Mr. Otter is recognized for an opening statement.

Mr. OTTER. I will pass.

Mr. WHITFIELD. Ms. Myrick.

Mrs. MYRICK. Thank you, Mr. Chairman. And thank you for the hearing today. All of us of course send all of our prayers and lots of other things that we can send to help to the Katrina victims. And we have been doing that, we will continue doing so, and I want to thank all the volunteers as well who have pitched in to help. Thank you to all the panelists who are sitting here patiently waiting. I will be brief. I just have two things I want to touch on today.

One of them is what I call price gouging, because in my own area of Charlotte, North Carolina, it was mind boggling how fast the prices rose at the pump. They no more than posted the high premium price and they were right back upping the regular. It was just a continual circle over and over again. And it is not that I don't want people to make a fair profit. Of course, that is what we are all about in America. That is not the point. I just want to make sure that people aren't arbitrarily raising their prices. And it is a serious issue that we need to examine.

Second is the oil and gas futures market. I have been concerned about this for some time because I think we can reel this in in a way that is going to have an effect on prices in the near future, not like the long-term remedies of building refineries which we also need to look at. But I have had concerns for many months that some speculators have been driving prices of gas higher than the factors of supply and demand really warrant. And I am particularly concerned about the over-the-counter market for energy derivatives which is subject to very limited oversight under the Commodities Future Trading Commission, the way I understand it.

I know there are many factors involved in the final price of gas in our neighborhood stations, such as the taxes and the refinery costs and the distribution costs and the profits, which I said before need to happen. But we need to examine what is going on here, because it appears to me that it is abusive and manipulative trading in some cases.

And so I thank you again, Mr. Chairman, for this hearing. I truly believe this gives us an opportunity to look closely at what we need to be doing for the future, because the global situation is not going to change and, as was stated before, our committee has jurisdiction over a lot of the health issues that are going to be coming up and we need to be doing those, too. And I yield my time.

Chairman BARTON. The gentlelady yields back. The gentlelady from California, Ms. Capps.

Mrs. CAPPS. Thank you, Mr. Chairman, and to our witnesses for being here today. We have all been moved by the tragedy on the gulf coast and our thoughts and prayers are with the thousands of Americans, fellow citizens of ours so painfully and personally affected. We also are thankful and need to keep thanking folks for the countless acts of heroism and selflessness, from both the area's residents and from people across the country responding to this tragedy. And now Congress has a critical role to play here in the aftermath of Katrina. I believe there are two significant areas in which Congress has major responsibility.

First, we have to provide the financial support for the people affected by Katrina. I am glad we have moved the emergency funding bill last week to start this process and there will be more funding requests coming. We are going to have to do a lot more to help these folks put their lives back together, and I hope we will work in a bipartisan fashion to do so.

And the second thing we must do is to figure out what went wrong with the Federal response and why so that it never happens in this way again. And I believe we need to do this in a bipartisan way as well. The Federal response, as has been said over and over again, was late and it was ineffective. This administration utterly failed in its responsibility to help prepare for the disaster ahead of time and to help in its aftermath. There are disasters waiting to come, so we must do this work. Hundreds of thousands of gulf coast residents have paid a very high price for our failure. The administration's actions or inactions were an insult to all Americans and simply inexcusable. I believe that Congress has an important job in investigating these shortcomings, and I hope this committee will be vigilant in pursuing this inquiry, and I am thankful that this hearing will start this process. The lives of Americans will be affected by how well we do our job and by how well the administration does its job and the private sector as well, this time and the next time. There will be a next time.

So I hope, Mr. Chairman, that this hearing is only the first of many that we can be holding, because studying the lessons of Katrina should help us to avoid similar problems in the future.

Finally, Mr. Chairman, I know there are many calls now for congressional action to address the high gas prices. There are things we should do and things we shouldn't do. For example, resuscitating the ill-conceived refinery legislation is one we shouldn't do. We do need more refineries. But as has been noted, environmental regulations aren't the problem here. So you don't need to waive them to get a refinery bill. The problem is that the refining industry makes a lot more money with a tight refining capacity. The industry doesn't want to build more refineries because it makes too much money the way things are. On the other hand, if we had cut down on some of our demand over the last decade or so, we wouldn't be in such a predicament right now. Demand reduction works, even as the President now belatedly recognized, evidenced by his call last week for conservation.

Mr. Chairman, you scheduled this hearing long before Katrina, and I would remind committee members that record gas prices were here long before Katrina hit and they will be here long after the effects of Katrina are dealt with. If we don't do something about our insatiable appetite for fossil fuel, shoving more tax breaks to industries making record profits and gutting the laws that protect our environment are simply uncalled for. It should be rejected. And I do yield back.

Chairman BARTON. The gentlelady yields back. Mr. Sullivan to make an opening statement?

Mr. SULLIVAN. Saving my time.

Chairman BARTON. The gentleman defers. Does Dr. Burgess wish to make an opening statement?

Mr. BURGESS. I will defer.

Chairman BARTON. All right. Mr. Walden?

Mr. WALDEN. Mr. Chairman, I will defer as well and save my time for the witnesses.

Chairman BARTON. All right. Mr. Otter? He defers. I think we have deferred on the Republican side, so we go to the gentleman from Pennsylvania.

Mr. DOYLE. And I will also defer.

Chairman BARTON. We have got a string going here. Mr. Allen.

Mr. ALLEN. I am afraid I am going to break the string, Mr. Chairman.

Chairman BARTON. The gentleman is recognized for 3 minutes.

Mr. ALLEN. Thank you, Mr. Chairman, for convening this hearing. The victims of Hurricane Katrina remain in our thoughts and prayers. When the 1998 ice storm crippled Maine, the Nation rallied to our aid. Maine is prepared to do the same for the people of the gulf coast in their hour of need. We are a nation that draws strength from shared adversity, and I hope that, working together, we will emerge from this terrible tragedy a stronger and more united people.

The Federal Government's response to this crisis has been, in a word, pathetic. But that response should be the subject of another hearing. Today we will be focusing on Hurricane Katrina's effect on energy prices, but let us not deceive ourselves or our constituents: Gas prices, heating oil futures, and oil company profits were at record highs before Katrina struck. We cannot blame high gas prices on Katrina alone.

From 1977 through 2002, the number of refineries in the U.S. decreased from 282 to 153. During this period of time, gasoline demand rose 27 percent. Refiners in the last decade have spent $47 billion to expand existing capacity by 13 percent, but demand has grown even faster.

Why not more refineries? The answer is profit margin. Fewer refineries mean higher profits. The strategy has worked; oil profits have soared into the billions. That may be all well and good for ExxonMobil, and for others, but what about everyone else? The increased profit margins for the oil companies are driving my constituents out of business. Small businesses in Maine are being crushed by increased gas prices, not to mention the spike that is coming in their heating oil bills.

Maine's large fleet of independent truckers are suffering and at grave risk of going out of business. Maine's fishing fleet is suffering as well.

In 1962, facing a similar threat to the Nation's economy due to the pricing practices of the Nation's steel manufacturers, President Kennedy summoned steel barons to the White House and demanded that they reduce prices. They backed down. This President, President Bush, needs to call oil company CEOs to the White House and demand sacrifice on their part. That may seem like fantasy, but it is the kind of leadership that we need today.

I would just add one other point. On Thursday, Valero's chief executive Bill Greehey, commenting on the FTC's decision last week to authorize Valero's $8 billion purchase of Premcor, said that: We are in a new era for refining where I believe you will continue to see higher highs and higher lows, among other things, for product margins.

That is what has been going on in that industry, and Mr. Chairman, that is what we need to investigate here. I yield back.

Chairman BARTON. The gentleman yields back. The gentlelady Ms. Schakowsky wish to make an opening statement?

Ms. SCHAKOWSKY. Yes, Mr. Chairman. And thank you for calling this timely hearing. Americans have been riveted to their televisions watching with shock and shame, not shock and awe, as the Federal Government failed in its primary mission, providing for the safety and security of its citizens. As reporters and camera crews brought images that look like they came from another country instead of the superpower of the world. As they were able to make it to the Superdome and convention center, Americans watched and waited in disbelief for help to arrive. For many, help came too late. This predictable and predicted catastrophe, as the Sun Times editorialized, exposed the plight of the Nation's have nots, all those Americans, not refugees from another country, but the millions of American citizens who are not part of the ownership society. Now we know what that means. If you own a car, you can escape disaster. If you own a tank of gasoline or enough money to buy a hotel room, you might survive in this ownership society.

Make no mistake, millions of Americans are angry, millions of Americans are ashamed. And, yes, no matter how they and we may be scolded for doing so, they blame the Federal Government, they blame this administration for failing to do its job, failing to prepare for this crisis, and failing promptly to deal with it. Many Americans shook their heads and asked: Is this my country? Newt Gingrich said, quote: As a test of Homeland Security, this was a failure. He said this is not a moment to defend inadequacy. End quote.

Other crises and potential crises are now looming, and we in Congress have responsibility as well to face up to that fact and deal with it. One of those is an energy crisis. The question is, are we going to act now to prevent a catastrophic energy crisis, or will we wait to scramble to pick up the pieces in the aftermath? This time, the President and the Congress have to anticipate a breach in the levees. In my view, we already squandered an opportunity to look ahead and mitigate an energy crisis that leaves our country at the mercy of hurricanes and vulnerable oil rigs and oil refineries and foreign countries when this committee and this Congress passed an energy bill that the President's own experts said could increase prices at the pump.

Days before Katrina struck, the price of a barrel of crude was $66, double what it was in January 2004. In Chicago the price was already nearly $3 a gallon, the highest in the country. Katrina exacerbated a preexisting condition. Now we must assure that immediate needs are met and that we look ahead at the cost and availability not only of gasoline, but, as the cold weather approaches, heating oil and natural gas. How are the poor, several of whom because of Katrina now have to face, going to stay warm. And what about middle-class families, small businesses, and farmers? Our constituents can't afford $1,000 monthly heating bills.

Can we look that far ahead and plan? In the aftershocks of Katrina, can we leave Americans out in the cold while energy companies are left with money to burn?

I hope that no member has the audacity to suggest that weakening environmental standards or drilling in the Arctic wilderness or any other transparent political fix will alleviate this energy crisis. The only way to mitigate this pending catastrophe is for Congress, with this great committee taking the lead, to be bold enough to enact laws that will hold down costs, prevent profiteering off the backs of the American people, and protect those who are hit hardest by increases in energy costs.

Thank you, Mr. Chairman.

Chairman BARTON. We thank the gentlelady. Mr. Radanovich, would you like a statement?

Mr. RADANOVICH. Waive.

Chairman BARTON. Okay. The gentlelady from California, Ms. Solis.

Ms. SOLIS. Thank you, Mr. Chairman. Yes, I have a statement.

As we sort through the issues surrounding recovering from Katrina, it is important for us to remember many of the communities that are still suffering right at this moment. First responders and other emergency personnel volunteers and even firemen from the local D.C. Area are training and working to continue finding survivors and evacuating the rest.

I am glad that we are here today to begin to address this issue. As we begin to learn, these evacuees and emergency responders are at increased risk for disease and infection caused by the mix of contaminants in the water they are wading through, particularly to those engaging in rescue and recovery missions, but also to those who lived in the Superdome and struggled through the water to escape the city. All, regardless of race, income, ethnicity, and country of origin, must receive adequate health care and treatment and counseling, mental counseling. I hope, Chairman Barton, that we will have a hearing to better understand the health implications of this hurricane. And I am also extremely concerned about the environmental and drinking water infrastructure implications of Hurricane Katrina.

On Sunday, on Meet the Press, Secretary Chertoff commented: We are going to have to clean probably the greatest environmental mess we have ever seen in this country.

Today's Washington Post identified just the beginning of the environmental problems the gulf coast will be facing. These include contaminated water which will likely be undrinkable for many years to come, unknown damage to the drinking water infrastructure, toxic fumes from fires which continue to burn. State authorities announced a litany of contaminants which are likely to be found in the flood waters, including tens of millions of pounds of concrete, lumber, cars, and animal carcasses. Sewage treatment plants were destroyed. Two major spills sent 78,000 barrels of oil into a local lake there, and fuel from 2,000 fuel tanks and leaking gasoline from flooded cars and boats also coated the city.

As ranking Democrat on the subcommittee with jurisdiction over these environmental hazards, I call on the Chair of the subcommittee, Mr. Gillmor, and Chairman Barton to begin hearings on the environmental implications of Hurricane Katrina. It is critical that as we move forward to clean up we rebuild New Orleans, that it be done in a manner which will protect the health and safety of our communities. So I encourage our colleagues not to disregard public health and environmental regulations.

And, last, with respect to the gasoline prices, we do have to have a thorough investigation here. In California for the last 3 months we have experienced high rates of gasoline prices far beyond the $3 mark. We need to do something now. We need to call in all, all resources that we can to look at what kind of price gouging has gone on.

I also would like to submit that there are several refineries that are dormant right now in our country. We should probably be going back and looking at those current refineries and trying to provide assistance there so we can startup and provide the kind of assistance that our consumers are waiting for.

Thank you very much, Mr. Chairman.

Chairman BARTON. We thank the gentlelady.

Does Mr. Fossella wish to make an opening statement? The gentleman defers. Does Mr. Gonzalez wish to make an opening statement?

Mr. GONZALEZ. Yes, Mr. Chairman. And I will be quick. I think what we see today as far as any shortages, price increases, and such, and the crisis that we face truly are just symptoms of underlying policies that have been inadequate and unrealistic. I think we need to start off with a firm understanding, if we are going to do something that is realistic and substantive, that there are no quick fixes, first of all; that there should not be any sacred cows. All of us represent a sacred cow or two. And, of course, it is not going to be pain-free. And that means for the industry and for the consumer. And if we believe we can get away with any kind of substantive policy changes that will address these problems without what I have just said, and that is the sacred cows and foregoing some of those interests, and that there is not going to be some pain felt by every American, then we will not accomplish what needs to be accomplished.

I think the American public will grasp certain concepts that we will discuss here today and that witnesses will touch on, such as production capacity on the domestic side. The location of where we have our facilities, they will understand that. And, again, just on the capacity side. But will they really understand other things that really come into this mix and I think have already been referred to by Congressman Shadegg? And we are talking about the futures market. How many Americans understand the futures market, the oil futures market? Or hedging? What does all that mean to them? What it means is exactly what is happening to them today when it comes to the volatility of the marketplace.

And with that in, Mr. Chairman, I hope that we will realistically promote policies that take all of this into account. Thank you.

Chairman BARTON. Does the gentlelady from Wisconsin wish to make an opening statement? The gentlelady is recognized for 3 minutes.

Ms. BALDWIN. Yes. Thank you, Mr. Chairman. My thoughts are with all of those who are suffering the effects of Katrina and also with those who are suffering the consequences of a painfully slow and uncoordinated response to Katrina.

I keep asking myself how a country that has spent the last 4 years planning for catastrophe found itself so ill-prepared for this catastrophe. There is a huge public call to assign blame to the planners and to name the stunning vacuum of leadership from this President and his FEMA Director immediately following this disaster. I know there is also an effort to subdue congressional critique and inquiry at least while the rescue and relief operation is still ongoing.

We have been urged to focus on the present and the future. But how can we do that properly without understanding the past? Our history? And which decisions, both recent and in the more distant past, have exacerbated and intensified last week's natural disaster?

Last week showed us and all of America, and in fact the world, many things, among them that our social safety net has been badly neglected. It showed us that we have been inadequate stewards of the environment, whether it is our failure to fight poverty and provide health care to all in America or our failure to protect the natural buffers, the coastal wetlands, the barrier islands which serve as Mother Nature's shock absorbers, the failure to make proper and adequate investments in infrastructure, including our emergency communication infrastructure, our failure to listen to scientists long warning us of climate change, or our failure to embark upon a path that decreases rather than increases our dependence on finite resources so that future generations won't experience the fear and anxiety that grips all of our constituents when fuel becomes unaffordable. All of this was stunningly revealed last week.

Let us not ignore what was exposed. I have talked about the public calls for blaming the planners. In a real way, we on this committee and in this Congress are planners, planners for the future. This time, let us seize the opportunity to work for the common good, to help those with the least, not just those with the most, and to make good upon the social compact.

Mr. Chairman, I look forward to working with you on these very big challenges.

Chairman BARTON. I thank the gentlelady.

Does Mr. Ross wish to make an opening statement?

Mr. ROSS. Yes, sir, Mr. Chairman. In fact, I just left a conference call which I will soon be joining again with our Governor of Arkansas, who is housing about 60,000 of our neighbors from Mississippi and Louisiana. As you can imagine, we have a lot of challenges that we want to meet, and we want to be there for them and lend a helping hand.

I have grave concerns about the response time in the aftermath of this hurricane and subsequent flooding and levee failures as it relates to FEMA, and I believe that we need to make FEMA a cabinet level position and remove it from Homeland Security. We have some short- and long-term needs that are going to have to be met for the people of Mississippi, Louisiana, and Alabama. I believe that we must have a bipartisan commission, much like the 9/11 Commission, to figure out what went wrong and how to avoid this from happening in the future. But there is time for those things. Right now is the time I believe to try and restore order in New Orleans, to help the people of these three States get their lives back together, and obviously the challenge of recovering the bodies that remain in the devastation of this hurricane.

But today, this hearing before the Energy and Commerce Committee is, quite frankly, about dealing with the aftermath of Katrina as it relates to energy, to gasoline supply and prices, and so let me say this: That over the August district work period I traveled the fourth district, in fact about 8,000 miles worth of traveling in my district, listening to the concerns of constituents about rising gas and diesel prices. I heard this before the hurricane. Obviously, it was compounded by the hurricane. I witnessed firsthand already inflated gas prices jump from $2.45 a gallon to $3.25 a gallon in communities throughout Arkansas. These are the very towns and communities our neighbors from Mississippi and Louisiana and Alabama have traveled to seeking shelter. Many citizens in my rural congressional district commute over 100 miles round trip for work each day. Many farmers in my district face hardships in operating the necessary equipment, especially in this drought, to harvest their crops due to high diesel prices.

These citizens, as well as those impacted by the hurricane in Mississippi and Louisiana and Alabama, simply cannot afford these drastic increases in fuel prices.

We need to ensure the people of this country that oil market manipulation and price gouging are not occurring; and, if the Federal Trade Commission's ongoing investigations do find manipulations, we need to move swiftly and effectively to punish those taking advantage of this situation. Oil production platforms, import terminals, pipelines, and refineries were all affected as a result of Hurricane Katrina. The full impact that Hurricane Katrina will have on oil markets will depend on how quickly these facilities will be able to recover to pre-hurricane status.

And, Mr. Chairman, finally let me just encourage this committee to work to do all it can in a bipartisan way to bring down the high cost of gasoline, to maintain an adequate supply while also meeting the needs and challenges of the people that have been directly impacted by this horrible natural disaster. And, with that, I yield back the balance of my time.

Chairman BARTON. We thank the gentleman.

Does the gentleman from Massachusetts, Mr. Markey, wish to make an opening statement?

Mr. MARKEY. I do, Mr. Chairman.

Chairman BARTON. The gentleman is recognized.

Mr. MARKEY. Thank you, Mr. Chairman. When the price of gasoline is $2.50 at the pump and 3 days later $3.50 is what is being charged to consumers as they are tipped upside down in gas stations across America and having money shaken out of their pockets, then there is profiteering, there is price gouging which happens.

The President should have announced that he was deploying the Strategic Petroleum Reserve on the first day of the crisis last week, at the very beginning, not 5 days later after the oil speculators were able to take advantage of consumers all across our country. The President was at least 5 days late in deploying the Strategic Petroleum Reserve and asking for help from our allies around the world. We only have 3 percent of the oil reserves in the world. God put most of the reserves under certain Middle Eastern countries, but we put 70 percent of that oil into our gasoline tanks.

The Republican energy bill which was passed and signed by a Republican President was an historic failure. It did not deal with the issue of fuel economy standards for SUVs and automobiles, it did not have a renewable portfolio standard so that all utilities in America increased dramatically their use of renewables. I believe that what we should be doing right now is suspending all royalty relief for oil and gas companies across America, and giving that relief to the victims of Hurricane Katrina.

Here is the oil company profits over the last 3 years. It has just skyrocketed. And there are estimates that ExxonMobil can make upwards of $40 billion this year. We just gave $10 billion for relief down in the gulf area. $40 billion for one company. And the prices, the prices at the pump have just skyrocketed over the last very brief period of time. And no relief, no answer from this administration. In fact, they admit that their bill does nothing.

So we continue to pollute our air, we continue to have increases in climate change, we continue to see our wetlands disappear, we continue to turn a blind eye to human rights abuses by our OPEC suppliers, we continue to not really complain about these incubators of terrorism over in the Middle East, and we continue to argue—this administration continues to argue that we need to give more royalty relief to oil and gas companies.

This hearing is a very important first step. But what we need is every CEO of every oil company, and I would also recommend OPEC ministers, come in here and that they be requested—we can't make them, but we request them to testify as to what they are going to do in order to ensure that there is an adequate supply of oil for our country.

Chairman BARTON. The gentleman's time has expired. We thank the gentleman. Does the gentleman from New Hampshire wish to make an opening statement?

Mr. BASS. I will be brief.

I want to thank you for holding the hearing. I understand that subject matter has been broadened from just talking about gasoline and energy to talking about elements relating to the latest catastrophe that has beset our Nation.

I just want to say that it is obvious that our Nation has been lulled into complacency with respect to the availability of cheap gasoline in the past; and we really haven't planned for a perfect storm like that which we are experiencing today with tight supplies, massive global demand and political unrest and then, of course, Katrina. Maybe we could never have been prepared for this kind of an event.

But I just want to say that last week I set up a special Web site that would allow constituents in the Granite State to fill out a form if they—which would be e-mailed to me directly which would outline instances of—specifically about price changes: the date, the reason, the period of time it occurred between one price and another price. And I have notified these citizens—I would say there have been well over 100 responses in the last 7 days—that I will turn all of this information over to the Department of Energy and the Federal Trade Commission so that the appropriate action be taken, if it is justified.

You know, I think it is appropriate to look into the processes whereby oil is extracted from the ground, transported, marketed, delivered and so forth and see if there are areas that require policy attention. But I personally find it difficult to define the term "price gouging," and I will be interested to hear what our witnesses have to say about that.

Mr. Chairman, I will stop my statement here so that we can get on with the important testimony that we are about to hear from the people who are here today. Thank you.

Chairman BARTON. Does the gentleman from Ohio wish to make an opening statement?

Mr. STRICKLAND. Yes, sir.

Mr. Chairman, I was reading in the New York Times comments from an engineering professor at the State University of Louisiana who had served as a consultant on the Louisiana State evacuation plan. He said that little attention has been paid to the evacuation of New Orleans's low mobility population: the elderly, the infirm and the poor without cars or other means of fleeing the city, about 100,000 people.

Mr. Chairman, we knew this disaster was upon us days before it reached our shore. In fact, the President went on television and urged people to evacuate the city. We saw the TV pictures of cars lining the freeways as they were heading northward out of harm's way. But apparently there were many in New Orleans and elsewhere along the Hurricane's path that did not have cars, that did not have credit cards. They had no means of renting an automobile for transportation. They could not afford a bus ticket. They simply were left behind. They were the poorest among the region's population.

Then the flooding came; and these—the sickest, the poorest, the oldest, along with children—have died. And the sad truth is that many have died unnecessarily. Many have died simply because they lacked for water, for food; they lacked for timely medical attention.

Mr. Chairman, we are the greatest and most advanced Nation on the face of the earth. We have at our disposal every resource that is known to mankind. Yet when disaster hit our own country, when our own citizens were without food, water and medical care, we did not respond in a timely manner. So many were lost. And those who lost their lives were primarily black, and they were primarily poor, and that should strike at the conscience of every one of us.

Mr. Chairman, one of the things that must be done is for us to examine ourselves as a Federal Government and as a people why is it, why is it that it is the poor, the minority, the child, the elderly and the infirm who are most likely to suffer in times of disaster?

But we are here today to talk about our Nation's energy. I noticed that the President said a few days ago he had zero tolerance for looters. This Nation is waiting for the President to speak so strongly about gougers. Will he tell us and will he tell the oil company executives that he has zero tolerance for gouging at the gas pumps?

I yield back the remainder of my time.

Chairman BARTON. The gentleman yields back 3 seconds. We appreciate it.

Does Mr. Terry wish to make an opening statement?

Mr. TERRY. No.

Chairman BARTON. Does Mrs. Blackburn wish to make an opening statement.

Mrs. BLACKBURN. Mr. Chairman, having been in Mississippi with my family in the middle of this for the last few weeks I will submit my statement and look forward to hearing from our panel.

[The prepared statement of Hon. Marsha Blackburn follows:]

PREPARED STATEMENT OF HON. MARSHA BLACKBURN, A REPRESENTATIVE IN CONGRESS FROM THE STATE OF TENNESSEE

Mr. Chairman, during the past decade Congress waged debate over whether to increase domestic oil exploration and encourage construction of new refineries.

Hurricane Katrina made it abundantly clear that this nation can no longer engage in seemingly endless debate, but must actively work to discover and harvest American oil. We must encourage construction of new refineries. This is not to say that we should end our alternative fuel research and development efforts, but that we must have a realistic view of our current consumption needs.

Over the past few decades, environmental groups and some of my colleagues across the aisle have been very successful in their efforts to stymie domestic exploration. The National Energy Policy Act which we passed this July after several years of effort took steps to ease the regulatory red tape that has prevented us from accessing domestic oil supplies and constructing refineries. In light of what we've learned from Hurricane Katrina, I'd suggest we build on that legislation and taking immediate steps to open ANWR.

Today we know that had we been able to pass that legislation years ago we'd very likely be less reliant on the gulf region's oil industries and the current price increases and periodic supply shortages would not be nearly as painful.

This is not as complicated a problem as some would have us believe. We need more domestic oil and we must increase our refining capacity. Both of those needs are within our power to address. Our energy security absolutely cannot remain so vulnerable to a single although significant natural disaster like Hurricane Katrina.

Mr. Chairman, I thank you for holding this important hearing today and I look forward to taking the steps necessary to strengthen our energy infrastructure.

Chairman BARTON. Seeing no other member present who wishes to make an opening statement, the Chair asks unanimous consent to put into the record at this point in time the statement from the distinguished chairman of the Small Business Committee, Congressman Manzullo.

Hearing no objection, so ordered.

[The prepared statement of Hon. Donald A. Manzullo follows:]

PREPARED STATEMENT OF HON. DONALD A. MANZULLO, CHAIRMAN, HOUSE COMMITTEE ON SMALL BUSINESS

As the Chairman of the Small Business Committee, I hear everyday how the price of energy affects entrepreneurs. It costs more everyday to simply turn on the lights of a business when it opens its doors. It costs more to ship merchandise, both raw material and finished products. Every consumer sees prices rising on the most basic products. Every time people add gasoline to their car, there is less money in their pocket for other purchases.

In the northern Illinois Congressional district I am proud to represent, many of my constituents have already been hurt by the increase in energy prices. Richard Beuth is a farmer in Seward, Illinois and he has told me how every facet of his operation rising fuel prices has impacted. He explains that the cost of fertilizer has doubled over the past year. A year ago at this time, fertilizer cost $250 a ton and when it increased to $350 a ton he wondered how he would absorb this increase. Now, in the span of a year, it has increased to $500. Also, he tells me there is a scarcity of fertilizer on the market for purchase.

Additionally shipping costs are hurting his farming operation. It costs more to buy seeds, fertilizer and other products have increased because of shipping costs. Farmers also get hit with shipping costs as they send their crops to market. Richard explained to me that the cost to farm an acre has increased between 15 and 20 percent. He explains that because the price of corn is down, he is operating at a loss.

Richard Todd runs Todd Transfer Trucking of Rockford, Illinois. Richard tells me that the cost of fuel is tipping the balance of the scales to where he is operating without any profits. His company normally runs on a margin of two to three percent and fuel increases are eating into that margin. He says that his fuel prices are up $90,000 in the last six months, with all other factors remaining the same.

Additionally, his company is forced to absorb more of the costs because their biggest customers will not accept a fuel surcharge. He is only able to pass on his additional fuels costs or surcharges to his smaller customers, who themselves are struggling.

Another example is Bob Trojan of Rockford Linear Actuation, a hydraulic cylinder manufacturer. Bob explains that the cost to run his plant has increased between 20 and 30 percent because of soaring fuel prices. He says that he fully expects to the price of steel to rise, because of production costs, which will again increases his input costs.

Because he is a small manufacturer, trucking companies have passed surcharges on to his shipments. He ships his products all over the United States and Europe. Bob says that customers and vendors are more cautious in taking trips to see products because travel costs are so expensive.

Rising energy costs hurt every aspect of business. These costs are spiraling out of control. It is imperative that we find ways to curtail energy prices before businesses are forced to close their doors. The economic aftershocks of Hurricane Katrina affected communities all across this nation. Mr. Chairman, I trust that as oil refineries come back on-line in Louisiana and production is restored to pre-storm levels, we will see a decrease in energy costs so that our small businesses and small manufacturers will remain open. Any change in U.S. government policy to address the high cost of energy should keep in mind the perspective of small business because this sector of our economy generates most of the new jobs and economic growth in this country. Thank you, Mr. Chairman, for allowing me this opportunity.

[Additional statements submitted for the record follow:]

PREPARED STATEMENT OF HON. MICHAEL C. BURGESS, A REPRESENTATIVE IN CONGRESS FROM THE STATE OF TEXAS

First, I want to thank Chairman Barton for convening this hearing today. The Chairman has indicated that this will be the first of a series of hearings to examine impact of Hurricane Katrina.

In the wake of Hurricane Katrina, many of my constituents have contacted me with concerns about price gouging by gasoline retailers.

It is tempting to be led by emotion and make quick decisions in order to show that we are “doing something.” But I believe that the best thing to do in this situation is to study the issue as deliberately as possible.

In the Dallas-Fort Worth area gasoline prices increased by anywhere between 30-50 cents per gallon in the last week alone. I know Chairman Upton indicated that he received reports that one point that gasoline increased by a dollar per gallon overnight in Michigan.

At the same time, we know that Hurricane Katrina resulted in the suspension of 25 percent of U.S. oil production and took 25 percent of U.S. refining capacity offline. Since domestic oil and gas refineries have operated at nearly 100 percent capacity over the last few years, the loss of even one U.S. refinery would have reduced supply and increased prices at the pump.

We need to determine if the problem is inappropriate pricing or a problem with supply.

We need to make sure to fix the right problem. Trying to fix the wrong problem can only make things worse—we all remember the long lines at the gas pump in the 1970s.

If it is determined that illegal pricing has occurred, I will support prosecution of wrongdoers to the utmost of my ability. I think it is unconscionable that opportunists would take advantage of this national tragedy for financial gain.

But, it is important that we, as policy makers, avoid single synapse reactions which can translate into untenable public policy. We should examine the strategy in place for dealing with this type of emergency situation; and if no such strategy exists, we should work to develop one. We need to learn from this experience and determine how we can prevent this loss of supply in the future.

In conclusion, I'd like to again thank the Chairman for holding this hearing. I look forward to hearing from the witnesses who are appearing before us.

PREPARED STATEMENT OF HON. BART STUPAK, A REPRESENTATIVE IN CONGRESS FROM THE STATE OF MICHIGAN

Thank you Chairman Barton and Ranking Member Dingell for calling this hearing today and welcome to the witnesses. This hearing comes at a time when our southern states are struggling with the ramifications of a devastating natural disaster and the rest of our Nation is being hit with gas and oil prices that are the highest this Country has ever seen.

I hope that the witnesses here today will provide information that will help this Committee understand why these high prices are occurring and when the American people can expect a reprieve.

As every member of the Committee knows first hand, our constituents are angry about high gas prices—and they have a right to be!

More than 5,000 complaints have been logged in at the Energy Department's gas-price gouging hotline, and there have been reports of pump prices hitting $6 per gallon in some parts of the country where gas prices were in the range of $2.50 per gallon. Furthermore, oil prices have reached as high as $70 per barrel.

My home state has been hit particularly hard by these prices, especially in my Northern Michigan district. With tourism as the largest industry in the region during the summer months, gas prices have taken a toll on small businesses this year even before Katrina.

Many Northern Michigan residents who must commute to distant communities for work find themselves putting in the first hour or two just to pay for the gas to and from their job.

The fact is we rely on oil to fuel our cars, homes and economy, so we're forced to pay the price. Like my constituents, I find myself at the pump feeling completely helpless to stop the seemingly endless rise in cost.

Many weeks prior to Hurricane Katrina's wrath, gas prices had been increasing at an alarming rate. These prices have been pinching the pockets of the middle class for well over a year.

During this time, the Administration did nothing to curb these rising prices despite the urging of myself and other Members of Congress to defer shipments of oil and tap into the Strategic Petroleum Reserve.

Instead, the Administration chose to wait until Hurricane Katrina put the country in dire straights before releasing this desperately needed oil. Had they immediately released this oil, the situation after Katrina might not have been so grim.

However, supply of oil is not the only factor that is affecting gas prices. The United States refining capacity is concentrated in the southern states and some of those states have unfortunately been devastated by Katrina.

While many existing refineries have expanded operations, it has barely kept pace with demand. Most U.S. refineries are operating at or near 94 percent capacity, and the United States now has to import 10 percent of already refined gasoline.

For the current refiners, this limited capacity keeps gasoline prices high and profits up. But for the consumer, and the overall health of the American economy, it's a potential disaster.

Hurricane Katrina is not the only cause of these independent gas prices. This Administration's foreign policy has also directly influenced the price of gas we are paying today. Poorly planned foreign policy and mismanaged international diplomacy have created major instability in the oil rich regions of the world, including Venezuela which has been one of our largest suppliers.

This pervasive instability in the Middle East and the Iraq war has lead to a commodity futures market where an additional $10 to $15 "risk premium" is added to each crude oil contract sold on the market. This directly correlates to higher prices at the pump for consumers and only increases with concerns over natural disasters such as Hurricane Katrina.

I hope the witnesses here today will address these issues. Thank you.

PREPARED STATEMENT OF HON. ALBERT R. WYNN, A REPRESENTATIVE IN CONGRESS FROM THE STATE OF MARYLAND

Chairman Barton and Ranking Member Dingell, thank you for holding this hearing at such a critical point during this state of national crisis. Hurricane Katrina has ravaged the Gulf Coast region, leaving thousands without a place to call home, crippling and separating families, and the death toll continues to rise as we press forward with relief efforts.

The Administration, state, and local officials' sluggish response and incompetence in addressing this catastrophic natural disaster are unacceptable. The lack of coordination, limited military presence, and insufficient supplies reveal a shocking ineptitude in planning for emergency situations, such as witnessed along the Gulf Coast.

In addition to the human tragedy, in the context of this committee's jurisdiction, Hurricane Katrina resulted in a significant loss of refining capacity, intensifying already high gas prices. At least 20 percent of the nation's refining capacity ceased operations or reduced runs as a direct result of the disaster. No new refineries have been built in this country since 1976, yet over the past 20 years U.S. demand for gasoline has increased over 20 percent. Correspondingly, refining capacity has decreased by ten percent over the same time period. This is an unsustainable situation and we must work towards increasing domestic refining capacity.

The current price of gasoline, when adjusted for inflation, is as high as gas prices during the 1979 crisis! What makes matters worse is that even before the hurricane, crude oil and gas prices were inching towards unprecedented heights.

Unfortunately, in the aftermath of Hurricane Katrina, the issue of price gouging has been illuminated. In the witnesses' testimony they do not address the jurisdiction of the federal government over price gouging; nor do they specify what specifically defines price gouging. These are matters that must be addressed given the current crisis. We need federal authority over price gouging; the American people should not be subjected to artificially high prices without an effective means of recourse.

We should also remember that the reduced refining capacity and supply will impact the price of home heating oil this winter. While it is currently a warm day in D.C., in a couple of months the weather will cool significantly. For low-income residents, this could mean the difference between putting food on the table and surviving the cold winter. Given the high prices in energy, we should ensure that our low income heating assistance is sufficient to support the inflated prices.

Adjustments in international supply and demand of oil, continually increasing refining capacity, speeding up our energy independence from foreign sources of oil, and ultimately shifting towards a hydrogen economy are necessary to address America's energy crisis.

I look forward to hearing the testimony of those present today and I thank you all for speaking with my colleagues and I about the aftermath of Hurricane Katrina and her impact on oil and gas production.

PREPARED STATEMENT OF HON. JIM DAVIS, A REPRESENTATIVE IN CONGRESS FROM THE STATE OF FLORIDA

Mr. Chairman and Members of the Committee, thank you for holding this hearing on the effect that Hurricane Katrina has had on the supply and prices of gasoline in the United States. Last month, Congress passed a massive energy bill that did nothing to address the price of gas at the pump. It is unfortunate that it took this natural disaster to finally get Congress to act on addressing the price consumers pay at the pump.

Today, we are here to specifically look at the economy of gasoline—a system that has proven to be fundamentally flawed, even before Katrina devastated our coasts.

It is my hope that we will uncover long term solutions to address the supply and pricing of gasoline in this process of Congressional hearings and investigations that are sure to come.

We have all heard the stories of price gauging at the pump—I urge the witnesses at this hearing today to heed the call of this Congress and keep in mind that America is watching your actions closely to ensure that suppliers of gasoline are not unfairly profiting in this time of national crisis.

Gasoline is unlike any other US commodity. The fuel that every American relies on in one way or another is impacted by a global market of supply and demand which alters the prices of products from Tupperware to jet fuel. The refining process is one of the steps in the pricing of gasoline. Some Members of the House of Representatives argue that relaxing the laws that protect consumers will assist in easing this severely bottlenecked market. I don't believe that approach will really address the problem or tell us why the years of bottlenecks and tight supplies have allowed the refineries to maintain and increase their profit margins. The Members of this Committee must be aware that opportunistic exploitation of consumer protections will not be tolerated by the American people.

Today, it is as clear as it has ever been that we are unable to drill our way out of oil and gas dependence. While solutions to short term disruption solutions are needed at this time to help address the impact of Hurricane Katrina, we must look to solutions that will affect the long term energy markets and keep in our minds the economic, environmental and homeland security of our children's and grandchildren's futures. I am eager to work with the Members of this committee towards the goal of finding real solutions to obstacles that will be outlined in the witness's testimonies.

I also look forward to hearing from the courageous Governors of Louisiana and Mississippi. All of America appreciates their leadership through these difficult and trying times for their state and our nation. Floridians are said to have PhD's in hurricane preparedness, and it is with this knowledge that we will help in the recovery process for Hurricane Katrina in any way we can.

Mr. Chairman, I urge the Committee to put partisan politics aside; we Floridians cannot wait through a year of hearings and investigations to find out what failures occurred in the preparation and response to Katrina as hurricane season is not yet over for us. We must get to the bottom of this as no state stands to gain or loose from learning the lessons of Katrina and the aftermath quickly.

Mr. Chairman, again, I thank you and the Members of this Committee for the opportunity to discuss gas prices today and look forward to working with you on this and many other issues in the future.

Chairman BARTON. The Chair now is going to recognize its first panel. We appreciate your patience, gentlemen.

I think it shows the importance of this hearing that, of the 56 members of the Energy and Commerce Committee, 45 have made an appearance and, of those 45, over 30 have made opening statements. That shows the concern the country has about what has happened and is worried and concerned about the response of the Federal Government to the catastrophe. So I do thank you again for your patience.

We are going to recognize David Garman, who is the Under Secretary for Energy, Science and Environment at the Department of Energy for 7 minutes; and then we will go to Mr. Caruso, Mr. Seesel and Mr. Moran. Thank you, gentlemen, for waiting.

Secretary Garman, you are recognized for 7 minutes.

STATEMENTS OF HON. DAVID K. GARMAN, UNDER SECRETARY FOR ENERGY, SCIENCE AND ENVIRONMENT, DEPARTMENT OF ENERGY; HON. GUY F. CARUSO, ADMINISTRATOR, ENERGY INFORMATION ADMINISTRATION; JOHN H. SEESEL, ASSOCIATE GENERAL COUNSEL FOR ENERGY, FEDERAL TRADE COMMISSION; AND KENNETH P. MORAN, ACTING DIRECTOR, OFFICE OF HOMELAND SECURITY, ENFORCEMENT BUREAU, FEDERAL COMMUNICATIONS COMMISSION

Mr. GARMAN. Mr. Chairman and members of the committee, apart from the human dimension, which weighs heavily on all of our minds, Hurricane Katrina had a devastating impact on energy infrastructure, prices and markets.

For example, Katrina shut in roughly 1.4 million barrels of crude oil production, roughly 95 percent of all U.S. Gulf production. Katrina halted 25 percent of all gulf coast refining, or approximately 2 million barrels per day. Ten refineries were totally shut down, and six were reduced in their runs. Some undamaged refineries suffered from crude oil supply shortfalls.

Approximately 2.7 million households were without electricity at one point, in addition to the loss of electricity to refineries and pipelines. Three major pipelines carrying crude and petroleum products to large portions of the Nation were out of service, with estimates of repairs ranging from days to weeks.

Mindful of these impacts, we did the following:

First, the Department of Energy, within 48 hours of receiving the first requests, approved loans of crude oil from the Strategic Petroleum Reserve to refineries, 12.6 million barrels as of yesterday afternoon. Some of that oil is being delivered as we speak.

Second, the President has authorized the release and sale of oil from the Strategic Petroleum Reserve. The Department has issued the formal notice of sale of an initial 30 million barrels of oil yesterday. Bids will be opened on Friday, assessed over the weekend, and we should be in a position to issue a notice of apparently successful offerers by Monday, September 12. Oil will flow from that release and sale as soon as the winning bidders provide for the oil's transportation.

Third, the Department has worked with the International Energy Agency to coordinate the release of an additional 33 million barrels of crude oil and refined product from reserves of our nations to provide additional supply to global markets.

Fourth, DOE's Office of Electricity Delivery and Energy Reliability began working with other Federal, State and local officials, utilities, municipalities, power marketing administrations and cooperatives even before the storm struck to accelerate the restoration of power. This was important not only to the affected populations but the Nation as a whole since the refineries and the product pipelines depend on this power to deliver gasoline, diesel and other petroleum products to demand.

Fifth, the Environmental Protection Agency issued a nationwide waiver to allow the use of winter blend reformulated gasoline in stock to increase the flow of refined products to consumers. EPA is also allowing the use of diesel fuel with sulfur content exceeding the 500 parts per million limitation.

Sixth, the Department of Homeland Security temporarily waived Jones Act restrictions on transportation fuel supplies by tanker.

Seventh, the Treasury Department announced that off-road diesel would be permitted for road use to bring more diesel into the market during this emergency.

Eighth, the Department of Transportation waived the restriction on hours that can be driven by truckers to keep goods and services and products, including energy, moving.

Ninth, the President and the Secretary have repeated their calls for all Americans to use energy wisely. Energy efficient practices, exercised by millions of American consumers, can have a substantial impact.

Tenth, the Navy and Coast Guard are surveying and, as necessary, clearing shipping channels of sunken obstructions that would affect gasoline or crude shipments.

While we still face a difficult situation, there are encouraging developments. Four hundred thousand barrels of the lost production in the gulf have already been restored. The latest assessments from the Department of the Interior suggest that 99 percent of prior platform production will eventually be restored.

Of the 10 refineries that were completely shut down, we expect four to be operational within the next week or so. Of the six that went to reduced runs, all are expected to be fully operational by tomorrow. I am informed that five more that were undamaged are now receiving supplies of oil from the Strategic Petroleum Reserve. Of the 2 million barrel a day of refining capacity lost, roughly half that capacity is now back on line.

Meanwhile, we understand that over 20 tankers carrying gasoline are currently en route to the United States from Europe. I am also informed that, as of this morning, power has been restored to over 1.8 million households of the 2.7 million households without power at peak impact. We are most grateful for the work of thousands of utility crews working nearly nonstop, many of whom who have been sleeping in their trucks while they have not been working to restore power.

Also of good news for all consumers around the Nation is the fact that all three of the major pipelines are back in service at full or nearly full capacity far sooner than most observers had predicted. In the near term, we need to get these production and refining facilities back on line while encouraging Americans to use energy wisely.

Again, millions of American consumers can have an impact by doing simple things such as consolidating trips, keeping their vehicles in tune, keeping tires at their proper inflation and by driving more slowly and smoothly.

In the longer term, new supplies of petroleum and alternatives to petroleum, including hydrogen fuel as the President proposed back in January, 2003, along with provisions of the just-passed energy bill can help us overcome these challenges.

I would be pleased to answer any questions the committee might have either today or in the future. Thank you, Mr. Chairman.

[The prepared statement of Hon. David Garman follows:]

PREPARED STATEMENT OF HON. DAVID GARMAN, UNDERSECRETARY, DEPARTMENT OF ENERGY

Mr. Chairman and Members of the Committee, thank you for the invitation to appear this morning on the subject of Hurricane Katrina and its effect on energy supply and prices.

Let me start by saying this is a tragedy of monumental proportions. Hurricane Katrina is one of the worst national disasters in our nation's history.

It has been responsible for an unknown number of deaths—possibly in the many thousands.

And for those who survived, it has utterly destroyed homes, schools, businesses and livelihoods. For them, it will be years before life returns to normal—if it ever can.

It is also the largest single disaster impacting the energy infrastructure of this country.

At the Department of Energy, our focus is on two aspects of the events in the Gulf of Mexico.

First, obviously, we are concerned about the direct impact of the storm on the residents of Louisiana, Mississippi, Alabama, Florida and other affected states.

And because the Gulf Coast plays such a critical role in supplying much of the nation's energy needs, we are also concerned about the hurricane's broader effect on the country as a who le and on international markets.

I want you to know at the outset that Secretary Bodman has committed the Department of Energy to doing everything in its power to meet the immediate needs of those affected by Hurricane Katrina—both on the Gulf Coast and throughout the rest of the country—and we have marshaled all of our resources to fulfill that commitment.

Within the last week, the Department of Energy dispatched employees to emergency response centers throughout the southeastern United States to coordinate power restoration efforts. DOE staff are working closely with state and local officials, first responders, and power companies to begin restoring power and fuel supplies as quickly as possible, wherever possible.

In the immediate aftermath of Katrina, upwards of 2.7 million customers were without electric power in Louisiana, Mississippi, Alabama, Florida, Georgia, and Tennessee. One week ago, for instance, more than 90 percent of the residents in the state of Mississippi had no electricity—including those hundreds of miles from where the hurricane made landfall.

Power has been restored in many of these areas. As of 11 a.m. yesterday, fewer than a million customers remained without electric power due to Hurricane Katrina. In Louisiana and Mississippi, 971,360 were without power. Alabama has essentially restored all customers without electric power.

In some places—and not just New Orleans—it may be weeks, perhaps months, before power can be restored. In Biloxi, Gulfport, and elsewhere on the Coast, the electricity infrastructure of transmission, substations, and distribution has been damaged or destroyed. Those capabilities must be restored and rebuilt, and this cannot be done overnight. The publicly owned municipal and cooperative utilities in these states, with the help of other utilities and contractors from many states, are undertaking the massive job of restoring the system. But it will take time.

A number of challenges are hindering this effort. One is just the massive scope of the destruction, as we have all seen on television. Not to minimize the suffering caused by the hurricanes that battered Florida last year, but Hurricane Katrina's devastation is in an entirely different category. Upon seeing what Katrina had wrought on the Mississippi coast, Governor Barbour remarked that it is what Hiroshima must have looked like 60 years ago. I don't think anyone could accuse the Governor of hyperbole.

On top of the sheer devastation caused by the storm as it passed through, the subsequent flooding in New Orleans, Mobile, and elsewhere adds further huge complications.

Well over 10,000 crews have arrived throughout the affected region to work on electricity restoration. As they finish their work in certain places, they move on to the next ones. As Florida utilities have completed their work, crews from these companies and their contractors have moved to the Gulf Coast to support restoration work. Crews have come from many states and Canada to support utility restoration.

But this is a massive area we are talking about, and a number of factors are slowing progress. At this point inaccessibility and the extensive damage from flooding and saltwater are the biggest challenges. We have heard from Entergy that its single biggest problem to restoring power in the greater New Orleans area is the lack of food and water for its repair crews, who have literally been sleeping in their trucks.

The affected states face a massive challenge, but we will work with state and local leaders, with utilities and power companies, and with anyone else to try to restore power wherever possible as quickly as possible.

While the Department works with people on the ground to restore power, we are also monitoring the effects of the storm on the nation's energy markets.

Nine refineries that supply nearly 10 percent of the nation's gasoline were shuttered by the storm.

Thousands of energy industry workers in the Gulf Coast had to be evacuated.

Oil and Gas production rigs and other infrastructure were damaged.

The pipelines supplying Gulf Coast gasoline and natural gas to the Midwest and Eastern part of the country were affected, as well. However, damage was not as severe as we at first had feared.

One week after the storm, these are back at full or near full capacity. Meanwhile, 95 percent of the nation's refining capacity should be operating by mid-September.

Despite this news, it is not clear how long it will be before energy production and distribution in the Gulf is back to normal.

The Department of Energy and the Bush Administration are very concerned about the effects of this disaster on already tight markets.

And we are concerned about the impact of higher gasoline prices on the average American.

Accordingly, we have taken a number of steps to try to alleviate the situation. Last week, the Department of Energy entered into separate agreements with several energy companies to loan more than 12 million barrels of oil from the Strategic Petroleum Reserve in order to limit disruptions in crude supplies for refineries.

The crude oil will be loaned from the SPR under short-term contractual agreements and returned to the Reserve once supply conditions return to normal.

I want to point out we have taken very quick action in this regard. Oil was on the way to refineries within 48 hours of loan requests being made.

Further, in the aftermath of the storm, we outlined the impact on our energy sector for the members of the International Energy Agency to determine whether it was necessary to supply additional crude oil and gasoline products to the market. On Friday, the members of the IEA made a historic decision to provide crude oil from each member's strategic reserves.

Under this agreement, IEA member countries have agreed to make available 60 million barrels, or an average of 2 million barrels per day, for 30 days beginning immediately. This will consist of both oil and gasoline, with an emphasis on refined product.

The United States is a member of the International Energy Agency, of course, so to meet our obligations as a member of the IEA, we will be releasing 30 million barrels of crude oil from the United States Strategic Petroleum Reserve.

In addition to these efforts, I want to add that the Environmental Protection Agency has granted a nationwide waiver for fuel blends to make more gasoline and diesel fuel available throughout the country. The EPA action will permit the early use of wintertime gasoline blends and, we expect, will take some pressure off the price of gas.

On top of this, I want to point out that the President has made an appeal to the American people to conserve gasoline during this time of tightened supply. There are a number of things that people can do to reduce their use of gasoline, such as carpooling, driving slower, bundling errands together to make fewer trips, and telecommuting.

One final point I want to make concerns the anecdotal reports all of us have heard about price gouging in various parts of the country in the days after Katrina hit. Our Department and our Administration take the subject of excessive pricing very seriously. It is unconscionable that Americans would seek to exploit a tragedy for profit.

DOE has established a web site where Americans can report gasoline price gouging. All complaints registered with the Department of Energy will be collected and transmitted to the Federal Trade Commission, U.S. Department of Justice, and individual State Attorneys General for investigation and prosecution where appropriate.

Chairman Barton—members of the Committee—I want to thank you for the opportunity to come before you this morning to apprise you of our Department's efforts in the wake of Hurricane Katrina.

I would be happy to respond to any questions you and the other members may have.

Chairman BARTON. We thank you, Mr. Secretary.

We now want to hear from the Honorable Guy Caruso, who is the Administrator of the Energy Information Administration.

Welcome, Administrator Caruso. You are recognized for 7 minutes, also.

STATEMENT OF HON. GUY F. CARUSO

Mr. CARUSO. Thank you very much, Mr. Chairman, for this opportunity to present the Energy Information Administration's views and analysis of energy markets in the aftermath of Katrina.

As you know, EIA is the independent statistical and analytical agency in the Department of Energy; and we do not promote, formulate or take positions on policy issues.

As the Chairman has mentioned, even before the tragic hurricane, crude oil and gasoline prices were already at high levels. On August 29, average gasoline prices were $2.61 per gallon, and diesel prices were $2.59 per gallon. Crude oil prices on the futures market had increased by nearly 60 percent over the same period compared with last year, due in large part to substantial growth in world oil demand, which has used up much of the world's productive capacity. Refineries have been running at high levels of utilization in many parts of the world, including the United States; and the high production of distillate fuels and higher-than-average refinery outages this summer added to the tight gasoline markets.

Natural gas markets were also tight on the eve of the hurricane and futures prices were $10.85 per million Btus or more than double year earlier levels.

Hurricane Katrina has had a significant impact, particularly on gasoline, diesel fuel and natural gas prices. For example, EIA's survey data released yesterday showed that the national average price of regular gasoline prices rose 46 cents per gallon, to $3.07, between August 29 and Labor Day, while diesel prices rose 31 cents, to $2.90 per gallon. While prices rose throughout the country, the East Coast experienced the largest price increase in both fuels.

The near-term outlook for the oil and gas markets will depend on a number of factors, most importantly the timing and pace of the recovery of the infrastructure and operations in the Gulf.

Production of both oil and natural gas in the Gulf of Mexico has already recovered substantially from the peak impacts, as Mr. Garman has pointed out in his statement.

The infrastructure has been coming back more quickly than many had expected, as Mr. Garman has mentioned, and, fortunately for natural gas markets, we are in the shoulder season, between the period of high demand for electricity generation for air conditioning and the high demand for heating fuel.

The level of natural gas in storage remains above the 5-year average, but the disruption in operations due to Katrina is likely to reduce the amount put in storage during the remainder of the injection season.

Today, we released our September Short-Term Energy Outlook; and, as you can imagine, the uncertainty in this outlook, which goes out to the remainder of 2006, is greater than ever. Nevertheless, we consider three cases in this current outlook based on the speed of recovery from the effects of Hurricane Katrina. We include in that report a slow, a medium, and a fast recovery case. The fast recovery case assumes a very favorable set of circumstances for returning operations to normal, while the slow recovery case assumes that significant impacts on oil and natural gas production and delivery continue at least into November. In all cases, normal operations are achieved or nearly achieved by December. We assume that the loans and releases of crude oil and products from Government stocks will help to offset the price increases due to Katrina.

The WTI crude oil price averaged $65 per barrel in August and reached $70 in peak trading last week. Crude oil prices have retreated from those heights in recent days, and we expect they will trend downward in the fourth quarter of 2005, although staying above $60 for the remainder of the year and into 2006.

The national average price of unleaded gasoline was $2.49 per gallon in August, with prices generally rising throughout the month. Projected gasoline prices in the near term are very sensitive to the assumptions that I have mentioned in the three cases. Gasoline prices, however, should ease in the coming weeks as supply improves. We project $2.60 per gallon gasoline in the fourth quarter and an average price for 2006 of $2.40.

The heating oil prices, however, will show a substantial increase this winter compared with last year. Assuming normal winter weather, heating oil prices are expected to be about 30 percent higher this winter compared with last winter in the medium recovery case. Of course, this assumes a normal winter.

Natural gas prices probably will be impacted even more than heating oil and are likely to stay tight over the next couple of months as the heating season begins. We anticipate the September spot price for natural gas to average about $13 dollars per million Btus and about $11.50 per million Btus in the fourth quarter. Based on the present trends, natural gas price this winter are expected to be significantly higher than last winter.

In sum, Mr. Chairman, the impact of Katrina on energy prices has been to make a tight market situation for oil and natural gas even more challenging for the industry and for consumers.

This concludes my statement. I would be happy to answer any questions now or at any time that you deem appropriate, sir.

[The prepared statement of Hon. Guy Caruso follows:]

PREPARED STATEMENT OF HON. GUY CARUSO, ADMINISTRATOR, ENERGY INFORMATION ADMINISTRATION, U.S. DEPARTMENT OF ENERGY

Mr. Chairman and Members of the Committee: I appreciate the opportunity to appear before you today to discuss gasoline prices in the United States and recent developments in world oil markets.

The Energy Information Administration (EIA) is the independent statistical and analytical agency within the Department of Energy. We are charged with providing objective, timely, and relevant data, analysis, and projections for the Department of Energy, other government agencies, the U.S. Congress, and the public. We do not take positions on policy issues, but we do produce data and analysis reports that are meant to assist policymakers determine energy policy. Because the Department of Energy Organization Act gives EIA an element of independence with respect to the analyses that we conduct and publish, our views should not be construed as representing those of the Department of Energy or the Administration.

The devastation of Hurricane Katrina included offshore production, refineries, and loss of power to run pipelines and otherwise-working refineries. Damage assessments are ongoing but still incomplete. With the current tight global petroleum market, gasoline and distillate prices have risen sharply. How far and how long they remain elevated will depend on the severity of damage to petroleum facilities. Our understanding of the situation is rapidly evolving, and I will discuss this in my oral remarks. This written testimony focuses on events prior to the hurricane and challenges to gasoline markets following the recovery.

Even prior to Hurricane Katrina, petroleum prices, including gasoline, were setting new records as crude oil prices climbed. Gasoline prices as of August 29 were $2.61, which was 73 cents per gallon higher than a year ago, and, on average for the month, were 58 cents per gallon higher. Yesterday's prices, which will be released late this afternoon, will undoubtedly be—much higher given the significant disruptions experienced due to Hurricane Katrina. A consumer who drives about 1,000 miles per month in a car that gets about 20 miles per gallon paid almost $30 more for that car's fuel during August this year than last August. Businesses and government budgets are also affected, as it costs more to fill their vehicle fleets.

The remainder of this testimony describes the fundamentals affecting petroleum prices, focusing on crude oil and gasoline. The underlying market situation today, even before Katrina, is one in which the spare crude oil production, refinery, and tanker capacities that existed for more than a decade prior to 2003 were reduced more quickly than EIA or other analysts anticipated. Little spare capacity, both upstream and downstream, not only supports higher prices, but they also add to price volatility, since any upset to supply/demand balances regionally cannot be resolved quickly. Restoring spare capacity will not be easy or rapid, because an increase in capacity takes time and investment, and growing demand will require capacity increases just to maintain current cushions, which suggests that high prices and potential volatility will be with us for some time.

Changes in the gasoline price at the pump are driven mainly by changes in crude oil prices and changes in wholesale gasoline prices. Crude oil cost represented nearly 60 percent of the gasoline price this summer and explains much of the variation in gasoline price. Crude oil prices are driven and set by international markets. The wholesale price of gasoline or its spot price is influenced first by crude oil but also by seasonal demand variations and by regional refinery and distribution supply and demand balances. Retail price changes generally lag behind wholesale price changes.

INTERNATIONAL CRUDE OIL MARKETS

Turning to crude oil prices first, Figure 1 shows that the current crude price increase began in 2004, when crude oil prices almost doubled from 2003 levels, rising from about $30 per barrel at the end of 2003 to peak at $56.37 on October 26, 2004. After falling back briefly, prices then continued to rise in 2005.

This is a significant change from what we experienced during much of the 1980s and 1990s. For most of the time since the early 1980s, we have lived in a market in which spare crude oil production, refining, and delivery system capacity existed. Crude oil suppliers outside of the Organization of Petroleum Exporting Countries (OPEC) produce at maximum rates (i.e., no surplus production capacity) for economic reasons, thus, the world's surplus crude oil production capacity resides in OPEC (mainly Saudi Arabia). The large growth in non-OPEC capacity and production in areas like the North Sea and Alaskan North Slope, along with softening demand from high prices, led to major cuts in OPEC production in the 1980s, creating large capacity surpluses. As demand grew through the 1990s, OPEC production increased, but new productive capacity was not added. Short-term imbalances between supply and demand occurred and we experienced some price swings, but those imbalances did not last long, as capacity generally existed to remedy the situation within a year.

During most of the 1990s, the West Texas Intermediate (WTI) crude oil price averaged close to $20 per barrel, but plunged to almost $10 per barrel in late 1998 as a result of the Asian financial crisis slowing demand growth, at the same time as extra supply from Iraq was entering the market for the first time since the Gulf War. OPEC producers reacted by reducing production, and crude oil prices not only recovered, but increased to about $30 per barrel as demand grew in the face of OPEC production discipline.

Beginning in 2004, world oil demand growth accelerated significantly. For the 10 years prior to 2004, world oil demand growth had averaged 1.2 million barrels per day. But in 2004, world demand jumped by 2.6 million barrels per day, led by an unprecedented increase in demand from China of about 1 million barrels per day, compared to that country's increase of 0.4 million barrels from 2002 to 2003. This unusually rapid demand growth along with growth in the United States and the rest of the world, quickly used up much of OPEC's available surplus crude oil production capacity (Figure 2). As the world balance between supply and demand tightened considerably, ongoing supply uncertainties associated with Russia, Iraq, and Nigeria added to market concerns over the availability of crude oil, and prices rose. In 2005, Iran, Ecuador, and Venezuela added new uncertainties.

Global oil demand is expected to grow more slowly during 2005 and 2006, increasing by about 1.7 to 1.8 million barrels per day. China's demand is projected to increase by 0.5 million barrels per day and U.S. demand by 0.4 million barrels per day in 2006. Together, these two areas are projected to account for about 50 percent of the world's petroleum demand growth next year.

Crude oil production capacity increases are expected to keep up with these demand increases. Production increases from OPEC members are projected to represent almost one-third of the world production growth next year, and the former Soviet Union is expected to provide an additional 40 percent of the increase. Other areas such as the United States and other non-OPEC countries will provide additional production volumes. However, EIA is not projecting much increase in the surplus capacity cushion any time soon. Spare capacity is projected to remain at or below 1.2 million barrels per day in 2005.

We are facing tight crude oil markets for a number of years. EIA's *Short-Term Energy Outlook* is projecting WTI crude oil prices to remain above $55 through 2006. Even if demand softens or capacity is developed faster than anticipated, statements from OPEC members indicate an intention to keep prices from falling below $50 per barrel. While high relative to recent years, the price of crude oil, adjusted for inflation, is still below the levels seen in the early 1980s.

This tight balance results in different behavior and price implications than exhibited by the short-term market imbalances seen for the past 20 years. Instead of high prices being accompanied by low inventories and expectations for prices to be falling quickly in the future, today, in both crude oil and product markets, we see high prices with high inventories. Consumers exhibit similar behavior when they expect to experience higher prices in the near future. For example, consumers top off their gasoline tanks before a bad storm that could limit supplies and drive prices up in their region.

Prior to Hurricane Katrina, crude oil prices increased about 39 cents per gallon in summer 2005 over summer 2004, while gasoline prices only increased 34 cents per gallon (Figure 3). Although refinery and distribution and marketing contributions to gasoline prices were on average lower this summer on average than last summer, seasonal and local supply conditions affected these refinery contributions to price gasoline more strongly at the end of the summer, as described next.

U.S. PRODUCT MARKETS

Tightening in other parts of the supply chain beyond crude oil exacerbated product price increases in the United States and in the rest of the world. World refining capacity utilization increased from 85 percent to 87 percent from 2003 to 2004, driven in large part by increases in demand and utilization in areas like China and India. While adequate refining capacity is available to meet demand today, the refining system cannot shift quickly to meet unexpected needs. With refinery capacity running at high utilization levels in many parts of the world, including the United States, product balancing is frequently done through international trade, which means products must travel long distances, stretching out the time it takes to resolve imbalances. This sluggish response puts additional pressure on product prices beyond the effect of high crude oil prices and can result in price spikes if a regional shortage evolves.

Product markets in the United States provide an example of various supply and demand balancing effects on price. In the United States, the spread between wholesale product prices and crude oil prices is often higher in spring and summer than during the rest of the year. Gasoline is the highest volume product refineries produce, and spring and summer are when gasoline demand is typically the highest. Gasoline spreads typically increase at this time of year, lifting overall refinery margins to their highest seasonal level. Distillate product (diesel and heating oil) spreads are usually lower in spring and summer, but they represent only about half as much volume as gasoline production.

U.S. petroleum product price spreads were very unusual in spring and summer 2005. Wholesale gasoline price spreads through July were slightly above the average for the past 5 years, but lower than spreads seen in 2004. Heating oil and diesel spreads were unprecedented, exceeding gasoline spreads from April through July. This unusual distillate market was seen throughout the world as distillate demand grew rapidly and ultra-low sulfur diesel demand in Europe pulled on tight supplies. Distillate prices remained above gasoline prices in Europe as well as Asia. This unusual distillate market ultimately affected gasoline.

Gasoline and distillate products are produced together at the same refineries. In the spring, the U.S. inventories for gasoline were high and prices were lower than for distillates. Distillate inventories were low, and the price incentives caused refiners to respond by producing unusually high yields of distillate, which resulted in reduced gasoline yields. The consequence was that U.S. distillate inventories rose from below normal to above normal, and gasoline inventories fell from above normal to normal into July.

In addition to the switch in yield patterns, unplanned refinery outages in July and August added to the tightening gasoline market. The high demand summer season is when U.S refiners run close to or at full utilization rates, but outages always occur. The degree of outages varies, and preliminary data indicate a higher level than average occurred in July and August of this year. Had refineries been able to run at the same utilizations as last year, they would have run about 200 thousand barrels per day more crude oil, and the gasoline inventories in the July/August period would now be in the middle of their seasonal range, even with the higher-than-usual distillate yields.

The loss of supply and rapid decline in gasoline inventories starting in July resulted in an increase in gasoline price spreads (Figure 4). Higher gasoline spreads encourage more gasoline imports, and some refiners may have shifted yields to produce more gasoline, but with the peak summer driving season at an end, and winter heating needs ahead, we would expect a continued focus on maximizing production of distillates.

The high level of refinery outages in July and August increased pressure on gasoline prices, adding possibly 8 to 15 cents per gallon. Wholesale prices were poised to decline as some of the refinery problems were being resolved, but then the Gulf Coast was hit by Hurricane Katrina. Both spot market prices and near-month futures prices for gasoline and distillate products have risen dramatically in the days following the hurricane. Retail prices, which follow wholesale prices with a lag, are also rising. We expect that prices will begin to fall back as production and refining capacity are restored, although the pace of restoration is at present highly uncertain. While the gasoline price and supply situation will also be helped by the seasonal decline in U.S. gasoline demand after Labor Day, seasonal trends in crude oil markets will work in the opposite direction as world crude oil demand begins to increase in the fall with the onset of the Northern Hemisphere heating season.

Looking ahead to next summer, high crude oil prices are expected to continue to support high prices for all petroleum products, including gasoline. In addition, gasoline prices may see some additional pressure since the industry is moving quickly to eliminate methyl tertiary butyl ether (MTBE). While the removal of the oxygen content requirement in the recently-enacted Energy Policy Act of 2005, without some accompanying liability protection, may have hastened companies' decisions to remove MTBE, companies were moving in that direction anyway. Removing the oxygen content requirement will help consumers in the long run by providing more supply options for refiners and blenders. In the short run, however, the loss of gasoline production capability and some potential sources of gasoline imports that will occur when phasing out MTBE cannot be made up easily. The distribution system will also have to adjust, depending on how the industry shifts. The result is that we may see increased volatility during the transition, as we have seen with other fuel specification transitions.

In addition to potential supply problems due to removal of MTBE, the United States will begin the ultra-low sulfur diesel program. In June 2006, suppliers will begin providing diesel fuel to the on-road market that contains less than 15 parts per million sulfur. Following a full recovery from Katrina, production capability to produce ultra-low sulfur diesel is felt to be adequate, but the industry is still struggling to determine how to deliver the product through its pipeline and storage tank system without contamination. Many issues remain to be resolved, implying this transition may also add pressure to the system, and can be expected to affect gasoline as well as distillate prices.

Next year is also the first year of the renewable fuel standard established under the new energy bill, and while meeting the total volumes of ethanol required under this standard should not be difficult, a credit trading program must be in place and operating smoothly to enable each gasoline supplier to meet its obligation. It is our understanding that Environmental Protection Agency (EPA) and the industry are working towards this goal, but little time exists for EPA and the industry to get everything prepared.

One more specification change slated for 2006 is the final phase of the Tier 2 low-sulfur gasoline program for refiners and importers, who will be providing gasoline with an average sulfur content of 30 parts per million or less, which is less than one-tenth the average sulfur content before the program began. With many refiners already producing gasoline at 30 parts per million, this last phase may be less challenging than the removal of MTBE and the start of ultra-low sulfur diesel. It is one more additional strain on the supply system, however. For example, if a refinery loses a desulfurization unit, the stricter specifications may result in no production of gasoline, whereas, in the past, the refinery might have been able to produce more volumes at higher sulfur levels for a longer time.

CONCLUSION

In conclusion, the world is experiencing an underlying change in petroleum markets with the development of tight supplies that will not likely change quickly. Hurricane Katrina has significantly exacerbated the near-term supply tightness, especially in the U.S. market for gasoline and diesel fuel. Even after production and refinery operations fully recover from the effects of Katrina, capacity increases will be needed throughout the supply chain to keep up with demand. Until the world returns to more spare capacity, particularly in crude oil supply, crude oil and petroleum product prices will remain high. Even if the balance should relax unexpectedly, OPEC members have expressed an interest to maintain prices well above their prior target range. While the system currently can meet demand, it cannot respond quickly to unexpected changes. We will see shifts in imbalances from one region of the world to another and from one product to another, as we saw with gasoline and distillate in the United States. The gasoline market in the United States is subject not only to the higher crude oil prices and generally tight market conditions, but also to volatility from continuing specification changes down the road, with next summer presenting a number of such specification challenges.

This completes my testimony, Mr. Chairman. I would be glad to respond to any questions you and the other Committee members may have.

Figure 1. Crude Oil Prices Rose Rapidly in 2004 & 2005

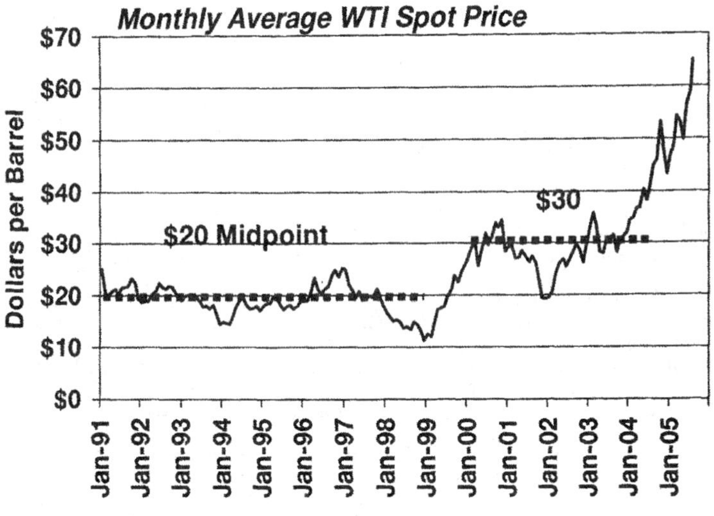

Source: Reuters

Figure 2. OPEC Spare Capacity Is Extremely Tight

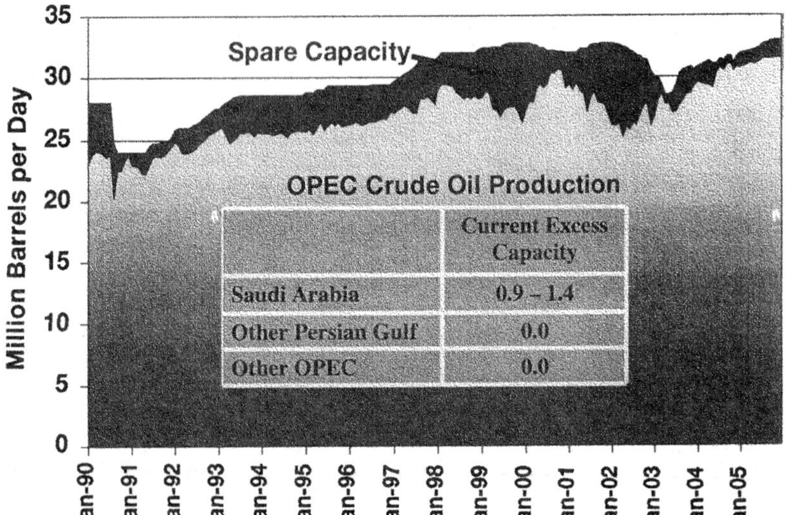

Source: Energy Information Administration estimates.

Figure 3. Crude Oil is Largest Contributor to Gasoline Price Increase Since Last Year

Components of Retail Gasoline Prices
U.S. Average, May-August

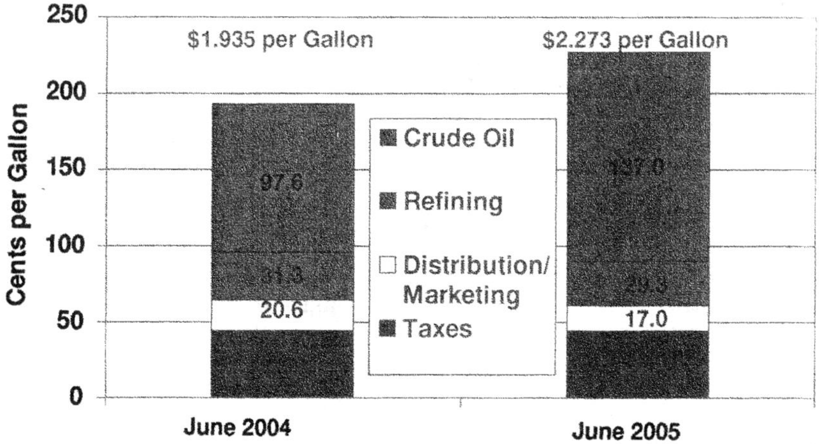

Source: Energy Information Administration

Figure 4. Gasoline Spreads High in 2004 and Soften Some in 2005

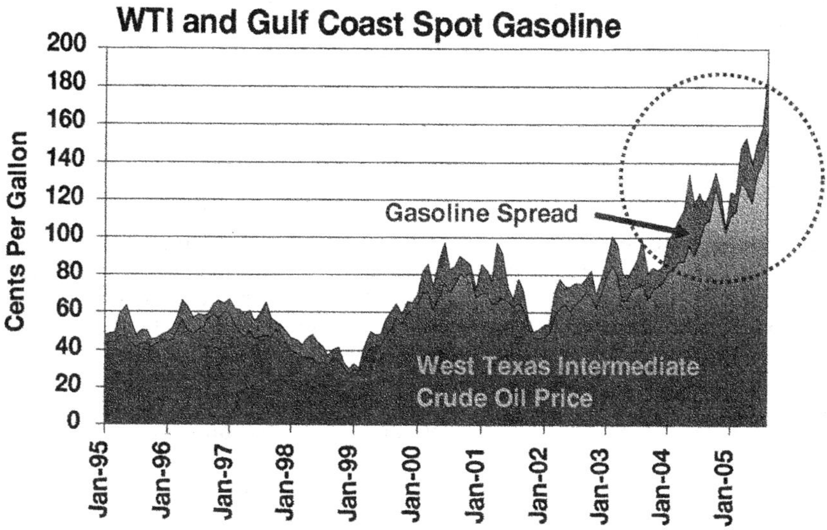

Source: Reuters

Chairman BARTON. Thank you, Mr. Director.

We now want to hear Mr. John Seesel, who is the Associate General Counsel for Energy at the Federal Trade Commission.

Welcome, Mr. Seesel; and you are recognized for 7 minutes.

STATEMENT OF JOHN H. SEESEL

Mr. SEESEL. Thank you, Mr. Chairman.

Good afternoon, Mr. Chairman and members of the committee. I am John Seesel, the Associate General Counsel for Energy at the Federal Trade Commission. I am pleased to have this opportunity to discuss the FTC's actions to promote competition in the petroleum industry and to protect consumers who use gasoline, diesel fuel and the other petroleum products so vital to our Nation's economy.

The Nation, indeed the world, continues to witness the heart-rending destruction and misery that Hurricane Katrina left in its wake. The Commission mourns the loss of life and the many other tragedies that have unfolded in the past 10 days in the States along the gulf coast.

Today's hearing focuses on one of Katrina's most important economic consequences, the storm's impact on the Nation's gasoline supply and on gasoline prices. I want to assure this committee that the FTC is acutely aware of the pain that high gasoline prices that we have experienced recently has caused American families and businesses, and we are continuing our intense scrutiny of conduct in the petroleum industry in the aftermath of Katrina.

The FTC will proceed aggressively against any violations of the antitrust and consumer protection laws that it enforces. We are on high alert.

The Commission is committed to maintaining competitive markets in refined petroleum products. We achieve this objective through a three-pronged approach: vigorous law enforcement against anti-competitive mergers and business behavior, careful study of various developments with competitive implications for the petroleum industry, and an ongoing project to monitor gasoline and diesel prices in order to detect unusual price movements.

A significant recent development in the FTC's law enforcement program is the issuance of dual consent orders in late July designed to remedy the anti-competitive effects of Unocal's allegedly deceptive conduct in connection with the development of reformulated gasoline in California, as well as the alleged anti-competitive effects that were anticipated from Chevron's acquisition of Unocal.

The Commission's first complaint alleged that Unocal had deceived the California Air Resources Board, CARB for short, in developing standards for reformulated gasoline. The Commission challenged Unocal's misrepresentation that certain technology was in the public domain while it pursued patents on that technology to enable it to charge substantial royalties. The proposed merger between Chevron and Unocal raised the concern that if Chevron had acquired Unocal's patents, Chevron could have obtained sensitive information and thus could have used this information and power to facilitate coordination among competitors to raise gasoline prices.

The two consent orders embodying Chevron's commitment not to enforce the Unocal patents provided a significant victory for consumers. The Commission has estimated that the main relief provided by these orders could save California gasoline consumers around $500 million dollars per year. The FTC will continue its energetic enforcement of the antitrust laws against collusive and monopolistic practices in this country.

In aid of its extensive law enforcement work, the FTC also conducts careful research on key competitive issues in the petroleum industry. I especially commend our recent report on gasoline price changes to the committee's attention. The report sets forth in detail the numerous supply, demand and competitive factors that influence gasoline prices or cause gasoline price spikes.

The report shows that the market for gasoline functions as any other market is expected to when supply is significantly constrained and demand keeps rising. As important, the report also shows that market forces in the form of changes in how much gasoline can producers supply and consumers demand can ameliorate price increases.

A related FTC study issued last year was our staff report on mergers, structural change and antitrust enforcement in the petroleum industry over the last 20 years.

The third prong of our approach is a continuous effort by our staff to identify unusual gasoline and diesel price movements. Our economists monitor daily pricing data from 20 wholesale and approximately 360 retail areas across the Nation. If the statistical model that they apply detects any unusual price movement that cannot be explained by refinery outage, a pipeline break or another business related cause, the FTC staff, in consultation with other Federal and State officials, will examine whether a law violation has occurred.

The Commission is acutely aware of the escalating prices that consumers pay for gasoline, and we will examine any information that we receive about pricing to determine whether there is a basis for legal action under the anti-collusion and anti-monopoly statutes that the FTC enforces. For those complaints that are not a violation of Federal law, the State attorneys general appear to have begun major multi-State initiatives to pursue any such complaint under State statutes.

The energy industry, especially the petroleum sector, has been a centerpiece of FTC antitrust enforcement for decades, and the Commission expects to devote substantial resources to policing the competitiveness of the industry in this time of economic duress for many of our fellow citizens. Moreover, as it always does, the Commission will give State and local officials as much assistance as it can as those authorities carry out their responsibilities.

Thank you again for the opportunity to present the FTC's views, Mr. Chairman. I will be happy to answer any questions.

[The prepared statement of John H. Seesel follows:]

Prepared Statement of John H. Seesel, Associate General Counsel for Energy, Federal Trade Commission

I. INTRODUCTION

Mr. Chairman and members of the Committee, I am John Seesel, the Federal Trade Commission's Associate General Counsel for Energy. I am pleased to appear before you to present the Commission's testimony on FTC initiatives to protect competitive markets in the production, distribution, and sale of gasoline, and to discuss an important recent Commission study on the factors that affect gasoline prices.[1]

The petroleum industry plays a crucial role in our economy. Not only do changes in gasoline prices affect consumers directly, but the price and availability of gasoline also influence many other economic sectors. No other industry's performance is more deeply felt or carefully scrutinized.

Gasoline prices are among the most visible prices in our complex economy. Consumers closely follow gasoline prices, and in recent months these prices have experienced dramatic increases. In recent weeks, prices of gasoline have exceeded $3.00 a gallon in some markets. Despite higher prices, demand for gasoline continues to grow, increasing at a 1.6 percent rate over the most recent four-week period for which data are available (August 19), over that same period for last year. Gasoline inventories remain at the lower end of the average range. These rising prices command our attention.

On top of this tight market, Hurricane Katrina has temporarily disrupted an important source of crude oil and gasoline supply. At one point, over 95 percent of Gulf Coast crude oil production was shut in, and numerous refineries and pipelines were either damaged or without electricity.[2] Because of this massive supply disruption, price relief has been and will be delayed.

The FTC has been and remains vigilant regarding anticompetitive conduct in this industry. Recent activity includes, on June 10, 2005, the acceptance of two consent orders that resolved the competitive concerns relating to Chevron's acquisition of Unocal and settled the FTC's 2003 monopolization complaint against Unocal. The Unocal settlement alone has the potential of saving billions of dollars for consumers nationwide in future years. In addition, in early July 2005, the Commission published its study of the factors that affect gasoline prices.[3] This study grew out of conferences of industry, consumer, academic, and government participants held by the Commission over the past four years, as well as years of research and experience, and sheds light on how gasoline prices are set.

In 2004, the FTC staff published a study reviewing the petroleum industry's mergers and structural changes as well as the antitrust enforcement actions the FTC has taken.[4] Commission enforcement statistics show that the agency has taken action against proposed mergers in this industry at concentration levels lower than in other industries. Since 1981, the FTC has filed complaints against 19 large petroleum mergers. In 13 of these cases, the FTC obtained significant divestitures. Of the six other matters, the parties in four cases abandoned the transactions altogether after our respective antitrust challenges; one case resulted in a remedy requiring the acquiring firm to provide the Commission with advance notice of its intent to acquire or merge with another entity; and the sixth case is ongoing.

In addition to litigation and industry studies, the Commission also protects consumers through other initiatives. The Commission actively monitors wholesale and retail prices of gasoline.[5] Three years ago, the FTC launched an initiative to monitor gasoline prices to identify "unusual" movements in prices[6] and then examine whether any such movements might result from anticompetitive conduct that violates Section 5 of the FTC Act. FTC economists developed a statistical model for identifying

---

[1] This written statement represents the views of the Federal Trade Commission. My oral presentation and responses to questions are my own and do not necessarily represent the views of the Commission or any Commissioner.

[2] *See* Minerals Mgmt. Serv., U.S. Dep't of the Interior, Release No. 3328, *Hurricane Katrina Evacuation and Production Shut-in Statistics Report as of Tuesday, August 30, 2005* (2005), *at* http://www.mms.gov/ooc/press/2005/press0830.htm.

[3] Federal Trade Commission, Gasoline Price Changes: The Dynamic of Supply, Demand, and Competition (2005) [hereinafter Gasoline Price Changes], *available at* http://www.ftc.gov/reports/gasprices05/050705gaspricesrpt.pdf.

[4] Bureau of Economics, Federal Trade Commission, The Petroleum Industry: Mergers, Structural Change, and Antitrust Enforcement (2004) [hereinafter Petroleum Merger Report], *available at* http://www.ftc.gov/os/2004/08/040813mergersinpetrolberpt.pdf.

[5] *See* FTC, *Oil and Gas Industry Initiatives, at* http://www.ftc.gov/ftc/oilgas/index.html.

[6] An "unusual" price movement in a given area is a price that is significantly out of line with the historical relationship between the price of gasoline in that area and the gasoline prices prevailing in other areas.

such movements. The agency's economists daily scrutinize price movements in 20 wholesale and approximately 360 retail markets across the country. In no other industry does the Commission so closely monitor prices.

This gasoline monitoring and investigation initiative focuses on the timely identification of unusual movements in gasoline prices (compared to historical trends) to determine if a law enforcement investigation is warranted. If the FTC staff detects unusual price movements in an area, it researches the possible causes, including consultation, if appropriate, with the state Attorneys General, state energy agencies, and the Department of Energy's ("DOE") Energy Information Administration. The FTC staff also monitors DOE's gasoline price "hotline" complaints. If the staff concludes that the unusual price movement likely results from a "natural" cause (i.e., a cause unrelated to anticompetitive conduct), absent other evidence of potential anticompetitive conduct, it does not investigate further (although it continues to monitor).[7] The Commission's experience from its past investigations and the current monitoring initiative indicate that unusual movements in gasoline prices typically have a natural cause. FTC staff further investigates unusual price movements that do not appear to be explained by "natural" causes to determine whether anticompetitive conduct may be a cause. Cooperation with state law enforcement officials is an important element of such investigations.

The Commission's testimony today addresses the Committee's inquiries in two parts. It first reviews the basic tools that the Commission uses to promote competition in the petroleum industry: challenging potentially anticompetitive mergers, prosecuting nonmerger antitrust violations, monitoring industry behavior to detect possible anticompetitive conduct, and researching petroleum sector developments. This review of the Commission's petroleum industry agenda highlights the FTC's contributions to promoting and maintaining competition in the industry. The Commission places a premium on careful research, industry monitoring, and investigations to understand current petroleum industry developments and to identify accurately obstacles to competition, whether arising from private behavior or from public policies. The petroleum industry's performance is shaped by the interaction of extraordinarily complex, fast-changing commercial arrangements and an elaborate set of public regulatory commands. A well-informed understanding of these factors is essential if FTC actions are to benefit consumers.

The second part of this testimony reviews the learning the Commission has derived from its conferences and research and its review of recent gasoline price changes. Among other findings, this discussion highlights the paramount role that crude oil prices play in determining both the levels and the volatility of gasoline prices in the United States. Changes in crude oil prices account for approximately 85 percent of the variability of gasoline prices.[8] When crude oil prices rise, so do gasoline prices. Crude oil prices are determined by supply and demand conditions worldwide. The supply of crude is strongly influenced by production levels set by members of the Organization of Petroleum Exporting Countries ("OPEC"). Demand has increased substantially over the past few years, both in the United States and in the developing economies of China and India. When worldwide supply and demand conditions result in crude oil prices in the range of $70 per barrel, it is not surprising that we see higher gasoline prices nationwide.

II. FTC ACTIVITIES TO MAINTAIN AND PROMOTE COMPETITION IN THE PETROLEUM INDUSTRY

A. Merger Enforcement in the Petroleum Industry

The Commission has gained much of its antitrust enforcement experience in the petroleum industry by analyzing proposed mergers and challenging transactions that likely would reduce competition, thus resulting in higher prices.[9] In 2004, the Commission released data on all horizontal merger investigations and enforcement actions from 1996 to 2003.[10] These data show that the Commission has brought more merger cases at lower levels of concentration in the petroleum industry than

---

[7] Natural causes include movements in crude oil prices, supply outages (e.g., from refinery fires or pipeline disruptions), or changes in and/or transitions to new fuel requirements imposed by air quality standards.

[8] *See* GASOLINE PRICE CHANGES, *supra* note 3, at 13.

[9] Section 7 of the Clayton Act prohibits acquisitions where the anticompetitive effects may occur "in any line of commerce or in any activity affecting commerce in any section of the country." 15 U.S.C. § 18.

[10] Federal Trade Commission Horizontal Merger Investigation Data, Fiscal Years 1996-2003 (Feb. 2, 2004), Table 3.1, et seq.; FTC Horizontal Merger Investigations Post-Merger HHI and Change in HHI for Oil Markets, FY 1996 through FY 2003 (May 27, 2004), *available at* http://www.ftc.gov/opa/2004/05/040527petrolactionsHHIdeltachart.pdf.

in other industries. Unlike in other industries, the Commission has obtained merger relief in moderately concentrated petroleum markets.

Several recent merger investigations illustrate the FTC's approach to merger analysis in the petroleum industry. The most recently completed case involved Chevron's acquisition of the Union Oil Company of California ("Unocal"). When the merger investigation began, the Commission was in the middle of an ongoing monopolization case against Unocal that would have been affected by the merger. Thus, the Commission settled both the merger and the monopolization matters with separate consent orders that preserved competition in all relevant merger markets and obtained complete relief on the monopolization claim.[11] The nonmerger case is discussed below.

Another recent merger case that resulted in a divestiture order resolved a complaint concerning the acquisition of Kaneb Services and Kaneb Pipe Line Partners, companies that engaged in petroleum transportation and terminaling in a number of markets, by Valero L.P., the largest petroleum terminal operator and second largest operator of liquid petroleum pipelines in the United States.[12] The complaint alleged that the acquisition had the potential to increase prices in bulk gasoline and diesel markets.[13]

The FTC's consent order requires the parties to divest assets sufficient to maintain premerger competition, including certain Kaneb Philadelphia-area terminals, Kaneb's West pipeline system in Colorado's Front Range, and Kaneb's Martinez and Richmond terminals in Northern California.[14] In addition, the order forbids Valero L.P. from discriminating in favor of or otherwise preferring its Valero Energy affiliate in bulk ethanol terminaling services, and requires Valero to maintain customer confidentiality at the Selby and Stockton terminals in Northern California. The order succeeds in maintaining import possibilities for wholesale customers in Northern California, Denver, and greater Philadelphia and precludes the merging parties from undertaking an anticompetitive price increase.

Most recently, the Commission filed a complaint on July 27, 2005, in federal district court in Hawaii, alleging that Aloha Petroleum's proposed acquisition of Trustreet Properties' half interest in an import-capable terminal and retail gasoline assets on the island of Oahu would reduce the number of gasoline marketers and could lead to higher gasoline prices for Hawaii consumers.[15] Because this matter is currently in litigation, this testimony will not discuss it in any more detail.

In the past few years, the Commission has brought a number of other important merger cases. One of these involved the merger of Chevron and Texaco,[16] which combined assets located throughout the United States. Following an investigation in which 12 states participated, the Commission issued a consent order against the merging parties requiring numerous divestitures to maintain competition in particular relevant markets, primarily in the western and southern United States.[17] Among other requirements, the consent order compelled Texaco to (a) divest to Shell and/or Saudi Refining, Inc., all of its interests in two joint ventures—Equilon[18] and Motiva[19]—through which Texaco had been competing with Chevron in gasoline marketing in the western and southern United States; (b) divest all assets relating to the refining, bulk supply, and marketing of gasoline satisfying California's environmental quality standards; (c) divest assets relating to the refining and bulk supply of gasoline and jet fuel in the Pacific Northwest; and (d) divest various pipelines used to transport petroleum products.

---

[11] *Chevron Corp.,* FTC Docket No. C-4144 (July 27, 2005) (consent order), *at* http://www.ftc.gov/os/caselist/0510125/050802do0510125.pdf; *Union Oil Co. of California,* FTC Docket No. 9305 (July 27, 2005) (consent order), at http://www.ftc.gov/os/adjpro/d9305/050802do.pdf.

[12] *Valero L.P.,* FTC Docket No. C-4141 (June 14, 2005) (complaint), *at* http://www.ftc.gov/os/caselist/0510022/050615comp0510022.pdf.

[13] *Id.*

[14] *Valero L. P.,* FTC Docket No. C-4141 (July 22, 2005) (consent order), *at* http://www.ftc.gov/os/caselist/0510022/050726do0510022.pdf.

[15] *Aloha Petroleum Ltd.,* FTC File No. 051 0131 (July 27, 2005) (complaint), *at* http://www.ftc.gov/os/caselist/1510131/050728comp1510131.pdf .

[16] *Chevron Corp.,* FTC Docket No. C-4023 (Jan. 2, 2002) (consent order), *at* http://www.ftc.gov/os/2002/01/chevronorder.pdf.

[17] *Id.*

[18] Shell and Texaco jointly controlled the Equilon venture, whose major assets included full or partial ownership in four refineries, about 65 terminals, and various pipelines. Equilon marketed gasoline through approximately 9,700 branded gas stations nationwide.

[19] Motiva, jointly controlled by Texaco, Shell, and Saudi Refining, consisted of their eastern and Gulf Coast refining and marketing businesses. Its major assets included full or partial ownership in four refineries and about 50 terminals, with the companies' products marketed through about 14,000 branded gas stations nationwide.

Another petroleum industry transaction that the Commission challenged successfully was the $6 billion merger between Valero Energy Corp. ("Valero") and Ultramar Diamond Shamrock Corp. ("Ultramar").[20] Both Valero and Ultramar were leading refiners and marketers of gasoline that met the specifications of the California Air Resources Board ("CARB") and were the only significant suppliers to independent stations in California. The Commission's complaint alleged competitive concerns in both the refining and bulk supply of CARB gasoline in two separate geographic markets, the state of California and Northern California, and the Commission contended that the merger could raise the cost to California consumers by at least $150 million annually for every one-cent-per-gallon price increase at retail.[21] To remedy the alleged violations, the consent order settling the case required Valero to divest: (a) an Ultramar refinery in Avon, California; (b) all bulk gasoline supply contracts associated with that refinery; and (c) 70 Ultramar retail stations in Northern California.[22]

A final example is the Commission's 2002 challenge to the merger of Phillips Petroleum Company and Conoco Inc., alleging that the transaction would harm competition in the Midwest and Rocky Mountain regions of the United States. To resolve that challenge, the Commission required the divestiture of: (a) the Phillips refinery in Woods Cross, Utah, and all of the Phillips-related marketing assets served by that refinery; (b) Conoco's refinery in Commerce City, Colorado (near Denver), and all of the Phillips marketing assets in Eastern Colorado; and (c) the Phillips light petroleum products terminal in Spokane, Washington.[23] The Commission's order ensured that competition would not be lost and that gasoline prices would not increase as a result of the merger.

B. Nonmerger Investigations into Gasoline Pricing

In addition to scrutinizing mergers, the Commission aggressively polices anticompetitive conduct. When it appears that higher prices might result from collusive activity or from anticompetitive unilateral activity by a firm with market power, the agency investigates to determine whether unfair methods of competition have been used. If the facts warrant, the Commission challenges the anticompetitive behavior, usually by issuing an administrative complaint.

Several recent petroleum investigations are illustrative. On March 4, 2003, the Commission issued the administrative complaint referred to above, stating that it had reason to believe that Unocal had violated Section 5 of the FTC Act.[24] The Commission alleged that Unocal deceived the California Air Resources Board ("CARB") in connection with regulatory proceedings to develop the reformulated gasoline ("RFG") standards that CARB adopted. Unocal allegedly misrepresented that certain technology was non-proprietary and in the public domain, while at the same time it pursued patents that would enable it to charge substantial royalties if CARB mandated the use of Unocal's technology in the refining of CARB-compliant summertime RFG. The Commission alleged that, as a result of these activities, Unocal illegally acquired monopoly power in the technology market for producing the new CARB-compliant summertime RFG, thus undermining competition and harming

---

[20] *Valero Energy Corp.*, FTC Docket No. C-4031 (Feb. 19, 2002) (consent order), at http://www.ftc.gov/os/2002/02/valerodo.pdf.

[21] *Valero Energy Corp,* FTC. Docket No. C-4031 (Dec. 18, 2001), at http://www.ftc.gov/os/2001/12/valerocmp.pdf.

[22] *Valero Energy Corp., supra* note 20.

[23] *Conoco Inc. and Phillips Petroleum Corp.,* FTC Docket No. C-4058 (Aug. 30, 2002) (Analysis of Proposed Consent Order to Aid Public Comment), *at* http://www.ftc.gov/os/2002/08/conocophillipsan.htm. Not all oil industry merger activity raises competitive concerns. For example, in 2003, the Commission closed its investigation of Sunoco's acquisition of the Coastal Eagle Point refinery in the Philadelphia area without requiring relief. The Commission noted that the acquisition would have no anticompetitive effects and seemed likely to yield substantial efficiencies that would benefit consumers. *Sunoco Inc. / Coastal Eagle Point Oil Co.,* FTC File No. 031 0139 (Dec. 29, 2003) (Statement of the Commission), *at* http://www.ftc.gov/os/caselist/0310139/031229stmt0310139.pdf. The FTC also considered the likely competitive effects of Phillips Petroleum's proposed acquisition of Tosco. After careful scrutiny, the Commission declined to challenge the acquisition. A statement issued in connection with the closing of the investigation set forth the FTC's reasoning in detail. *Phillips Petroleum Corp.,* FTC File No. 011 0095 (Sept. 17, 2001) (Statement of the Commission), *at* http://www.ftc.gov/os/2001/09/phillipstoscostmt.htm.

Acquisitions of firms operating mainly in oil or natural gas exploration and production are unlikely to raise antitrust concerns, because that segment of the industry is generally unconcentrated. Acquisitions involving firms with de minimis market shares, or with production capacity or operations that do not overlap geographically, are also unlikely to raise antitrust concerns.

[24] *Union Oil Co. of California,* FTC Docket No. 9305 (Mar. 4, 2003) (complaint), *at* http://www.ftc.gov/os/2003/03/unocalcmp.htm.

consumers in the downstream product market for CARB-compliant summertime RFG in California. The Commission estimated that Unocal's enforcement of its patents could potentially result in over $500 million of additional consumer costs each year.

The proposed merger between Chevron and Unocal raised additional concerns. Although Unocal had no horizontal refining or retailing overlaps with Chevron, it had claimed the right to collect patent royalties from companies that had refining and retailing assets (including Chevron). If Chevron had unconditionally inherited these patents by acquisition, it would have been in a position to obtain sensitive information and to claim royalties from its own horizontal downstream competitors. Chevron, the Commission alleged, could have used this information and this power to facilitate coordinated interaction and detect any deviations.

The Commission resolved both the Chevron/Unocal merger investigation and the monopolization case against Unocal with consent orders. The key element in these settlements is Chevron's agreement not to enforce the Unocal patents.[25] The FTC's settlement of these two matters is thus a double victory for California consumers. The Commission's monopolization case against Unocal was complex and, with possible appeals, could have taken years to resolve, with substantial royalties to Unocal—and higher consumer prices—in the interim. The settlement provides the full relief sought in the monopolization case and also resolves the only competitive issue raised by the proposed merger. With the settlement, consumers will benefit immediately from the elimination of royalty payments on the Unocal patents, and potential merger efficiencies could result in additional savings at the pump.

The FTC undertook another major nonmerger investigation during 1998-2001, examining the major oil refiners' marketing and distribution practices in Arizona, California, Nevada, Oregon, and Washington (the "Western States" investigation).[26] The agency initiated the Western States investigation out of concern that differences in gasoline prices in Los Angeles, San Francisco, and San Diego might be due partly to anticompetitive activities. The Commission's staff examined over 300 boxes of documents, conducted 100 interviews, held over 30 investigational hearings, and analyzed a substantial amount of pricing data. The investigation uncovered no basis to allege an antitrust violation. Specifically, the investigation detected no evidence of a horizontal agreement on price or output or the adoption of any illegal vertical distribution practice at any level of supply. The investigation also found no evidence that any refiner had the unilateral ability to raise prices profitably in any market or reduce output at the wholesale level. Accordingly, the Commission closed the investigation in May 2001.

In conducting these and other inquiries, the Commission makes the important distinction between short-term and long-term effects. While a refinery outage on the West Coast could significantly affect short-term prices, the FTC did not find that it would be profitable in the long run for a refiner to restrict its output to raise the level of prices in the market. For example, absent planned maintenance or unplanned outages, refineries on the West Coast (and in the rest of the country) generally run at full (or nearly full) capacity. If gasoline is in short supply in a locality due to refinery or pipeline outages, and there are no immediate alternatives, a market participant may find that it can profitably increase prices by reducing its refinery output—generally only for a short time, until the outage is fixed or alternative supply becomes available. This transient power over price—which occurs infrequently and lasts only as long as the shortage—should not be confused with the durable power over price that is the hallmark of market power in antitrust law.

In addition to the *Unocal* and West Coast pricing investigations, the Commission conducted a nine-month investigation into the causes of gasoline price spikes in local markets in the Midwest in the spring and early summer of 2000.[27] As ex-

---

[25] *Union Oil Co. of California, supra* note 11.

[26] FTC Press Release, *FTC Closes Western States Gasoline Investigation* (May 7, 2001), *available at* http://www.ftc.gov/opa/2001/05/westerngas.htm. In part, this investigation focused on "zone pricing" and "redlining." See *Statement of Commissioners Sheila F. Anthony, Orson Swindle and Thomas B. Leary, available at* http://www.ftc.gov/os/2001/05/wsgpiswindle.htm, and *Statement of Commissioner Mozelle W. Thompson, available at* http://www.ftc.gov/os/2001/05/wsgpithompson.htm, for a more detailed discussion of these practices and the Commission's findings. *See also* Cary A. Deck & Bart J. Wilson, *Experimental Gasoline Markets,* Federal Trade Commission, Bureau of Economics Working Paper (Aug. 2003), *available at* http://www.ftc.gov/be/workpapers/wp263.pdf, and David W. Meyer & Jeffrey H. Fischer, *The Economics of Price Zones and Territorial Restrictions in Gasoline Marketing,* Federal Trade Commission, Bureau of Economics Working Paper (Mar. 2004), *available at* http://www.ftc.gov/be/workpapers/wp271.pdf.

[27] Midwest Gasoline Price Investigation, Final Report of the Federal Trade Commission (Mar. 29, 2001), *available at* http://www.ftc.gov/os/2001/03/mwgasrpt.htm; *see also* Remarks of Jeremy

plained in a 2001 report, the Commission found that a variety of factors contributed in different degrees to the price spikes. Primary factors included refinery production problems (e.g., refinery breakdowns and unexpected difficulties in producing the new summer-grade RFG gasoline required for use in Chicago and Milwaukee), pipeline disruptions, and low inventories. Secondary factors included high crude oil prices that contributed to low inventory levels, the unavailability of substitutes for certain environmentally required gasoline formulations, increased demand for gasoline in the Midwest, and *ad valorem* taxes in certain states. The industry responded quickly to the price spike. Within three or four weeks, an increased supply of product had been delivered to the Midwest areas suffering from the supply disruption. By mid-July 2000, prices had receded to pre-spike or even lower levels.

The Commission's merger investigations also are relevant to the detection of non-merger antitrust violations. FTC oil and gas merger investigations during the past decade uniformly have been major undertakings that have reviewed all pertinent facets of the relevant petroleum markets. These investigations have involved the review of thousands of boxes of documents in discovery, examination of witnesses under oath, and exhaustive questioning of outside experts. The FTC staff, therefore, have learned information that also could assist in detecting and investigating potentially anticompetitive conduct.

III. COMMISSION REPORT ON FACTORS THAT AFFECT THE PRICE OF GASOLINE

What are the causes of high gasoline prices and gasoline price spikes? These important questions require a thorough and accurate analysis of the factors—supply, demand, and competition, as well as federal, state, and local regulations—that drive gasoline prices, so that policymakers can evaluate and choose strategies likely to succeed in addressing high gasoline prices.

The Commission addressed these issues by conducting extensive research concerning gasoline price fluctuations, analyzing specific instances of apparent gasoline price anomalies, and holding a series of conferences[28] on the factors that affect gasoline prices, leading to the publication of a report[29] that draws on what the Commission has learned about the factors that can influence gasoline prices or cause gasoline price spikes. We discuss the findings of our study, but first set out three basic lessons that emerge from our collective work.

First, in general, the price of gasoline reflects producers' costs and consumers' willingness to pay. Gasoline prices rise if it costs more to produce and supply gasoline, or if people wish to buy more gasoline at the current price—that is, when demand is greater than supply. Gasoline prices fall if it costs less to produce and supply gasoline, or if people wish to buy less gasoline at the current price—that is, when supply is greater than demand. Gasoline prices will stop rising or falling when they reach the level at which the quantity consumers demand matches the quantity that producers will supply.

Second, how consumers respond to price changes will affect how high prices rise and how low they fall. Limited substitutes for gasoline restrict the options available to consumers to respond to price increases in the short run. Because gasoline consumers typically do not reduce their purchases substantially in response to price increases, they are vulnerable to substantial price increases.

Third, producers' responses to price changes will affect how high prices rise, and how low they fall. In general, when there is not enough gasoline to meet consumers' demands at current prices, higher prices will signal a potential profit opportunity and may bring additional supply into the market. Additional supply will be available to the extent that an increase in price exceeds the producers' cost of expanding output.

The vast majority of the Commission's investigations and studies have revealed market factors as the primary drivers of both price increases and price spikes. There is a complex landscape of market forces that affect gasoline prices in the United States.

A. Worldwide Supply, Demand, and Competition for Crude Oil Are the Most Important Factors in the National Average Price of Gasoline in the United States

Crude oil is a commodity that is traded on world markets, and the world price of crude oil is the most important factor in the price of gasoline in the United States

---

Bulow, Director, Bureau of Economics, *The Midwest Gasoline Investigation, available at* http://www.ftc.gov/speeches/other/midwestgas.htm.

[28] FTC Press Release, *FTC to Hold Second Public Conference on the U.S. Oil and Gasoline Industry in May 2002* (Dec. 21, 2001), *available at* http://www.ftc.gov/opa/2001/12/gasconf.htm.

[29] GASOLINE PRICE CHANGES, *supra* note 3.

and all other markets. Over the past 20 years, changes in crude oil prices have explained approximately 85 percent of the changes in the price of gasoline.[30] United States refiners compete with refiners all around the world to obtain crude, and the United States now imports more than 60 percent of its crude from foreign sources.

If world crude prices rise, then U.S. refiners must pay higher prices for the crude they buy. Facing higher input costs from crude, refiners charge more for the gasoline they sell at wholesale. This requires retail stations to pay more for their gasoline. In turn, retail stations, facing higher input costs, charge consumers more at the pump. In short, when crude oil prices rise, gasoline prices rise because gasoline becomes more costly to produce.

Crude oil prices are not wholly market-determined. Since 1973, decisions by OPEC have been a significant factor in the prices that refiners pay for crude oil. Over time, OPEC has met with varying degrees of success in raising crude oil prices. (For example, OPEC members can be tempted to "cheat" and sometimes sell more crude oil than specified by OPEC limits.) Higher world crude prices due to OPEC's actions, however, increased the incentives to search for oil in other areas, and crude supplies from non-OPEC members such as Canada, the United Kingdom, and Norway have increased significantly. Nonetheless, OPEC still produces a large enough share of world crude oil to exert market power and strongly influence the price of crude oil when its members adhere to their assigned production quotas. Especially when demand surges unexpectedly, as in 2004, OPEC decisions on whether to increase supply to meet demand can have a significant impact on world crude oil prices.

Crude oil consumption has fallen during some periods over the past 30 years, partially in reaction to higher prices and partially in response to federal laws, such as requirements to increase the fuel efficiency of cars. Gasoline consumption in the United States fell significantly between 1978 and 1982, and remained lower during the 1980s than it had been at the beginning of 1978.[31] Overall, however, the long-run trend is toward significantly increased demand for crude oil. Over the last 20 years, United States consumption of all refined petroleum products increased on average by 1.4 percent per year, leading to a total increase of nearly 30 percent.[32]

Crude oil prices have been increasing rapidly in recent months. Demand has remained high in the United States, and large demand increases from rapidly industrializing countries, particularly China and India, have made supplies much tighter than expected.[33]

B. Gasoline Supply, Demand, and Competition Produced Relatively Low and Stable Prices From 1984 Until 2004, Despite Substantial Increases in United States Gasoline Consumption

Consumer demand for gasoline in the United States has risen substantially, especially since 1990.[34] In 1978, U.S. gasoline consumption was about 7.4 million barrels per day. By 1981, in the face of sharply escalating crude oil and gasoline prices and a recession, U.S. gasoline consumption had fallen to approximately 6.5 million barrels per day.[35] As gasoline prices began to fall in the 1980s, U.S. consumption of gasoline began to rise once again. By 1993, consumption rose above 1978 levels, and

---

[30] A simple regression of the monthly average national price of gasoline on the monthly average price of West Texas Intermediate crude oil shows that the variation in the price of crude oil—based on data for the period January 1984 to October 2003—explains approximately 85 percent of the variation in the price of gasoline. This is similar to the range of effects given in United States Department of Energy/Energy Information Administration, *Price Changes in the Gasoline Market: Are Midwestern Gasoline Prices Downward Sticky?*, DOE/EIA-0626 (Feb. 1999). More complex regression analysis and more disaggregated data may give somewhat different estimates, but the latter estimates are likely to be of the same general magnitude.

This percentage may vary across states or regions. See Prepared Statement of Justine Hastings before the Committee on the Judiciary, Subcommittee on Antitrust, Competition Policy and Consumer Rights, United States Senate, *Crude Oil: The Source of Higher Gas Prices* (Apr. 7, 2004). Dr. Hastings found a range from approximately 70 percent for California to 91 percent for South Carolina. South Carolina uses only conventional gasoline and is supplied largely by major product pipelines that pass through the state on their way north from the large refinery centers on the Gulf Coast. California, with its unique fuel specifications and its relative isolation from refinery centers in other parts of the United States, historically has been more susceptible to supply disruptions that can cause major gasoline price changes, independent of crude oil price changes.

[31] GASOLINE PRICE CHANGES, *supra* note 3, at 43-45.

[32] *Id.* at 19.

[33] This phenomenon was not limited to crude oil: other commodities that form the basis for expanded growth in developing economies, such as steel and lumber, also saw unexpectedly rapid growth in demand, along with higher prices. *Id.* at 27.

[34] *Id.* at 48.

[35] *Id.*

it has continued to increase at a fairly steady rate since then. In 2004, U.S. gasoline consumption averaged about 9 million barrels per day, and the EIA's forecast is for 9.2 million barrels per day this year.[36]

Despite high gasoline prices across the nation, demand has not fallen off in 2005. Gasoline demand this summer driving season has been above last year's record driving-season demand and well above the average for the previous four years. Average daily demand of finished gasoline for May was 9.3 millions barrels per day, an increase of 1.2 percent over May of 2004, and 5.5 percent higher than the average demand for the previous four summers. Similarly, June's demand was up 2.8 percent over last June (up 5.4 percent from the average of the previous four years) and July's demand increase was up 3.2 percent over July of 2004 (up 4.6 percent from average of the last four years). Gasoline demand for the last four weeks ending August 26 was level with the demand for the same period last year, despite much higher prices.[37]

In spite of these substantial demand increases, increased supply from U.S. refineries and imports have kept gasoline prices relatively steady until 2004. A comparison of "real" average annual retail gasoline prices and average annual retail gasoline consumption in the United States from 1978 through 2004 shows that, in general, gasoline prices remained relatively stable despite significantly increased demand.[38] Indeed, over the very long run in the 84-year period between 1919 and 2003, real annual average retail gasoline prices in the United States did not increase at all. The data show that, from 1986 through 2003, real national average retail prices for gasoline, including taxes, generally were below $2.00 per gallon (in 2004 dollars). By contrast, between 1919 and 1985, real national average retail gasoline prices were above $2.00 per gallon (in 2004 dollars) more often than not.[39]

Average U.S. retail prices have been increasing since 2003, however, from an average of $1.56 in 2003 to an average of $2.04 in the first five months of 2005.[40] In the last two months, the prices have moved even higher. It is difficult to predict whether these increases represent the beginning of a longer-term trend or are merely normal market fluctuations caused by unexpectedly strong short-term worldwide demand for crude oil, as well as reflecting the effects of instability in such producing areas as the Middle East and Venezuela.

One of the reasons why long-term real prices have been relatively contained is that United States refiners have taken advantage of economies of scale and adopted more efficient technologies and business strategies. Between 1985 and 2005, U.S. refineries increased their total capacity to refine crude oil into various refined petroleum products by 8.9 percent, moving from 15.7 million barrels per day in 1985 to 17.133 million barrels per day as of August 2005.[41] This increase—approximately 1.4 million barrels per day—is roughly equivalent to adding approximately 10 to 12 average-sized refineries to industry supply. Yet U.S. refiners did not build any new refineries during this time. Rather, they added this capacity through the expansion of existing refineries. They also have adopted processing methods that broaden the range of crude oils that they can process and allow them to produce more refined product for each barrel of crude processed. In addition, they have lowered inventory holdings, thereby lowering inventory costs (although lower inventory holdings may

---

[36] *See id.* at 49; EIA, DOE/EIA-0202, Short-term Energy Outlook, Apr. 2005, app. at 5 tbl.A5, *at* http://www.eia.doe.gov/pub/forecasting/steo/oldsteos/apr05.pdf.

[37] EIA, DOE/EIA-0208(2005-34), WEEKLY PETROLEUM STATUS REPORT, August 31, 2005, *at 17,* tbl.11, at http://www.eia.doe.gov/pub/oil_gas/petroleum/data_publications/weekly_petroleum_status_report/historical/2005/2005_08_31/pdf/wpsrall.pdf.

[38] "Real" prices are adjusted for inflation and therefore reflect the different values of a dollar at different times; they provide more accurate comparisons of prices in different time periods. "Nominal" prices are the literal prices shown at the time of purchase.

[39] *See* GASOLINE PRICE CHANGES, *supra* note 3, at 43-47.

[40] The higher prices in 2005 appear to be the result of market factors that have uniformly affected the entire country. At least for the part of this year that preceded Hurricane Katrina, the FTC's Gasoline Price Monitoring Project has detected no evidence of significant unusual local or regional gasoline pricing anywhere in the United States during this summer driving season. This contrasts with the past two summers, during which various regional supply shocks, such as the Arizona pipeline shutdown and Northeast blackouts of August 2003, and the several unanticipated regional refinery outages and late summer hurricanes during the summer of 2004, significantly increased prices in some areas above levels that might be expected based on historical price patterns.

[41] PETROLEUM MERGER REPORT, *supra* note 4, at 196, tbl.7-1; EIA, DOE/EIA-0340(04)/1, 1 PETROLEUM SUPPLY ANNUAL 2004, *at* 78, tbl.36 (2005), at http://www.eia.doe.gov/pub/oil_gas/petroleum/data_publications/petroleum_supply_annual/psa_volume1/current/pdf/volume1_all.pdf. EIA, DOE/EIA-0208(2005-33), WEEKLY PETROLEUM STATUS REPORT, August 24, 2005, *at* http://www.eia.doe.gov/pub/oil_gas/petroleum/data_publications/weekly_petroleum_status_report/historical/2005/2005_08_24/pdf/wpsrall.pdf.

also make an area more susceptible to short-term price spikes when there is a disruption in supply).

Offsetting some of the observed efficiency gains, increased environmental requirements since 1992 have likely raised the retail price of gasoline by a few cents per gallon in some areas. Because gasoline use is a major factor in air pollution in the United States, the U.S. Environmental Protection Agency—under the Clean Air Act[42]—requires various gasoline blends for particular geographic areas that have not met certain air quality standards. While available information shows that the air quality in the United States has improved due to the Clean Air Act,[43] as with any regulatory program, costs come with the benefits. Environmental laws and regulations have required substantial and expensive refinery upgrades, particularly over the past 15 years. It costs more to produce cleaner gasoline than to produce conventional gasoline. Estimates of the increased costs of environmentally mandated gasoline range from $0.03 to $0.11 per gallon.[44]

Our studies indicate that higher retail prices are not caused by excess oil company profits. Although recent oil company profits may be high in absolute terms, industry profits have varied widely over time, as well as over industry segments and among firms.

EIA's Financial Reporting System ("FRS") tracks the financial performance of the 28 major energy producers currently operating in the United States. In 2003, these firms did have a return on capital employed of 12.8 percent, as compared to the 10 percent return on capital employed for the overall Standard & Poors ("S&P") Industrials. Between 1973 and 2003, however, the annual average return on equity for FRS companies was 12.6 percent, while it was 13.1 percent for the S&P Industrials.[45] High absolute profits do not contradict numbers showing that oil companies may at times earn less (as a percentage of capital or equity) than other industrial firms. This simply reflects the large amount of capital necessary to find, refine, and distribute petroleum products.

The rates of return on equity for FRS companies have varied widely over the years, ranging from as low as 1.1 percent to as high as 21.1 percent during the period from 1974 to 2003.[46] Returns on equity vary across firms as well. Crude oil exploration and production operations typically generate much higher and more volatile returns than refining and marketing. In essence, companies with exploration and production operations now find themselves in a position analogous to that of a homeowner who bought a house in a popular area just before increased demand for housing caused real estate prices to escalate. Like the homeowner, crude oil producers can charge higher prices due to increased demand. If high prices and high profits are expected to continue, they may draw greater investments over time into the oil industry—in particular, to crude exploration and production. Over the long run, these investments are likely to elicit more crude supply, which would exert a downward pressure on prices.

C. Other Factors, Such as Retail Station Density, New Retail Formats, and State and Local Regulations, Also Can Affect Retail Gasoline Prices

The interaction of supply and demand and industry efficiency are not the only factors that impact retail gasoline prices. State and local taxes can be a significant component of the final price of gasoline. In 2004, the average state sales tax was

---

[42] Beginning with the Clean Air Act Amendments of 1970 (Pub. L. No. 91-604, 84 Stat. 1698) and continuing with further amendments in 1990 (Pub. L. No. 101-549, 104 Stat. 2468) and the Energy Policy Act of 1992 (Pub. L. No. 102-486, 106 Stat. 2776), Congress has mandated substantial changes in the quality of gasoline, as well as diesel, that can be sold in the United States.

[43] Robert Larson, Acting Director of the Transportation and Regional Programs, Environmental Protection Agency, Remarks at the FTC Conference on Factors that Affect Prices of Refined Petroleum Products 79-80 (May 8, 2002).

[44] *See* EIA, *1995 Reformulated Gasoline Market Affected Refiners Differently,* in DOE/EIA-0380(1996/01), PETROLEUM MARKETING MONTHLY (1996), and studies cited therein. Environmental mandates are not the same in all areas of the country. The EPA requires particular gasoline blends for certain geographic areas, but it sometimes allows variations on those blends. Differing fuel specifications in different areas can limit the ability of gasoline wholesalers to find adequate substitutes in the event of a supply shortage. Thus, boutique fuels may exacerbate price variability in areas, such as California, that are not interconnected with large refining centers in other areas.

[45] *See* GASOLINE PRICE CHANGES, *supra* note 3, at 61.

[46] *Id.*

$0.225 per gallon, with the highest state tax at $0.334 per gallon (New York).[47] Some local governments also impose gasoline taxes.[48]

Local regulations may also have an impact on retail gasoline prices. For example, bans on self-service sales or below-cost sales appear to raise gasoline prices. New Jersey and Oregon ban self-service sales, thus requiring consumers to buy gasoline bundled with services that increase costs—that is, having staff available to pump the gasoline.[49] Some experts have estimated that self-service bans cost consumers between $0.02 to $0.05 per gallon.[50] In addition, some 11 states have laws banning below-cost sales, so that a gas station is required to charge a minimum amount above its wholesale gasoline price.[51] These laws harm consumers by depriving them of the lower prices that more efficient (*e.g.*, high-volume) stations can charge.

Not surprisingly, retail gasoline prices are likely to be lower when consumers can choose—and can switch their purchases—among a greater number of retail stations. A small number of empirical studies have examined gasoline station density in relation to prices. One study found that stations in Southern California that imposed a 1 percent price increase lost different amounts of sales, depending on how many competitors were close by.[52] Those with a large number of nearby competitors (27 or more within 2 miles) lost 4.4 percent of sales in response to a 1 percent price increase; those with a smaller number of nearby competitors (fewer than 19 within 2 miles) lost only 1.5 percent of sales.[53] With all else equal, stations that face greater lost sales from raising prices will likely have lower retail prices than stations that lose fewer sales from raising prices.

Station density depends on cost conditions in an area. For example, the size and density of a market will influence how many stations can operate and cover their fixed costs. Fixed costs will depend on the costs of land and of building a station. Zoning regulations also may limit the number of stations in an area below what market conditions indicate the area could profitably sustain. Studies suggest that entry by new gasoline competitors tends to be more difficult in areas with high land prices and strict zoning regulations.[54]

One of the biggest changes in retail sales of gasoline in the past three decades has been the development of such new formats as convenience stores and high-volume operations. These new formats appear to lower retail gasoline prices. The number of traditional gasoline-pump-and-repair-bay outlets has dwindled for a number of years, as brand-name gasoline retailers have moved toward a convenience store format. Independent gasoline/convenience stores—such as RaceTrac, Sheetz, QuikTrip, and Wawa—typically feature large convenience stores with multiple fuel islands and multi-product dispensers. They are sometimes called "pumpers" because of their large-volume fuel sales. By 1999, the latest year for which data are available, brand-name and independent convenience store and pumper stations accounted for almost 67 percent of the volume of U.S. retail gasoline sales.[55]

Another change to the retail gasoline market that appears to have helped keep gasoline prices lower is the entry of hypermarkets. Hypermarkets are large retailers of general merchandise and grocery items, such as Wal-Mart and Safeway, that have begun to sell gasoline. Hypermarket sites typically sell even larger volumes of gasoline than pumper stations—sometimes 4 to 8 times larger.[56] Hypermarkets' substantial economies of scale generally enable them to sell significantly greater volumes of gasoline at lower prices.

---

[47] See GASOLINE PRICE CHANGES, *supra* note 3, at 111—(noting that the other four states with the highest average taxes on gasoline in 2004 were Wisconsin ($0.33 per gallon), Connecticut ($0.325 per gallon), Rhode Island ($0.306 per gallon), and California ($0.301 per gallon)).

[48] *Id.* For example, all areas in Florida also have a local tax between $0.099 and $0.178 per gallon. Similarly, Honolulu has a local tax of $0.165 per gallon.

[49] *See, e.g.*, OREGON REV. STAT., ch. 480, § 480.315.

[50] *See* Michael G. Vita, Regulatory Restrictions on Vertical Integration and Control: The Competitive Impact of Gasoline Divorcement Policies, 18 J. Reg. Econ. 217 (2000); *see also* Ronald N. Johnson & Charles J. Romeo, The Impact of Self-Service Bans in the Retail Gasoline Market, 82 REV. ECON. & STAT. 625 (2000); Donald Vandegrift & Joseph A. Bisti, The Economic Effect of New Jersey's Self-Service Operations Ban on Retail Gasoline Markets, 24 J. Consumer Pol'y 63 (2001).

[51] See GASOLINE PRICE CHANGES, *supra* note 3, at 113.

[52] JOHN M. BARRON ET AL., CONSUMER AND COMPETITOR REACTIONS: EVIDENCE FROM A RETAIL-GASOLINE FIELD EXPERIMENT (Mar. 2004), *at* http://ssrn.com/abstract'616761.

[53] *Id.* at 13, 15, 30-31.

[54] *See id.* at 30-31; GOV'T ACCOUNTABILITY OFFICE (GAO), GAO/RCED-00-121, MOTORFUELS: CALIFORNIA GASOLINE PRICE BEHAVIOR 20 (2000), *available at* http://www.gao.gov/new/items/rc00121.pdf.

[55] PETROLEUM MERGER REPORT, *supra* note 4, at 246 tbl.9-5.

[56] *Id.* at 239.

The list of factors that have an impact on retail gasoline prices is not exhaustive, but it shows that prices are set by a complex array of market and regulatory forces working throughout the economy. In the long run, these forces have combined to produce remarkably stable prices in the face of consistently growing demand. Short-run variations, while sometimes painful to consumers, are unavoidable in an industry that depends on the demand and supply decisions of literally billions of people.

IV. CONCLUSION

The Federal Trade Commission has an aggressive program to enforce the antitrust laws in the petroleum industry. The Commission has taken action whenever a merger or nonmerger conduct has violated the law and threatened the welfare of consumers or competition in the industry. The Commission continues to study this industry in detail, to monitor wholesale and retail gasoline prices, and to search for instances of illegal mergers or anticompetitive conduct.

Thank you for this opportunity to present the FTC's views on this important topic. I would be glad to answer any questions that the Committee may have.

Chairman BARTON. Thank you, Mr. Seesel.

We now want to hear from Mr. Kenneth Moran, who is the Acting Director of the Office of Homeland Security Enforcement Bureau at the Federal Communications Commission.

Welcome, Mr. Moran, and you are recognized for 7 minutes.

STATEMENT OF KENNETH P. MORAN

Mr. MORAN. Thank you, Mr. Chairman.

Good afternoon, Mr. Chairman and distinguished members of the committee. My name is Ken Moran, and I serve as the Director of the Federal Communication's Commissions Office of Homeland Security. In that role, I am primarily responsible for coordinating the Commission's support of the Hurricane Katrina disaster relief efforts.

In my written testimony, I describe some of the damage to the communications industry resulting from Hurricane Katrina and the Commission's efforts to assist consumers, the industries that the agency regulates, and other Federal agencies during this difficult crisis. I ask that my written testimony be submitted into the record, and I will summarize those comments now.

Hurricane Katrina caused catastrophic damage and massive flooding in areas of Louisiana, Mississippi and Alabama. The loss of life and damage to property is astounding, and our thoughts and prayers go out to those affected by the disaster. Most of the communications industry sustained tremendous damage to their facilitates in the affected area, hampering rescue operations or emergency responders and affecting the communications of those still struggling with the affects of the hurricane.

Hurricane Katrina knocked out more than 3 million customer phone lines. Wireline communications networks sustained enormous damage both to the switching centers that route the calls and the lines that connect the buildings and the customers to the network.

Local wireless networks also sustained considerable damage. More than 1,000 cell sites were knocked out of service by the hurricane. During this disaster, millions of telephone calls simply have not been able to get through.

Also, of the 41 broadcast radio stations located in New Orleans and the surrounding area, only two AM and two FM stations remained on the air in the wake of the hurricane.

We know that extraordinary efforts are being made to maintain and restore service in the disaster zone. Broadcasters are getting some stations on the air, albeit at significantly reduced power, to provide survivors with important information. Wireline and wireless carriers have crews working around the clock to repair switching centers, customer lines and cell towers. Satellite service providers have helped bridge some of the gaps left by the outages by, for example, providing satellite phones and video links to law enforcement officials and medical personnel, emergency relief personnel and news outlets. Even with these efforts, though, many of the communications services in the affected areas remain down.

Today, we estimate that 1 million customer lines remain out of service. Six 911 centers still remain out of service, and approximately 30 percent of wireless telecom cell sites are not operational. Also, more than 50 radio and TV stations remain off the air. Many of these sites that are operational are dependent on the back-up energy supplies.

On August 30, the Commission established an internal task force consisting of senior executives and management from within the Commission. Chairman Martin directed the task force to coordinate the FCC's hurricane response activities which fall into two categories: regulatory relief; and industry outreach and coordination with other Federal agencies. The task force has been working on these assignments continuously since August 30, and the Commission was open throughout the Labor Day weekend to continue the work. To date, nearly 200 FCC employees have assisted in this effort.

The Commission has taken a number of steps to facilitate resumption of communication services in the affected areas and to authorize use of temporary communication services for use by disaster relief personnel and evacuees in shelters.

At the start of this disaster, the Commission notified communications providers that it would provide streamlined treatment for requests for special temporary authority in order to aid them in resuming and maintaining operations in the areas impacted by Hurricane Katrina. The FCC has received 22 special temporary authority requests and 77 requests for temporary frequency assignments. The Commission has also received a number of requests for temporary waiver of its rules. The Commission has granted most of these requests within 4 hours of their receipt and all requests within 24 hours. In addition, the Commission has released several public notices and quickly adopted orders to provide temporary relief.

My written testimony provides many examples of the Commission action to date.

The Commission has been working closely with industry as well as FEMA and the National Communications System consistent with procedures established in the National Response Plan. The Commission continues to reach out to the communications companies serving the affected areas to assess their operational status and determine the resources that they need to resume full operations.

The FCC provides the critical information about resources that communications providers need to restore and maintain service in the area to both FEMA and the NCS, who are responsible for ensuring that priority needs are met. The agency updates FEMA and NCS daily on the evolving needs.

The Commission is also responsible for providing the National Coordinating Center for Telecommunications with information on communications companies' operational status for incorporation into the governmentwide situation reports. Again, the agency gathers and submits this data daily.

In addition, the Commission has worked closely with the communications industry to help identify resources for use by disaster response personnel. The agency both transmits this information to the NCS and facilitates the industry's communication with other Federal officials.

Finally, the Commission has been coordinating with the Interagency Coordination Council on Individuals with Disabilities to ensure that the needs of the disability community are addressed in the coordinated Federal relief efforts.

In conclusion, the Commission is continuing to work with key Federal agencies and the communications industry to determine what additional actions can be taken to assist in the disaster relief and restoration effort. More information about these efforts is available on our Web site.

The Commission stands ready to work with the Congress, our colleagues at the State, Federal and local agencies and the American public to do whatever we can to help with disaster relief and restoration efforts. I would be pleased to respond to your comments. Thank you.

[The prepared statement of Kenneth P. Moran follows:]

PREPARED STATEMENT OF KENNETH P. MORAN, DIRECTOR, OFFICE OF HOMELAND SECURITY, ENFORCEMENT BUREAU, FEDERAL COMMUNICATIONS COMMISSION

Good morning, Mr. Chairman and distinguished members of the Committee. My name is Ken Moran and I serve as the Director of the Federal Communications Commission's Office of Homeland Security. In that role, I am primarily responsible for coordinating the Commission's support of the Hurricane Katrina disaster relief efforts.

In my testimony today, I will describe some of the damage wrought by Hurricane Katrina to the communications industry and the Commission's efforts to assist consumers, the industries the agency regulates, and other Federal Agencies during this difficult crisis.

Hurricane Katrina caused catastrophic damage and massive flooding in areas of Louisiana, Mississippi, and Alabama. The loss of life and damage to property is astounding, and our thoughts and prayers go out to those people affected by this disaster. As I am sure you are aware, most of the communications industry sustained tremendous damage to their facilities in the affected area, and the damage has had a significant impact. The damage to the communications infrastructure hampered the rescue operations of emergency responders. Relief efforts and survivors are still struggling with the effects of the hurricane. Survivors lack information about relief efforts. People displaced from their homes do not have the means to contact their loved ones to let them know they are safe. And of course, survivors remaining in the affected area lack a reliable means of contacting the authorities and getting help in life threatening situations.

STATUS OF COMMUNICATIONS NETWORKS

Hurricane Katrina knocked out more than 3 million customer phone lines in the Louisiana, Mississippi, and Alabama area. The wireline telecommunications network sustained enormous damage both to the switching centers that route calls and to the lines used to connect buildings and customers to the network. Local wireless networks also sustained considerable damage—more than a thousand cell sites were knocked out of service by the hurricane. During this disaster, millions of telephone calls simply have not been able to get through. Of the 41 broadcast radio stations located in New Orleans and the surrounding area, only two AM and two FM stations remained on the air in the wake of the hurricane.

Through network outage reports filed in accordance with the Commission's rules, and through data given to us voluntary by the industry, we understand that an extreme effort is being made to maintain and restore service in the disaster zone. Broadcasters are making every effort to get stations on-the-air, even at significantly reduced power, to provide survivors with important information. Wireline and wireless carriers have crews working to repair switching centers, customer lines, and cell towers. Satellite service providers have helped bridge some of the gaps left by the outages by, for instance, providing satellite phones and video links to law enforcement officials, medical personnel, emergency relief personnel, and news outlets.

Even with these efforts, given the enormity of the disaster, many of the communications services in the affected areas remain down. Today, we understand that more than one million customer lines and over 20 switching centers remain out of service. Approximately 1700 DS-3 interoffice facilities remain down. Six public safety answering points remain out of service. Approximately thirty percent of cell sites are not operational. Fifty to 100 radio and television stations remain off the air. Many of the sites that are operational are dependent on back-up energy supplies.

COMMISSION ACTIONS

On August 30th, Chairman Martin established an internal Task Force consisting of senior executives and management from within the Commission. Chairman Martin directed the Task Force to coordinate the FCC's hurricane response efforts, which fall into two categories: (1) regulatory relief; and (2) industry outreach and coordination with other federal agencies. The Task Force has been working on these assignments continuously since August 30th, and the Commission was open throughout the Labor Day weekend to continue the work. To date, nearly 200 FCC employees have assisted in this effort.

Regulatory Relief

The Commission has taken a number of steps to facilitate the resumption of communications services in the affected areas and to authorize the use of temporary communications services for use by disaster relief personnel and evacuees in shelters.

At the start of the disaster, the Commission notified communications providers that it would provide streamlined treatment for requests for special temporary authority (STA) in order to aid them in resuming and maintaining operations in areas impacted by Hurricane Katrina. The FCC has received at least 22 STA requests and 77 requests for temporary frequency assignments. The Commission also has received a number of requests for temporary waiver of its rules. The Commission has granted each of these requests within 4 hours of receipt of all necessary information from the requestor, except in instances requiring coordination with other government agencies. Even in those cases, requests have been granted within 24 hours. In addition, the Commission has released several public notices and quickly adopted orders to provide temporary relief.

Examples of the many steps the Commission has taken to assist disaster relief efforts and affected providers are listed in the attached appendix.

Industry Outreach and Coordination with Other Federal Agencies

The Commission has been working closely with industry as well as the Federal Emergency Management Agency (FEMA) and the National Communications System (NCS) pursuant to the procedures established in the National Response Plan. The Commission is continuously reaching out to communications companies serving the affected area—wireline and wireless network providers, broadcasters, cable providers, satellite providers—and to trade associations for these providers to assess the companies' status and determine what they need to resume operations. These efforts include Commission staff contacting each of the approximately 160 broadcast stations in the affected region.

The FCC provides the critical information about resources that communications providers need to restore and maintain service in the affected area to FEMA and NCS, who are responsible for ensuring that priority needs are met. For instance, the Commission identified wireline central offices and radio and television broadcasters that could be operational if provided fuel to power on-site generators. The agency updates FEMA and NCS daily on evolving needs.

The Commission also is responsible for providing the National Coordinating Center (NCC) with information on communications companies' operational status for incorporation into the government-wide situation reports. Again, the agency gathers and submits this data daily.

In addition, the FCC has worked closely with the communications industry to help identify resources for use by disaster response personnel. The agency both transmits this information to NCC and facilitates industry's communication with other federal officials. For example, Commission staff coordinated discussions between FEMA and a major Direct Broadcast Satellite (DBS) provider to set up free televisions at disaster relief facilities and to provide a nationwide channel for disaster emergency services programming. Staff also worked with a wide range of providers—including those offering competitive facilities-based telecommunications, satellite, wireless, wireless internet access and WI FI services—to identify those providers capable of offering facilities and services that can assist those in the affected area.

Finally, the Commission has been coordinating with the Interagency Coordinating Council on Individuals with Disabilities, organized by the Department of Homeland Security, to ensure that the needs of the disability community are addressed in the coordinated federal relief efforts.

CONCLUSION

FCC Chairman Kevin Martin, Commissioners Kathleen Abernathy, Michael Copps and Jonathan Adelstein, along with the FCC staff, commend the industry and the tremendous efforts it has made to begin to repair the infrastructure and restore communications service to the Gulf Coast. These extraordinary efforts to restore communications services are being performed by employees of the communications industry—many of whom may be personally impacted by this tragedy.

The Commission is continuing to work with other Federal agencies and the communications industry to determine what additional actions can be taken to assist in the disaster relief and restoration effort. More information about these efforts is—and will continue to be—available on the Commission's web site: http://www.fcc.gov/cgb/katrina/.

The Commission also will continue its important work in reaching out, and responding to, consumers affected by this tragedy. Since the hurricane struck, including over the Labor Day weekend, the Commission manned its toll-free consumer line to help individuals get access to critical information about telecommunications and broadcast services in the affected area. The agency will continue these and other efforts to address consumer concerns, in coordination with other government agencies, relief organizations, consumer groups and industry.

The damage wrought by Hurricane Katrina is tremendous and its effects will be felt for months and possibly years to come. The Commission stands ready to work with Congress, our colleagues at federal, state, and local agencies, and the American public to do whatever we can to help with the disaster relief and restoration efforts. I would be pleased to respond to your questions.

APPENDIX

Since Hurricane Katrina struck the Gulf Coast, the Commission has taken a number of steps to help the industry resume service and to assist the communications needs of disaster relief personnel and evacuees in shelters. Following are some examples of Commission actions:

- On September 2nd, the Commission granted STAs to operate ultra-wide band services "through-the-wall" imaging systems to locate survivors.
- On September 5th, the Commission temporarily authorized the Department of Defense to conduct ship-to-ship, ground, and air-to-ground operations in the affected area.
- Over the past week, the Commission granted STAs and temporary frequency authorizations to parties working to support relief efforts and to utilities working to restore phone and electric service in the affected area.
- Over the Labor Day weekend, the Commission granted a temporary waiver of its "slamming rules," which require carriers to ensure subscribers are notified before their long distance service is switched. This temporary waiver will permit carriers to temporarily transfer customers to long distance carriers with working facilities while restoration efforts are under way.
- On September 2nd, the Commission acted upon the request of the American Red Cross and temporarily reassigned the toll free 800 number "1-800-RED-CROSS" to the National Chapter of the American Red Cross. This action will facilitate the disaster relief operations and fundraising efforts of the American Red Cross—the only non-governmental agency with a specified lead role in the National Response Plan—by providing an easily-recognizable centralized telephonic point of contact for this important organization.
- Also on September 2nd, the Commission suspended its rules in order to permit noncommercial educational (NCE) radio and television stations in New Orleans

to rebroadcast programming, including commercial matter, received from commercial broadcast stations. This special relief is designed to bring immediate life saving and other important program information to the residents of New Orleans in the most expeditious manner possible.

- Between September 2nd and September 4th, the Commission granted STAs to provide Internet connectivity to more than 200 shelters operated by the American Red Cross.
- On September 1st and 2nd, the Commission temporarily waived certain rules applicable to NCE television and radio stations, allowing those stations to air fundraising programming to aid disaster relief efforts.
- On September 5th, the Commission granted experimental authorizations to permit the use of 3 FM signals to broadcast emergency information to the approximately 24,000 evacuees in the Houston Astrodome.
- Over the Labor Day weekend, the Commission granted a waiver of its numbering rules that require carriers to return certain unused telephone numbers. This action will permit carriers in the affected area to retain telephone numbers that are not in use for longer than 90 days in order to allow consumers returning to the affected area continued use of their telephone numbers.
- On September 1st, the Commission waived its rules in order to permit wireline and wireless carriers to port telephone numbers geographically outside of rate centers during this period of service disruption. This action is intended to help consumers keep using their telephone numbers during the crisis, to the extent facilities are available.
- Also on September 1st, the Commission granted an equipment authorization for a new digital microwave radio system. One of the major wireless carriers will use this equipment to replace equipment in Baton Rouge and southern Louisiana that was destroyed by the hurricane.
- On September 2nd, the Commission granted a request from the 800 MHz Transition Administrator to move Louisiana from Wave 2, which begins relocation negotiations in October 2005, to Wave 3. This action enables public safety entities in Louisiana to focus on more immediate public safety needs.
- On September 1st, the Commission issued informal guidance to amateur radio operators that they have authority to make transmissions necessary to meet essential communication needs and facilitate relief actions, and that prior Commission approval is not required for such transmissions.

FCC granted STA to the California Highway Patrol to operate portable and mobile radios in support of other law enforcement and relief agencies in Louisiana and Mississippi (9/6/05).

FCC granted STA to LifeCom/Air Methods to set up a control center with mobile radio communications in the 460 MHz band in the New Orleans area for disaster relief (9/6/05).

FCC granted an STA for stations licensed to American Family Association in Mississippi and Louisiana that ceased operations on August 28, 2005 to remain silent (9/6/05).

FCC granted an STA for WFMM(FM), Telesouth Communications, Inc., Sumrall, Mississippi, to remain silent after it went silent on 8/29/05 (9/6/05).

Chairman BARTON. Thank you, Mr. Moran.

The Chair now recognizes himself for the first 5 minutes of questions.

Secretary Garman, I want to commend you, the Deputy Secretary, and the full Secretary for your expeditious work on the SPR. I made a phone call to Secretary Bodman on Monday. I sent him a letter Monday afternoon asking that the SPR be utilized, and oil was released from the SPR on Thursday.

Crude prices on world markets have actually—they went up to over $70 a barrel briefly, but as of late yesterday they were down in the $65 to $66 range, and I do not know what the market is today. But one thing that the President and the Secretary of Energy have done right in the last week is use the SPR, and I want to publicly commend the Secretary and you for that decision.

I do have a question for you on refinery needs and this may be Mr. Caruso, also. Do you, Mr. Garman, and you—especially Mr.

Garman as a policymaker—think that we should have 100 percent refining capacity for our demand in this country?

Mr. GARMAN. Clearly, the margins that we have suffered under with inadequate refinery capacity in this country has had consumer impacts; and if we want to address that consumer impact, if we do not want to be dependent on foreign sources for petroleum product as we are on foreign sources for crude, then, yes, we ought to have sufficient refining capacity in this country to serve our needs, in my view.

Chairman BARTON. Okay. Mr. Caruso, as the director at EIA, my information is that we are consuming around 21 million barrels of refined products per day in this country, but our refinery capacity before Katrina was less than 17 million barrels per day. Do those numbers conform with your official numbers?

Mr. CARUSO. Yes, Mr. Chairman. The consumption for 2005, we are estimating, is 20.8, so very close to 21; and primary distillation capacity is 17.1 million barrels per day as we speak, prior to Katrina.

Chairman BARTON. Now the information that our staff has prepared in the aftermath of Katrina, even given the amazing efforts to bring refinery capacity back on line, showed about a million barrels per day of refinery capacity is out indefinitely because of water damage or power damage or hurricane damage. That is primarily a big refinery in Pascagoula and two refineries that are in or near New Orleans. Does that million barrels per day again conform with what you officially think is going to be long term out of order, either Mr. Caruso or Mr. Garman?

Mr. GARMAN. Roughly, yes. Those refineries, with minor damage, minor flooding or lack of power, have largely been brought back on line.

Those refineries—by my count, there are currently six of them that are off line—will probably take a bit longer to bring on line. I do not have a good damage assessment to be able to estimate. Perhaps some of the witnesses later in the day do.

Chairman BARTON. Mr. Caruso, do you want to do add anything to that?

Mr. CARUSO. The only thing that I would add is that those four that appear to have suffered major damage will be off line for a matter of months. So I probably would not use the word “indefinitely,” depending on what you——

Chairman BARTON. I understand.

Now this is for Mr. Garman. We are seriously thinking about preparing a refinery revitalization bill. In the current law that the President just signed, we give Governors of the 50 States the authority to request the EPA to appoint a facilitator to help facilitate and coordinate the various permit applications for refineries in this country. The House had passed a more comprehensive Refinery Revitalization Act, but the Senate would not agree to that in conference. So, Mr. Garman, if it is the will of this committee to expeditiously move on a Refinery Revitalization Act, do you believe that the Bush Administration, the Secretary would be supportive of that?

Mr. GARMAN. Without knowing the specific provisions, of course, we could make no commitments. But, clearly, refinery capacity is an issue. I am not certain why investment dollars and capital flows are not going into this opportunity. I would imagine, as is the case with many very large capital projects, people do not want to put their money at risk for a long period of time awaiting return in the face of regulatory uncertainty. We have seen that in the nuclear plant business where utility executives do not want to make that commitment and that large capital up-front commitment. A refinery is very much the same story, with a very uncertain regulatory path and very uncertain timing associated with that permitting process.

We have watched with some amazement as a potential refinery project in Arizona has been attempted to be built for the better part of a decade, if I am not mistaken; and investors have come and gone, somewhat frustrated by the inability to get the project under way.

Clearly, something needs to be done; and we would take a look at whatever the committee—and would be happy to work with the committee in brainstorming some of the ideas that might be employed.

Chairman BARTON. Thank you.

My time has expired, but I want to ask Mr. Seesel one question. It is my understanding that, under current law, there is no specific Federal legislation directly on point against price gouging. That is primarily a State issue, not a Federal issue. My question to you, Mr. Seesel, is there a standard definition of just exactly what price gouging is?

Mr. SEESEL. Mr. Chairman, I think there probably is a wide variety of definitions for that term. As far as I have been able to determine, for example, the States that have various laws against price gouging or similar terminology such as unconscionably high prices apply a lot of different criteria for measuring it either quantitatively in terms of a percentage over the usual price level before a certain event occurred or just more general language in terms of unconscionability or shockingly high. There is a lot of variations just among the 23 or so States that have those statutes on the book; and I think they are probably—it would be hard to find a consensus among just people in their normal everyday parlance on what they mean by gouging. I think there is a sort of "I know it when I see it" sort of sense among a lot of people about very high prices, but I think there is no real thing close to uniformity.

Chairman BARTON. Well, is it your view or the Federal Trade Commission's view that the State laws are adequate to handle price gouging, however one defines it, or would it be the FTC's view that we need specific Federal legislation on price gouging preempting State law? Do you have a position one way or the other on that?

Mr. SEESEL. Mr. Chairman, I don't think I can give you what the official FTC position on that would be. That would really require a determination by the Commission. But I can tell you that the issue of a Federal law about price gouging really has—sort of, it raises several issues, and I will try to do this quickly.

But one, obviously, is this difficulty of ascertaining how to measure what price gouging is and the fact that a lot of people have very different views on what that is.

In addition, I think any kind of effort to establish a Federal prohibition of price gouging, given the sort of uncertainty about what the term means, could also possibly create a replication of the experience this country went through in the 1970's. Essentially a well-intentioned effort to sort of stop what many might consider unconscionable pricing could essentially turn into an effort to control prices and profits, which certainly in the 1970's' experience led to a lot of unhappy experiences for this country in terms of gas lines and rationing and stations running out of gasoline entirely.

I think there is generally a sense, though—and, again, I am not speaking for the Commission officially—that the States have adequate firepower in their statutes to deal with this issue; and, as I said in my statement, I think a lot of Governors and attorneys general around the country are vigorously addressing that issue in the last few days.

Chairman BARTON. Thank you. I apologize to the committee for extending my time. I do not normally do that.

The Chair is trying to see who is the senior Democrat in attendance at the gavel, and I think it is Mr. Green. Mr. Green is the senior member. Mr. Gonzalez is closest to the Chair, but he just kind of cheated and moved up. So we are going to go to Mr. Green.

Mr. GREEN. Thank you, Mr. Chairman.

I know that we need to have a separate health hearing. I would hope we would, because I have some concern, like I said in my opening, about States' responsibility for their portion of Medicaid for people who are guests and not necessarily residents of that State. Ultimately, they may be, though.

Let me ask a question about high prices. It seems to me that high prices were caused originally, before we even had Katrina, by the global demand, the tight global supply, the limited domestic production and infrastructure. If it were possible in this—I guess the AEI. If it were possible to produce off shore on the east and west coast and site more energy infrastructure there, would a tricoastal energy infrastructure mean strategic stability for our energy section and wouldn't we also be more resistant to shocks like this one? Because I have lived on the gulf coast my whole life in Houston and, sure as you know, July comes around—to the end of October, in some cases—we are going to have a hurricane or tropical storm, and we will have a problem. But if we actually had a tricoastal energy production, instead of just the Gulf of Mexico and Alaska?

Mr. CARUSO. Well, it certainly would be my opinion that that would help the situation. Obviously, it is not a silver bullet, but it definitely would be a movement in the right direction.

Mr. GREEN. Any other response from anyone?

Mr. GARMAN. I would agree with that. Clearly, why is this energy infrastructure located on the gulf coast? And the answer is that is where it has been allowed to develop. That is where refineries have been allowed to be built. It is a lesson for all of us as we consider the location of new LNG terminals, as we become more dependent on natural gas. If we once again concentrate all of our LNG facilities on the gulf coast, as we have concentrated—and thank goodness for Point Cove in Maryland—but if we continue to do that, then we will be vulnerable by putting all of our energy eggs in one basket, so to speak.

Mr. GREEN. That leads to my next question. If your strategic energy situation is vulnerable and energy shocks travel so fast in our economy, I believe a national oversight is proper to ensure a national policy is made. You mentioned the example in the energy bill that just passed FERC will do the siting now on LNG terminals, and that was in the energy bill that the President signed over the break. Beyond the limited steps we can take in these situations with current authority, what else can be done to improve our ability to generate the most robust energy sector in the medium term? Should we have a Federal coordinating of these permits so we have a really a tricoast strategy? Are there solutions on that? Should we use the example of the energy bill on LNG for other production?

Mr. GARMAN. My own view is that it is prudent for us to allow the States to undertake their regulatory authority but to have back-stop capability in those key areas where it is needed, and that back-stop capability was provided in the energy bill for both electricity transmission and LNG siting.

But I think it is probably prudent, and I think the States would object, understandably, if we were to seize the powers and authorities that have been vested in them from them inordinately. But clearly it is something that, if the infrastructure is not being built, we have to ask the tough questions, why is that the case?

Mr. GREEN. So in the energy bill we did electricity and LNG. Now because of the production predominately in the Gulf of Mexico—and maybe instead of—you know, maybe we need to look at it and our committee needs to look at, like you said, a back-stop, some type of frame frames, maybe, that if you lease off of California or Florida that, you know, the States would have a certain period of time to approve, disapprove or whatever.

But, again, because we are not just talking about States' issues, we are talking about a national policy and actually international, which is the reason the energy bill dealt with the LNG terminal sitings.

So, Mr. Chairman, I guess I have only 30 seconds. But I know the issue of price gouging—what have past price gouging investigations found? Is it usually the large or small scale and is it usually large or small operators? Is the problem with oil companies or with retailers and distributors? And has that been—what has the history been? Because this is something—we have repeated this, I guess, for the last 30 years or more.

Mr. SEESEL. Representative Green, I guess the Commission's activity in the pricing area has largely involved receiving complaints about pricing behavior that violates one of the statutes that we enforce. So, generally, the investigations we have undertaken have looked at pricing that might have been collusive pricing and other activities.

For example, we had investigations in both the Midwest after the price spike about 5 years ago and in the Western States, an investigation that really lasted several years, between 1998 and 2001. And those were both really designed to look at whether pricing going on there was the result of collusive and coordinated activity among—at firms at various levels of the industry. There were long, arduous, detailed investigations that resulted in public reports and so forth, but they didn't turn up evidence of collusive behavior. There hasn't really been an effort to look at individual firm pricing behavior because of the problem that the Commission runs up against, that essentially it is dealing with a statute that primarily goes after collusive behavior, not individual firm behavior.

Mr. GREEN. Thank you.

Mr. HALL [presiding]. Thank you, Mr. Green.

I will recognize myself for my time, and I will not use the full amount, paying back some that the Chairman took on his own to extract questions and answers from you.

Mr. Garman, you were asked about the provisions in the Energy Policy Act. As we search for these policies that will relieve some pressure on the energy sector, both short term and then long term, I guess I want to focus on that Energy Policy Act. You spoke of the reticence or the lack of any indication of investors. It is a problem because, I guess, of the length of the return of their capital; and there are some incentives, though, that the EPA can make certain concessions or give certain instructions.

I guess what I want to know is whether or not—I know the Governors of the States have read the Energy Act, and I know that you all have gone over it and back and forth and everything. But everybody is aware of the provision in there and the number of the provision. I do not have to list that even for the policy. But have any of the States moved forward on this? Have you had any correspondence with them or any discussions with them for the use of this particular section of this Act?

Mr. GARMAN. I have not personally, nor am I aware that any States have focused on this provision yet. We are just at the interagency level. In fact, there is another meeting tomorrow at the White House, if I am not mistaken, where we are gathering to ensure that we have all of the implementation bases covered on the Act, that we are meeting, that assignments are made, that timeframes are being laid out, that we are making sure that the interagency cooperation that is needed to implement this Act is indeed under way.

Mr. HALL. Mr. Caruso, do you have any suggestions or any—have you had any contact or any inquiries from any of the States?

Mr. CARUSO. I have not, Mr. Chairman.

Mr. HALL. I am optimistic that one of the States other than one of the producing States might move forward with this policy. It is available to all of them.

What are the drawbacks, other than lack of expectation of early return on investment?

Mr. GARMAN. I think history has shown us that anyone trying to put a new energy facility of almost any kind—generation, transmission, LNG facility—faces a lot of local opposition. And NIMBYism rears its head, issues of environmental justice are brought to bear, and so investors tread lightly into that realm.

Mr. HALL. Well, I think one of the reasons that Joe Barton was able to pass an energy act when no one has been able to in the last 10 years is that he couched it with a lot of research—R&D, really mostly an R&D act, rather than an energy act.

But with the additional provisions and additional incentives in there and the situation we find ourselves in, there is increasing on an increasing ratio the difficulty of dealing with the people who are selling us energy that we have to rely on, that we do not really trust, and they do not really trust us.

With that ability to fall back on that new technology, it looks like we could shorten the time and make it a little more appealing to the investing public. They want a return, but they also want to see us solve a problem that might keep our kids out of the war, and that is to solve the energy problems of this country.

Mr. GARMAN. Yes, sir; and I think it is very important that we score a quick success on this. I think if the first new refinery gets built in 30 years, that will be a message to investors that this is an area that is ripe for new investment and new capital flow, and we can hopefully buildup the thin margins of capacity that we have suffered under for some time and have more cushion there to protect consumers.

Mr. HALL. All right. If I yield back 33 seconds, I have kept the faith with the chairman. I yield back.

The Chair at this time recognizes Ms. DeGette from Colorado for 5 minutes.

Ms. DEGETTE. Thank you, Mr. Chairman.

Mr. Garman, I wanted to ask you, as I mentioned in my opening statement, last night at the members briefing, Secretary Bodman told us that, aside from some short-term disruptions in the natural gas and also gasoline supplies in the southeastern United States, pretty much there were minimal disruptions in fuel supply. So I guess I am wondering, if that was the case, why did prices jump so significantly? For example, in Denver, Colorado, where my district is, not even in the supply chain of the gulf, the prices were going up as early as Tuesday of last week. I am wondering if the Department has some sense of why that happened?

Mr. GARMAN. I will provide an answer and then turn to my colleague from the EIA who actually is closer to the price situation.

But let me first make the comment that I believe Secretary Bodman, in making the comment that he made last night, was referring to permanent damage.

Ms. DEGETTE. Actually, I do not think that was true. I was there. But if you want to clarify his remark that way.

Mr. GARMAN. Because—I do. As I point out, Katrina, had a devastating impact on energy.

Ms. DEGETTE. But did it have a devastating impact as early as Monday and Tuesday in areas of the country that aren't even supplied by that region?

Mr. GARMAN. Yes. Refineries were shut down prior to the storm striking on Monday, and petroleum product and crude oil is a fungible material that affects prices outside a direct supply chain. Guy might have more to say on that.

Mr. CARUSO. Yes, there are two comments I could add. One, the market impact started on the weekend, actually, because some refineries were shut as early as Saturday.

And the second thing is that markets react to uncertainty. When there was uncertainty as early as Monday as to how much damage there would be, and how long it would last, the NYMEX futures market already started to rise. In fact, European markets rose, too so there are people in Paris and London today paying more for gas.

Ms. DEGETTE. And I guess it is a fine line between market uncertainty and price gouging, in many of our minds, because people price gouge because consumers are uncertain, and they know gas costs are rising, correct?

Mr. CARUSO. Well, as Mr. Seesel pointed out, the definition of price gouging is very nebulous. But, nevertheless, collusive and anticompetitive behavior is certainly——

Ms. DEGETTE. Well, and I want to ask both of you gentlemen just briefly, with respect to the Department's registration of consumer complaints about gasoline price gouging through its Web site and telephone hotline, how many complaints have been registered with the DOE?

Mr. GARMAN. The last time that I had checked, as of yesterday morning, I believe we had received on the order of 7,000 calls, which we were distributing to appropriate authorities at the Federal Trade Commission and States' attorneys generals.

Ms. DEGETTE. And what is the DOE going to do with this information after they distribute it to those appropriate agencies?

Mr. GARMAN. We do not have any regulatory or enforcement authority. They are the parties that do. So we are a collector of information, and we provide it——

Ms. DEGETTE. You are just going to pass that on.

Mr. GARMAN. We provide that information to the appropriate parties with the enforcement authority.

Ms. DEGETTE. Are you going to provide it also to the Department of Justice?

Mr. GARMAN. Yes, we are. And, yes, we have.

Ms. DEGETTE. Mr. Seesel, I would like to ask you, I was intrigued by your testimony, and we all know about half of the States currently have antigouging authority on their books, correct?

Mr. SEESEL. That is my understanding.

Ms. DEGETTE. And I think you were saying is that it is your view of the Federal statutes that if there is no proof of anticompetitive practices, then the Federal Government does not have the authority to prevent and punish price gouging. Would that be accurate?

Mr. SEESEL. I believe that would be accurate.

Ms. DEGETTE. So if one oil company is not colluding with another one, they just decide to price gouge on their own, it is your agency's view that, really, the Federal Government can't take any role in prosecuting that action, right?

Mr. SEESEL. Well, Representative DeGette, the antitrust laws that we enforce have historically consistently been interpreted not to give us the power to second-guess or sit in judgment on an individual firm's selection of a price.

Ms. DEGETTE. Exactly. So, really, State laws would be the only recourse for that kind of anticompetitive behavior, right?

Mr. SEESEL. That is correct.

Ms. DEGETTE. And for the half of the States that don't have that law on their books, there wouldn't be any recourse at all.

Mr. SEESEL. Well, I think there are a number of States have actually said they have more general consumer protection statutes that they are interpreting to deal with the gouging issue. Not the

States, the 23 or so States, that have explicit gouging prohibitions, but there are some other States that are starting to invoke their more general consumer protection laws.

Ms. DEGETTE. Just one last question. Do you think that it would be useful to give additional Federal authority to an agency like the FTC to be able to prosecute anticompetitive behavior in gas pricing?

Mr. SEESEL. Well, I certainly think that the FTC needs to deal with anticompetitive conduct in the sense that I have been talking about it, as the antitrust laws are historically interpreted.

Ms. DEGETTE. Right. But there is anticompetitive behavior that is not necessarily collusive behavior, Correct?

Mr. SEESEL. That is correct.

Ms. DEGETTE. And would you require additional Federal authority to do that?

Mr. SEESEL. I think in order for the FTC to deal with a unilateral price selection by a firm, we would need additional authority.

Ms. DEGETTE. Thank you, Mr. Chairman.

Mr. UPTON [presiding]. If I had known I was going to be chairman when it was time for my question, I might not have deferred on my opening statement but just stolen my 3 minutes. But I will now recognize myself for 8 minutes.

My first question has to deal with interoperability and public safety radio. As you know, this committee is poised to process a transition to digital bill, which would set a hard date for the broadcasters' return of their analog spectrum so it could be used for, among other things, public safety radio interoperability. And the tragic events of 9/11 underscored that dire need for a hard date so that we can clear the spectrum for public safety interoperability.

My understanding is that, in the wake of Katrina, public safety entities at all levels were not able to effectively communicate with each other, and I am hearing that there may have been a host of reasons for that, including the fact that police radio towers were knocked down, police radio batteries could not be recharged, no electricity. But can you tell me, Mr. Moran, about what, if anything, you have learned so far about the situation on the ground with respect to the lack of interoperability as a contributing factor to that situation?

Mr. MORAN. Yes, certainly. In the wake of the events in the gulf, the FCC works very closely with the National Communications System and other Federal agencies, so we do a lot of coordination, and we attempt to determine the status of the communications networks, which ones are working, which ones are not working. Also in those discussions there are—certain aspects of them, industry is directly involved. And the Commission, initially we went in there to see what we could do to expedite matters. And we would talk to the industry to see what their problems were, what systems were down, what could be done about it to quickly effect a good result there.

And I will tell you, the primary things that we were dealing with in the initial days were that the commercial power was out. Once it is out for a long period of time, backup batteries in the telecom and the communications systems will run down. Of course, many of the big installations have emergency generators; however, you have to get fuel to those generators to keep them running. And so what we were seeing in the first days after this was most of the infrastructure in the worst-hit areas were—communication infrastructure was not working because of power. But the biggest problems were as the carriers, as the communications carriers, were getting assets ready to get in there to see what they could fix, see what they could keep up, and try to assess the situation, the biggest problems we heard were they did not have security that would enable them to get their assets into the areas, and also transportation was really a serious problem trying to get fuel into these areas.

So the initial thing was it was a commercial power—it was a power issue, starting with commercial power and getting worse because facilities could not get in there.

Now, we dealt extensively with industry, and those were big issues that we were seeing and that we were dealing with FEMA and the NCS. We, of course, have all—we were watching this carefully, and we see the same videos that people around here are seeing where an emergency responder is using a couple of different communications devices because of an interoperability issue. And, of course, the underlying issue is that if some of the systems aren't even working, they might need more than one piece of equipment.

Clearly that is not an acceptable situation. The Commission—the effective communications for emergency responders is a priority for the Commission. We are looking into it; we will be working with the industry. And we have actually done a number of things in recent years to try to provide additional spectrum, for example, for the emergency service providers.

And so the initial problems we saw, we did see interoperability problems, but the biggest problems we saw initially were things that were needed to do to get the networks up. And that tended to be security issues, staging of personnel, to get them in there and have them secure, and also trying to get fuel into the areas until the power would come up.

Mr. UPTON. Mr. Caruso, in your statement you talked about the future particularly with home heating oil as well as natural gas. Now, you said that there were going to be significant increases in natural gas beginning this winter. Did Katrina seriously impact our natural gas supplies coming into the United States?

Mr. CARUSO. Yes.

Mr. UPTON. Tell me precisely what it was.

Mr. CARUSO. Yes. Initially—the Gulf of Mexico production of natural gas is about 21 percent of our national total.

Mr. UPTON. And where is that in terms of ramping back up to where it was?

Mr. CARUSO. It was 8.8 MMBtu per day shut in initially. As of yesterday, it was down to about 5 million Btu still shut in.

Mr. UPTON. And so it has come back to about 55 percent?

Mr. CARUSO. It is about 50 percent. And in addition to that, there are natural gas processing units onshore which were damaged, and that will affect the ability to——

Mr. UPTON. And do you think that will be a long-term problem, then, to get that back up to 100 percent where it was in natural gas?

Mr. CARUSO. In our preliminary assessment, we have it all back by end November. But certainly, until the investigators are able to get on the platforms and to actually test some of the pipeline infrastructure, we really don't know.

Mr. UPTON. I want Mr. Garman and Mr. Caruso and Mr. Seesel to respond to this question. Last week I drove probably every day a couple hundred miles in my district. And on Tuesday the gas price on average in our district in southwest Michigan went from 2.61 a gallon to some stations that day to 3.58, almost $1 increase within a couple of hours. When you talked about the price of oil per barrel going from $65 to $70, that is about a 7 percent increase. I think there are a lot of us that thought maybe gasoline would go up 7 to 10 percent, not literally $1 a gallon. Would you consider that type of an increase, knowing that we did have the supply—none of my stations in my district were out of gas. Would you consider that price gouging? Mr. Garman.

Mr. GARMAN. With that information alone, I couldn't make a judgment.

Mr. UPTON. Yes. The answer is yes. Mr. Caruso. I know you too well. Mr. Caruso.

Mr. CARUSO. I would have to say the same, but make the comment that the gasoline markets often times behave differently than crude markets, and vice versa, and it very much depends on the individual markets. For example, the NYMEX gasoline market, as I mentioned, went up sharply between Monday and Wednesday, much more sharply than crude, and many contracts are indexed to the NYMEX futures market. So part of that is explained by the wholesale and futures market rate.

Mr. UPTON. I look forward and I know we are going to have the witness from NYMEX in the second panel.

Mr. Seesel.

Mr. SEESEL. Again, Mr. Chairman, with the understanding of the sort of squishiness of the definition of the term "gouging," I would say that, again, I can't answer that, as Mr. Garman said, without information, but what you might be observing is a couple of things going on. One is a gasoline retailer hiking its price significantly because it expects to be paying a whole lot more for supplies that will be coming in the next day or the day after that, and generally an effort by the retailer sometimes to stay in business, at least to stay open and not put out a "no gas" sign, and essentially raising the price significantly in order to ration demand to accomplish that. It is—whether one calls it gouging, I don't know. It could be very well just fairly reasonable and expectable demand-and-supply responses to a situation of great shortage.

Mr. UPTON. Even though I am chairman for the moment, temporarily, my time has now expired. I yield to the gentleman from the great State of Michigan Mr. Stupak for 8 minutes, who also deferred.

Mr. STUPAK. Thank you.

Mr. Caruso, you are familiar with the term called "risk premium"; are you not?

Mr. CARUSO. Yes, Congressman.

Mr. STUPAK. And in layman's terms, risk premium is essentially the amount of money that is built into the future price of a good, in this case oil, that is above and beyond the amount of the normal price based on a number of factors that may impact the price of a good, such as terrorism, natural disasters, refinery problems, et cetera, right?

Mr. CARUSO. That is correct.

Mr. STUPAK. I have seen a number of articles that said that prior to the war in Iraq and prior to Hurricane Katrina, the risk premium on a barrel of oil, of crude, was in the neighborhood of about $2 to $4. Is that about right?

Mr. CARUSO. That is what a number of trade journals have said. It is not the position of EIA.

Mr. STUPAK. Right. Now, today, in the last couple of weeks here, the last couple of months, that terrorist premium risk, if you will, has gone up to $15 to $30 a barrel; is that correct?

Mr. CARUSO. I have not seen that large a number.

Mr. STUPAK. What would you think it is at right now, then, the risk premium on a barrel?

Mr. CARUSO. I think our models for crude oil indicate that you can explain most of the run-up in prices based on the lack of spare productive capacity for crude oil.

Mr. STUPAK. But to get back to the risk premium, though, As Kiplinger forecasts here, they say 15 to $20 a barrel, right? Would you disagree with that?

Mr. CARUSO. Yes.

Mr. STUPAK. What do you think it is then?

Mr. CARUSO. I think the risk premium for crude oil is very low. I think it——

Mr. STUPAK. Give me a number. What do you think it is?

Mr. CARUSO. I think it is probably only a few dollars.

Mr. STUPAK. You think it is still $2 to $4?

Mr. CARUSO. In that range. It hasn't changed that much for crude oil as a result of the recent event.

Mr. STUPAK. Here is another article that is within the last year, MSNBC, and this actually shows it about 15 to $30. And they talk about terrorism. Here is a—from Bloomberg.com which shows that the risk premium has substantially gone up, and this is actually from August 30, 2005, about $15 a barrel. So you are not familiar with any of these articles?

Mr. CARUSO. I am familiar with them. Yes. I disagree with them.

Mr. STUPAK. You disagree with them?

Mr. CARUSO. Yes, sir.

Mr. STUPAK. Well, let us assume that they are right and you are wrong. Okay? Apparently these authors in these articles disagree with you. So the price of oil right now per barrel is probably about 25 percent more, if we believe this risk premium is at $15, than what it should be, Right?

Mr. CARUSO. If you believe that, yes, sir.

Mr. STUPAK. How do you get your hands on this risk premium? I know you don't agree with us that it is $15 a barrel. You think it is much less than that. If it is not the risk premium, then what is making these prices fluctuate so much? And, really, the issue is $35, right? Now we are paying around $70, roughly.

Mr. CARUSO. That is right. It is 65 or——

Mr. STUPAK. Okay. So it has almost doubled in a year. If it is not a risk premium, then what is the factor that is causing it?

Mr. CARUSO. There are a number of factors. The first one is that world oil demand has grown rapidly in the last several years, putting upward pressure on price. There has not been sufficient productive capacity to meet that demand in the short term, there has been a lack of inventory to meet short-term demand, all of which means that we are operating an industry of 83 million barrels a day, on a global basis, at about 98 percent of capacity. So any short-term perturbations in either supply or demand, because of the low short-term elasticity of price or income, mean volatility and sharp price rises. And Katrina is a perfect example of that.

Mr. STUPAK. But actually once the administration finally heeded the advice of myself and others who have been for months saying release the SPR, didn't the price of oil go down after the barrels of oil were released from the SPR?

Mr. CARUSO. Yes, it did. As the Chairman mentioned, it had reached the peak of 70 in interday trading 1 day last week. It is now between 65 and 66 as of this morning.

Mr. STUPAK. So it has actually gone down. So isn't really the price that keeps it up is like instability in the world, such as like in Venezuela, one of our larger suppliers that we had some disagreements with, Iran that we have disagreements with over nuclear issues, Gulf, the war going on in Iraq, things like that?

Mr. CARUSO. Part of the reason that there is so much uncertainty and that refiners are willing to pay those prices is they don't know what is going to happen in places like those you have just mentioned.

Mr. STUPAK. Well, as these prices go up like this, whether it is risk premium or not, it has gone from 35 earlier this year to—peaked at 70. And who benefits? And take the case of Saudi Arabia; Aramco, right?

Mr. CARUSO. Yes, sir.

Mr. STUPAK. The people who purchased the oil from Saudi or Aramco, they would benefit, right? The refineries would benefit, Correct?

Mr. CARUSO. If they are able to sell it with enough of a profit margin, they benefit, yes, sir.

Mr. STUPAK. Sure. But the Saudis, if it takes $10 to get it out of the ground, and they sell it at whatever they sell it at, 40, there is a profit there. And they turn around and sell it somewhere else. And these futures are up to, what, 65 right now? So we have got some pretty good profits going right now, Right?

Mr. CARUSO. Absolutely. And——

Mr. STUPAK. In fact, every article I have seen, we have got record profits in parts of the industry, right?

Mr. CARUSO. That is correct.

Mr. STUPAK. Has the administration given any consideration to a windfall profits tax then? Mr. Garman, do you want to answer that?

Mr. GARMAN. I will, and say that, to my knowledge, no.

Mr. STUPAK. Is there—there is actually more—less refineries in the United States, but we are refining more oil into gas than ever before in the Nation's history, correct?

Mr. GARMAN. That is correct.

Mr. STUPAK. So everyone is making a pretty good profit here, and yet we are not doing anything to try and get control of the price of this oil other than release oil from this SPR, correct, trying to get your hands on this price's volatility. If it is not a risk premium, then, Mr. Caruso, if it is just supply and demand, but we have more, we are refining more, how do you account for those high prices then?

Mr. CARUSO. We don't have enough refining capacity. We are—as Mr. Garman mentioned, we are having to import more and more refined products from abroad.

Mr. STUPAK. We are importing about, what, 10 percent?

Mr. CARUSO. In refined products, a little over 2 million barrels a day on a net basis out of our 20-. So that is about right.

Mr. STUPAK. So what is the answer then to get control of these prices? More refineries?

Mr. GARMAN. There is a multitude of answers, and I don't think we should depend on any one answer. First, we should encourage new supply. In fact, let me make that second.

First, we should encourage conservation and efficient use of the supplies that we have. That is the quickest, cheapest, most dramatic effect that one can have in the short term, because it takes time to bring new production on line. New production is very important.

And then I think these two thin-capacity margins that Mr. Caruso have talked about are very, very important, both the thin-capacity margin on the production side. You know, for many years we had a production capacity margin of 3- to 5 million barrels a day. Now we are down to 1 million barrels a day, and most of that capacity margin exists in Saudi Arabia. It is in our interest to see capacity margins increased on the supply side upstream. It is also in our interest to see capacity margins increase down, on the downstream side, at the refinery side.

I think all of these are components, and I don't think we should hitch our wagon to any single effort. I think we have to take a comprehensive effort approach and urge Americans to conserve, urge producers to produce, urge refiners to invest in new refinery capability and capacity. We have to do all of those things if we expect to have a long-term impact on price.

Mr. STUPAK. So, in summation, when the President said in 2000 that he would just jawbone the Saudis into producing more crude, that really wasn't an answer or a correct answer to a complex problem.

Mr. GARMAN. The President, in my view, had a very comprehensive answer in his May 2001 energy plan that depended on both supply options and demand options. Roughly half, if I recollect correctly, of the recommendations in the President's original plan had to do with energy efficiency, renewable energy, and other alternatives to the status quo.

Mr. HALL [presiding]. The gentleman's time has expired.

The Chair recognizes the gentleman from Florida Mr. Stearns for 5 minutes.

Mr. STEARNS. Yes. Thank you, Mr. Chairman.

My colleague from Michigan, Mr. Upton, asked each of you a question about his experience driving around his district, and it appeared to me that we couldn't have a definition among you on what price gouging is. It seemed to me there is lots of factors. So when he tried to present you a case example, you really couldn't agree with him because you said there is other factors.

Now, if I describe cheating to you, I think we can all agree what cheating is. If I describe stealing, I think we could all agree what stealing is. It seems to me we just have to understand that gouging is something that we can discern and involves several components. I looked up the definition of it, and it comes from old Middle English which comes to sting. Now, following that definition is to basically—to cut or force out a rough cut of something.

I submit that gouging involves a couple things: It has a moral component; that is, it is a necessity that we need to have, so we are forced to buy it. Just before Valentine's Day, I notice that roses go up. But I don't necessarily have to have those roses, I can get carnations or something else. But if I have to get to work, I am going to need gasoline. So I think gouging involves a moral component.

Second, I think it involves a time factor, generally 1 or 2 days. If the price goes up, as Mr. Upton indicated, almost doubled in a period of 1 day, that is obviously price gouging.

And, last, I submit that the buyer is coerced and intimidated.

So I don't think—if you throw those three components in, it is not hard to discern and to see what is gouging.

Now, as I understand it, on the Federal level we do not directly have laws on gouging. Is that right, Mr. Seesel?

Mr. SEESEL. That is my understanding, Congressman.

Mr. STEARNS. But we do in the event of collusion. So, if companies work together, then we can step in and say there is collusion, but also price gouging; is that correct?

Mr. SEESEL. Well, collusion may take a number of forms. Traditionally, classically it would take the form of conspiracy to raise prices or reduce output. And so it might not take the form of gouging in the sense that you may be thinking of it going from 2.50 to 3.50 a gallon.

Mr. STEARNS. Do you think we should have Federal laws dealing with price gouging separate from the idea that you have to have collusion first?

Mr. SEESEL. As I said before, I think the Congress needs to tread quite carefully in this area because——

Mr. STEARNS. What about the idea that, even in the States where they have laws dealing with price gouging, it only generally applies in state of emergencies? So in this case we had four States declare a state of emergency. What about the other 46 States? How do you handle price gouging in those States?

Mr. SEESEL. I have seen quite a few media reports in the last few days, Congressman Stearns, of attorneys general and Governors in States quite far away from the gulf region that are applying either their specific price-gouging statutes or their more general consumer-protection statutes to deal with the gasoline pricing situation. And I think their sense is there is an economic emergency going on in their States. I think that is one of the rationales I have read. So that irrespective of the physical emergency of the hurricane, a lot of States have been able to proceed, begin investigations under the rubric of their general statutes.

Mr. STEARNS. In the State of Georgia, there was an example where gasoline was selling for $6.38 a gallon, yet that was not one of the States where there is an emergency. Texas, South Carolina, there was huge increases. Surely I would think from your standpoint that those would be areas you would investigate, because they are not in a state of emergency, yet it appears that the States are almost helpless to stop price gouging.

Mr. SEESEL. Well, Congressman, my understanding—and I may be wrong about one or two of the States you mentioned, but I think the attorneys general actually have announced they are looking at pricing issues for gasoline in those States. Even though—again, even though they are outside the Louisiana, Mississippi, Alabama, Texas area, they are still invoking their price-gouging or general consumer statutes to look at what has been going on.

Mr. STEARNS. Let me conclude before my time is out that you had a report that you issued, and this report came out before Katrina, before the hurricane, entitled: Commission Report on Factors that Affect the Price of Gasoline.

How has this report affected your ability to scrutinize the marketplace for collusion, for price gouging, for any things, even such things as State and local regulations that affect it? Maybe you can give me just a summary of what your report provided so that you could help us in the future on this matter.

Mr. SEESEL. Well, the report really—we have a fairly broad panoply of statutes that we enforce that we use in our law enforcement program. The report was really an attempt to set out in a very concise way all of the learning we have accomplished over the last 20 or 25 years in the petroleum industry on analyzing the various factors that will go into driving the price of gasoline, whether you are talking about supply factors, demand factors, competition for the various resources that go into the product such as crude oil and refining capacity and so forth. So what we attempted to set out here was a—I am sorry.

Mr. STEARNS. Mr. Chairman, could we allow him to finish his answer?

Mr. HALL. The gentleman will be allowed to finish his answer, if he wants.

Mr. STEARNS. Thank you, Mr. Chairman.

Mr. SEESEL. We essentially try to set out the entire spectrum of supply, demand, regulatory, and other factors that result in prices and volatility in gasoline.

Mr. STEARNS. Thank you, Mr. Chairman.

Mr. HALL. I thank you.

The Chair recognizes the gentleman New York, Mr. Engel, for 5 minutes.

Mr. ENGEL. Thank you, Mr. Chairman.

Gentlemen, with all due respect, and I think you have heard the frustrations of all of us, you know, if it looks like a rat and smells like a rat, it is a rat. The American people aren't stupid. And I remain totally unconvinced that 2 days after Hurricane Katrina happened, gasoline prices went sky high as a result of the catastrophe of that hurricane. It would certainly take much longer to have the hurricane's catastrophe translated into higher gas prices at the pump. There is no way other than price gouging that it could happen within 2 days. And, again, we all have seen that, when the price of oil drops, it takes several weeks, if not months, for that to be reflected at the pump with prices of gasoline dropping. So I just think that there is no way we can excuse it.

The American people aren't stupid. The Representatives on both sides of the aisle aren't stupid. We know that there was price gouging. And I just think it is absolutely unconscionable.

Last week I went to get gasoline right here on South Capitol Street, and there were two gasoline stations within a block from each other, and there was a 40-cent difference in the price of gasoline between those two stations. I have spoken with gasoline owners, owners of gas stations, who said that they were told by the companies to increase the prices.

I don't think this is something we can kind of talk away or just kind of have business as usual. The American people are sick and tired of it and want an explanation and don't want it to happen again. And, again, when prices sink next week or in a month or whenever it is to below $3 a gallon, we are not going to jump for joy, because as far as I am concerned, $2 and change is too much to pay.

I want to ask Mr. Garman, yesterday Senator Domenici said: We are too dependent on this part of the country. Congress must do something on offshore drilling because we need more diversity than what is out there.

I agree that we need to diversify our energy sources, but I respectfully disagree with the Senate chairman about how. To me, Hurricane Katrina has shown again that our Nation is overly dependent on oil itself, not gulf coast oil. If we are going to create a stable energy future for our country, we need to diversify away from our oil. The answer, in my opinion, is not opening drilling in Alaska or all along our Nation's coasts to increase oil supplies on the mere margins, but to aggressively promote technology such as advanced hybrid automobiles which will substantially reduce our demand for oil. If we were serious about energy policy in the wake of Katrina, we would significantly increase CAFE standards for passenger vehicles, not propose insignificant adjustments as the Bush administration recently did.

Next week I and a bipartisan group of Members will announce the founding of a new Oil and National Security Caucus. The purpose of our group will be to highlight bipartisan, common-sense ways to reduce our dangerous overreliance on oil. We will work with members of the committee and the administration to offer serious practical proposals to provide more balance in our energy mix.

Now, Mr. Secretary, in the wake of Hurricane Katrina, what policy adjustments are you and the administration proposing to diversify our Nation away from oil?

Mr. GARMAN. As you know—and admittedly this is a long-term approach—back in January 2003, the President in his State of the Union Address announced his hydrogen fuel initiative, which is a long-term effort to totally take our personal transportation off of petroleum, completely, through the use of hydrogen and fuel cell vehicles fueled by hydrogen fuel that can be produced here domestically from a variety of different energy, primary energy inputs. That is a long-term goal, admittedly, and we do not expect to see affordable hydrogen fuel cell vehicles that need no petroleum and emit no pollutants in an affordable fashion available to consumers prior to 2020. They are on the road today, but neither you nor I can probably afford them. So we have to bring down the cost.

In the energy bill that was just passed, a proposal that the President made back in January 2001, production tax—I am sorry, a consumer tax credit for hybrid vehicles to get more of these vehicles on the road is another very, very important component. We want to encourage that.

There are a lot of things in the energy bill that have not yet been implemented.

Mr. ENGEL. Let me just say, because I know my time is over, I am told that the FTC maintains a gasoline price monitoring project, and the DOE has a Web site for filing gasoline price-gouging complaints. What do you do with the Web site which permits the filing of complaints? Is it just for people to let off steam and feel good? What practical substance can we tell the American people that, if they feel they have been ripped off at the pump, that they can effectively do something and that government will move to make sure that it doesn't happen again?

Mr. SEESEL. Congressman, the Commission receives complaints from all kinds of sources. If we get a complaint that deals with gasoline pricing, to the extent the complaint spells out a law violation of the kind that we can go after, we will proceed vigorously against that. If it spells out an issue that the FTC really does not have authority to go after, that is the kind of thing we would refer to State attorneys general. It is not just a mechanism for people to let off steam; it is a way for us to learn information from consumers, some of which will be turned into law enforcement investigations that we can pursue.

Mr. WHITFIELD [presiding]. The gentleman's time has expired. It is my time to ask questions, so thank you.

You know, as elected representatives, and as people who depend upon being reelected to their position, obviously all of us are very much concerned and want to do everything we can to defend against higher gas prices. And our ability to do that will oftentimes depend upon whether or not we are reelected. But I was reading an article recently, on August 26, that indicated that in Amsterdam the price of gasoline was $7.13 per gallon. The price in Great Britain was $6.43 per gallon. The price in France was $6.90-some cents per gallon. The price in Belgium was in the $6 range. And the price in Greece was about $4.80 per gallon. And in the U.S., on August 26, it said the price was about $2.61 per gallon.

So the question I would ask: Is it realistic for the American people to expect low gas prices, maybe the lowest gas price in the world, when we have—is that not? What is the lowest gas price in the world?

Mr. CARUSO. The lowest is in places like Saudi Arabia and Iraq.

Mr. WHITFIELD. And how much are they paying per gallon?

Mr. CARUSO. Saudi Arabia, when I was there in May, they were paying about $1.50 a gallon. But they are much lower in Iran.

Mr. WHITFIELD. And how much are they paying in Iran?

Mr. CARUSO. Less than $1; 50 cents probably.

Mr. WHITFIELD. But is it realistic for a country like America where we have such a small amount of reserve. We do consume more than any other country in the world. Out of the 84 million barrels being consumed each day, we are consuming around 21 million barrels a day. Is it unrealistic for the American people to expect prices below $3 a gallon?

Mr. CARUSO. I mentioned that in our short-term outlook, prices are likely to come down below $3 in the coming days and weeks and to be back to about $2.60 in the fourth quarter. In the long run, the main difference between the European prices and the U.S. is taxes. They are a large component of the high price in the U.K.

Mr. WHITFIELD. It is my understanding that in Europe, I mean, the taxes might reflect 60 percent of the overall price of the gas, which might ask the question: Should we—I am not advocating this, but should we increase price on the gas for taxes?

There has been a lot of comments today about refineries and the lack of investment in refineries. And I read an article recently that said major oil companies are keeping a tight rein on their capital expenditures, and that they typically for any project will do a stress test for profitability at $20 a barrel or below; that they are making their decisions based on a price of $20 a barrel and below. Have you heard any comments about that, or do you think that is true? Or does that explain why we have not had a new refinery since 1976?

Mr. GARMAN. I am not familiar with the specific article that you read, but I am surmising that that $20 figure may relate to exploration and production investments by oil companies. They have been burned before when they have made their exploration and production investments in new finds expecting to find 25 or $30 a barrel oil, and then it fell to 18 or 19, and have been—had an unprofitable investment. So I have heard anecdotally that oil companies are now looking for 25 or even $30 investment, which is up from the past, in analyzing whether a new exploration and production investment is worthy of a payoff.

Mr. WHITFIELD. And I would be willing to stipulate that there has probably been some price gouging going on. My understanding, that at the retail level, that major oil companies own, what, like 10 percent of the retail outlets? Or is it more than that, or do you all have any idea?

Mr. CARUSO. I don't have that number off the top of my head, but I would certainly be able to provide it for the record.

Mr. WHITFIELD. Mr. Seesel, when was the last time a major oil company was successfully prosecuted for collusion or for monopolistic pricing of gasoline in the U.S.?

Mr. SEESEL. Mr. Chairman, I don't recall the last time. If what you are talking about was a criminal prosecution, which, of course, would be handled by the Justice Department, I don't recall the last instance of that, I am afraid.

Mr. WHITFIELD. What about from a civil standpoint?

Mr. SEESEL. From a civil standpoint? Could you hold on 1 second, sir? I am reminded, and I should have remembered, that the Federal Trade Commission's own administrative complaint against Unocal was essentially a monopolization complaint. So the allegation that Unocal had deceived the California Air Resources Board with regard to its patents on CARB gasoline was a monopolization case. That is the one that I mentioned in my opening statement.

Mr. WHITFIELD. What year was that?

Mr. SEESEL. The complaint was issued in 2003; the case was just settled about a month or so ago, about a month and a half ago.

Mr. WHITFIELD. And wasn't there some $20 million fine against some companies in Hawaii recently? Or was that a couple years ago? Or are you familiar with that?

Mr. SEESEL. I am not that familiar with that, Mr. Chairman.

Mr. WHITFIELD. Okay. Thank you.

My time has expired. I recognize the gentlelady from California Mrs. Capps for 5 minutes.

Mrs. CAPPS. Thank you, Mr. Chairman.

And I want to turn again to Mr. Moran, the FCC's topic. It is now 4 years after the tragic events of September 11 when this country made many promises and pledges to look carefully at what needed to be improved at that time. I want to ask you if you consider today's emergency alert system to be the most robust and effective mechanism for warning the American public of an emergency? And, if not, what steps has the FCC taken in the past 4 years to make the EAS more effective?

Mr. MORAN. Well, the EAS system is designed really to deliver the Presidential message to all Americans through the broadcast process when the Nation's security is at risk. It is available for use on a voluntary basis for State and local emergencies, and it can be used for that purpose, too.

The Commission—it is an operational system. It is tested all the time. We work with FEMA to ensure that this thing is operational. There haven't really been any changes in the EAS system over the last 4 years, if that is your question.

Mrs. CAPPS. Well, there have been no steps, no rulemaking?

Mr. MORAN. Yes. The Commission has a rulemaking—the Commission began a rulemaking a year ago. That rulemaking is in process. We are looking at a number of things. Among the things we are looking at, to see if we should make the use of the system by State and locals mandatory. Right now broadcasters have to agree to accept the State and local messages in order for it to work. We also are looking to make sure—the current system actually does not require digital broadcasters to be in the system, and we are looking at that——

Mrs. CAPPS. But I want to ask, when does the FCC plan to conclude this proceeding? And why is it taking so long?

Mr. MORAN. Well, we have an open proceeding. I really couldn't give you a timeframe as to exactly when it is going to happen, but the record has closed, and we are looking at—we are working on it actually right now. There is also, you may be aware that there is an executive branch—the executive branch began an overall review of all the emergency alert processes in recent months.

Mrs. CAPPS. In addition? That is a separate action from the FCC's action?

Mr. MORAN. Yes. There is a committee that has been established.

Mrs. CAPPS. So another committee. Another way of— Mr. MORAN. By the executive branch. But it is actually much broader than just EAS system. It is to look at such things as should we involve cellular phone systems in the system.

Mrs. CAPPS. Yes, I want to get to that, too. Was the emergency alert system activated with respect to Katrina and the breach of the levees?

Mr. MORAN. My understanding, it was not used in this instance.

Mrs. CAPPS. It wasn't used at all. And it has been 4 years. So nothing——

Mr. MORAN. Well, but the issue of is it used, that is an issue— the FCC, we have responsibilities to make sure we have an operational system, and we do, and we have testing rules, and we are involved in that.

Mrs. CAPPS. But you don't even know if it works because it wasn't used for Katrina.

Mr. MORAN. We do know that it works because there is a lot of testing procedures that we do to make sure that——

Mrs. CAPPS. Why wasn't it used?

Mr. MORAN. The decision as to if it is used or not even on the national message, that is a decision by the executive branch; but on the State and local side, it is the State and local government's decision as to whether or not to use it.

Mrs. CAPPS. Okay. Can I ask you how many times the Federal authorities have activated it since it has been put in place?

Mr. MORAN. The current EAS, I believe, was put in place in 1994, and it has not been used for the Presidential message; however, it has been used hundreds of times for State and local messages, for hurricane and tornadoes——

Mrs. CAPPS. But in this case, when the President activated the Federal Government's involvement after the hurricane, it still wasn't——

Mr. MORAN. It was not used by the Federal, and it is my understanding that State and local authorities have not used it either.

Mrs. CAPPS. They didn't use it either in this case?

Mr. MORAN. In this instance. I think the warning that the hurricanes were coming were so broadly known by everyone, it didn't probably need to be activated at that point.

Mrs. CAPPS. Well, that is a different question, isn't it, sir?

And then you started to—if I could just use the last few seconds. And, by the way, many States now have Amber Alerts. My State has one that works very well for child abductions all the time. So this technology is there.

But let me ask you finally to talk about the digital broadcasting. More and more people are watching digital television. That is a subject of concern to this committee as well. Does the FCC plan to institute that part of this? Will digital requirement become part of whatever is done?

Mr. MORAN. That is an issue in this proceeding, but I can't speak to the proceeding until the Commission decides how it is going to act in that. That is an issue before the Commission in this proceeding that I told you about. That is before the Commission, should we direct the digital broadcasters also to participate. That is before the proceeding, but it is an open proceeding.

Mrs. CAPPS. One final, because I have a lot more questions to ask. Who in the executive branch triggers the alert, if I could ask you that?

Mr. MORAN. It is in the White House. I think it might be the Office of Science and Technology Policy.

Mrs. CAPPS. But you don't know?

Mr. MORAN. I don't know for a fact.

Mr. SHADEGG [Presiding]. The gentlewoman's time has expired. The Chair recognizes the gentleman from Georgia for 5 minutes.

Mr. NORWOOD. Thank you very much, Mr. Chairman.

Mr. Garman, can you tell me once again how many refineries in this country are in operation?

Mr. GARMAN. There are a total of 10 refineries that are out of operation, that are shut down at this moment. However, we expect four of those to come back up relatively quickly, maybe one of them as soon as today.

Mr. HALL. But the total in the Nation?

Mr. GARMAN. We are hearing 136 from the folks behind us.

Mr. NORWOOD. And the percent of those in the gulf coast, Whether they are working or not?

Mr. GARMAN. There was 8 million barrels a day of refining capacity in the gulf coast; 2 million barrels a day were down at the height of the storm, and we have got about 1- of those back.

Mr. NORWOOD. But the percent at full production out of the Gulf of Mexico is what for the Nation?

Mr. GARMAN. Roughly 50 percent.

Mr. NORWOOD. That is what I thought.

Now, why is it all there? Why did so much accumulate in the gulf coast?

Mr. GARMAN. My theory is that, No. 1, that is where the production, most of the offshore production, virtually all of Nation's offshore production, is located there. That is where investors have been successful in building refineries.

Mr. NORWOOD. Because of why? Why were they successful there?

Mr. GARMAN. For a reason they found that closer to product, and they possibly found a more willing and obliging State regulatory regime.

Mr. NORWOOD. We haven't had a refinery in 30 years because of State regulation?

Mr. GARMAN. It is not solely State regulation.

Mr. NORWOOD. Okay. A good bit of it?

Mr. GARMAN. It is the willingness of the community to host a refinery.

Mr. NORWOOD. That is where I am sort of going. Are there many refineries in California?

Mr. GARMAN. I am told there are 21 refineries in California.

Mr. NORWOOD. And how about Massachusetts?

Mr. GARMAN. I am not aware of a refinery in Massachusetts.

Mr. NORWOOD. The problem is we don't have our refineries distributed around the country appropriately, I guess, and I think that is something that we significantly need to deal with, and I hope the energy bill does. And Chairman Barton says perhaps we go back and look at that again, but I view that as a major part of the problem, not just because of the hurricane which knocked out so much of our capacity but because it was in one spot.

One of you have said that the price of crude oil is the single largest factor in the price of gas. Am I correct in thinking that?

Mr. CARUSO. That is correct. It accounts for about 60 percent right now.

Mr. NORWOOD. And the price of crude oil then is determined by supply and demand?

Mr. SEESEL. Yes, sir.

Mr. NORWOOD. And a great deal of the supply today is from India—I mean, the demand is from India and China?

Mr. CARUSO. China is second largest.

Mr. NORWOOD. Are they making gasoline?

Mr. CARUSO. Not enough.

Mr. NORWOOD. Do they demand that much gasoline in China that the amount of crude that they are buying is going to gasoline?

Mr. CARUSO. In China there is a much smaller percentage of the total barrel going into gasoline because of their small passenger car fleet.

Mr. NORWOOD. So what do they do with that crude if they are not making gasoline?

Mr. CARUSO. Diesel fuel for generators and trucks and jet fuel, of course, and also diesel for railroads. And they do burn some for electric power, but not a lot.

Mr. NORWOOD. Well, the crude oil we purchase, what percent of it goes toward fuels, gasoline, diesel?

Mr. CARUSO. Approximately 70 percent of our oil.

Mr. NORWOOD. That is what I thought.

Is China anywhere near that?

Mr. CARUSO. No, sir. They are much lower. I would say probably about 25 percent.

Mr. NORWOOD. Let me just make one last comment, Mr. Chairman, to these gentlemen. And we thank you so much for your time. Mr. Upton made a point earlier asking, "was an increase of almost $1 or 75 cents over a period of 24 hours or 48 hours, is that called gouging?" And none of you could answer that or would answer that. And I think you are doing a disservice to your boss. The fact is it is gouging, and the reason it is gouging is because the American public perceive it as gouging.

I know there are other factors out there, there are other things, but the fact is most Americans don't understand and are not willing to understand why gas could go up 75 cents in 2 days. And we need to deal with that as we speak, I think, in an appropriate manner, and we, too, should agree that something is bad, wrong for that to go up 75 percent. Many people think that individual service stations are trying to make an extra buck—maybe they are, maybe they aren't—but I don't want you to deal with it, I don't want the Feds to deal with it. In my State of Georgia, as was pointed out by Mr. Stearns, some of them went up $6 a gallon. Our Governor took care of it: He sent the State Patrol over and arrested them. Now, that is who I want to deal with that. He called it gouging.

Most of us thought it was gouging. And he put a stop to it in Georgia very quickly.

And that is how I think we need to do it. We really don't need the Feds to help us with that. But I wish you would be careful of how you speak of the increased price of gasoline that has sometimes gone up in a most unreasonable manner.

With that, Mr. Chairman, I am happy to yield back.

Mr. HALL [presiding]. I thank the gentleman. I thank him for being so plainspoken. Did they torture that guy before they put him in jail?

Mr. NORWOOD. No. They even fed him, too. It stopped it, Mr. Chairman, I promise you.

Mr. HALL. I thank you.

The Chair recognizes the gentleman from Texas Mr. Gonzalez for 5 minutes.

Mr. GONZALEZ. Thank you very much, Mr. Chairman. I always wanted to be on this committee, got my wish, and of course the first thing we were dealing with was whether people's TVs would go dark after we converted to a digital system. It appears now we have to make a decision about gasoline, whether people are going to be able to afford gas to leave their homes; and, if they can't, they are going to stay home, and then when they turn their TVs on, they won't be able to get a picture once we convert over to digital.

But on the serious side, I had a question for Mr. Caruso. I think you are going to be the most quoted guy on Capitol Hill after today. What you are saying, in the fourth quarter gasoline prices will recede somewhere around to $2.60; is that correct?

Mr. CARUSO. That is correct.

Mr. GONZALEZ. And then next year, 2006, it will be around $2.40 nationwide average?

Mr. CARUSO. Yes, sir.

Mr. GONZALEZ. Because, believe me, our constituents are waiting for some sort of word about what they are going to be paying, and the fact that it is going to be coming down. For the first time, I think we have heard some discussion regarding efficiency and conservation.

And I am going to ask Mr. Caruso, and, of course, Mr. Garman, you can have your own opinion on this, this is from an article that appeared in the Wall Street Journal, And it said: Last month, the administration proposed a sweeping restructuring of the light truck fuel economy rules, claiming the new policy would save the country 10 billion gallons of gasoline over the lives of vehicles bought from 2008 to 2011. Critics say the administration's move wasn't enough. David Friedman of the Union of Concerned Scientists, environmental advocacy group, noted that 10 billion gallons of gasoline is the amount that the United States uses approximately every 25 days. Quote: It is meaningless, he said, of the administration's new fuel economy proposal. The administration is bragging about saving less than a month's worth of oil over two decades.

Would you agree with I guess it is Dr. Friedman's analysis of the administration's proposal, Mr. Caruso, as its real effectiveness?

Mr. CARUSO. I would not agree. I think that any savings is better than no savings. So the statistics that you have quoted I believe are accurate, but I would disagree with the characterization.

Mr. GONZALEZ. Well, something is better than nothing. I guess we can all agree on that. But sometimes something is not meaningful, and that is what I am getting at is we could do something that is more meaningful, or people will be staying home a lot longer. And if we don't do the digital conversion, right, then they won't even have a television either.

Mr. Garman, do you agree with David Friedman's analysis that it really isn't meaningful or substantive?

Mr. GARMAN. I would make a couple of observations. First, the administration previously raised CAFE standards on light-duty trucks in the 2005 to 2007 frame. I mean, we have raised—the administration has raised CAFE standards now twice, standards that had not been raised at all prior to or since the 1996 model year.

And I think it is also important to point out that consumers have the opportunity to choose high-mileage vehicles today. They don't. It is not a requirement that the government require manufacturers to make these vehicles. I drive a vehicle today, and have since 2001, that gets more than 50 miles per gallon, and that is a choice that I as a consumer can make and have made that choice.

So I think the question is consumers can buy vehicles today that deliver great efficiency, and we should encourage them to do so. The question is do we force markets and mandate consumer behavior and look to that as the answer? It is certainly an approach, and it is a tool that we have used in the administration, but we also want to encourage consumers to buy fuel-efficient vehicles not because the government is telling them they must, but because they realize that it is in their own self-interests.

Mr. GONZALEZ. And I understand the big argument about choice; consumers should be given choice. But, you know, policy and regulation and such is guidance so that you avoid a situation where maybe our oil industries and others aren't really prepared for suddenly a drastic shift in consumer choice. And that is what I am talking about. When you are talking about the price of gasoline reaching $3, $4, and $5, it definitely will cause a tremendous shift in the way the consumers will choose. And I do think we need responsible policies that will point us in the right direction. You are talking about hydrogen, we are talking about hybrids, we are talking about greater efficiencies. Those are realistic goals and policies that we should be instituting. And I think to simply say, well, we are going to deprive people of choice if we don't do these things, I don't think that that is really what is going on.

I have a real quick question for Mr. Moran, and my time is up, and I am hoping that they will allow you, and that is voice over Internet protocol. We know that you have policies, programs, regulations, and coordination of what is going on out there. But what about this new method or manner of providing phone service to households? That is not incorporated in any of your plans, is it?

Mr. MORAN. It is incorporated. We have done some things to allow voice over Internet to work and also to allow 911—to make it mandatory, actually, for voice over Internet providers to provide 911 services. But do you mean in the emergency alert?

Mr. GONZALEZ. In the emergency alert system, sure, Because I am just thinking that traditional providers come under the scheme. But I don't think voice over Internet protocol would.

Mr. MORAN. I think we may have had some questions in our proceeding. I would have to check, though. We cast a pretty wide net in our proceeding that is ongoing right now. I would have to ask to see if we asked questions about voice over Internet, and I don't recall if we did.

Mr. GONZALEZ. If you could get back to me on that.

Mr. MORAN. Sure.

Mr. GONZALEZ. Thank you very much, Mr. Chairman.

Mr. HALL. The gentleman's time has expired. Don't worry about us over in my area; we are still watching radio over there.

Mr. Shimkus, the gentleman from Illinois, is recognized for 5 minutes.

Mr. SHIMKUS. Thank you, Mr. Chairman. We have already mentioned about the inability to build a crude oil refinery, but it is not true that we haven't built refineries in this country; is that correct? I don't care who wants to jump up there. How about Mr. Garman?

Mr. GARMAN. Are you speaking of ethanol production?

Mr. SHIMKUS. Any.

Mr. GARMAN. Yes, sir. We have built—there are in excess of 70 ethanol production facilities in operation today, and that number is growing as the weeks go by.

Mr. SHIMKUS. And so I have always chided my friends that, for us in the Midwest, they don't want to move and build petroleum-based refineries; we will continue to build renewable-fuel refineries. In fact, I have got 90 ethanol plants that are on line today. There are 17 plants that are under construction as we speak. There are seven in Illinois, and 17 are planned for Illinois as we speak. That kind of talks to some of the benefits of what we did in the energy bill. We have to decrease our reliance on foreign oil, we did make great progress on this issue, and I would encourage other feedstocks to look at this as an alternative.

I also wanted to direct my colleagues, I do a lot of energy debates and discussions back in my district because of the expertise or lack thereof that you develop over the years. And one of the slides I had was China crude imports. In 1996, China demanded—and this is from the International Energy Agency, it is out of the National Journal—22,828 million tons of crude imports in 1996—that is a billion; 122—then there is 122,699 million tons in 2004, which is a sixfold increase in 6 years.

So for people to—so when I am asked by my public how—what is going on, I have to talk about that fuel tanker that is loaded with crude oil that is going to go to some port, it is either going to go to India or it is going to go to the United States, or it is going to go to China, what is going to determine where that tanker of crude oil is going to go, to what port? And, Mr. Garman, what would you say?

Mr. GARMAN. The willing buyer that will pay the highest price.

Mr. SHIMKUS. And that is what the public has to understand.

Now, I would like—Mr. Caruso, how much crude oil reserves do we have within the continental United States or off our Outer Continental Shelf?

Mr. CARUSO. We had proven reserves at the beginning of this year of about 28 billion barrels total, and that is in those areas that have been drilled.

Mr. SHIMKUS. How much do we not have access to?

Mr. CARUSO. I don't have that precise number, but there is a significant amount of resources that haven't been drilled and proven, but that the USGS believes are available to be discovered. And I could provide that number for you. But it is a significant—it will be a significant increment to the proved number I just gave you.

Mr. SHIMKUS. And so many of us are very frustrated by this debate. It is incomprehensible that we in this country are importing refined product. Just think of it from the job creation aspect. We allow someone else to get the crude oil in some other refinery, so they have got the tax base, they have the jobs, and we get the refined product? My constituents don't really understand that.

In 1998, my first term as a Member of Congress, myself and a colleague of mine, Karen McCarthy from the great State of Missouri, worked on changing Federal law under EPAct on which we were able to give biodiesel credit, which changed the debate from just vehicles purchased to fuel usage. And we have got a 50 percent credit now for fleets, and many fleets are moving to biodiesel, decreasing reliance on foreign oil. In fact, it has really hit now. Willie Nelson and all these stars are into soy diesel, and we are glad to have them on board.

Has the administration thought about working with us to move that credit to 100 percent?

Mr. GARMAN. We will certainly engage with the Treasury Department and have those discussions. To my knowledge, I have not been part of any such discussions.

Mr. SHIMKUS. If you can get back to me, I would appreciate it. And if we can be helpful, I would like to be.

Mr. Chairman, thank you. I yield back my time.

Mr. HALL. I think the gentleman's time has expired.

The Chair recognizes Mr. Inslee from the State of Washington for 5 minutes.

Mr. INSLEE. Thank you.

Last weekend I went down to the Astrodome to work with the evacuees, and I can tell you, I was so inspired talking to these people. The resilience, the graciousness, the appreciation of these people for what the country is doing for them, it was really inspiring.

On our way back, we were flying back from Houston to D.C., halfway through the flight one of the evacuees who was heading from Florida took out a razor blade and slashed his wrist to cut it and take his life. He had just simply had it. And the reason I mentioned, that we diverted our flight plan to Nashville to rescue him, and he is okay; we had two EMTs on the flight, and it was a happy ending of sorts. But we changed our flight plan to accommodate that.

And it is clear to me that we need to have some major flight plan changes in this country. We need a flight plan change on our fiscal policy to pay for this what is going to be close to a $100 billion disaster by the time we are done. We need a flight plan change to deal with global warming, which has the capacity to make these hurricanes more intense. And it is clear to me listening to you today we need a flight plan change on dealing with gouging and those who take advantage of these poor, not just evacuees, but all of us.

Now, what I would like to make real clear on, we have heard that the Federal Government does not have the ability to force these antigouging laws in the absence of collusion. I would like to know whether your administration is here asking the U.S. Congress to take action to give the administration more authority to prevent gouging from taking place. And the reason I ask you that is that this administration sat on its hands and let Enron abuse us to the tune of $1 billion and did nothing in the State of Washington. We don't want that to reoccur. So the question to you gentlemen: Are you asking us to do something to prevent gouging?

Mr. GARMAN. No, sir, I am not here today for that purpose.

Mr. INSLEE. Anyone else?

Mr. SEESEL. Congressman, as I said to some of your colleagues, I would need to check with the full Federal Trade Commission on what it is asking for, but I don't believe the Commission——

Mr. INSLEE. Well, I have to tell you, that is extremely disappointing. You know, when we heard the President say that no one could have anticipated that these levees would have been breached, and as a result we had a pathetically indifferent Federal response to this terrible tragedy, and those people ended up on the Astrodome floor, to see another pathetically indifferent response to gouging is very, very disappointing. We on this side will be introducing legislation which will call for giving you authority to prevent gouging in the case of natural disaster, taking into consideration the real prices, taking into consideration the amount of ramp-up. And when that happens, I hope you will come back to us and support that legislation to show a little more aggression dealing with this problem.

I want to go to the second issue. Mr. Garman, have you read the National Oceanographic and Atmospheric Agency report on global warming and its impact on hurricanes?

Mr. GARMAN. No, sir I have not.

Mr. ROSS. Let me help you out. The National Geographic Pollution Dynamics Agency of the administration—not a bunch of granola eaters in Berkeley—the administration says, and I quote, "the strongest hurricane in the present climate may be upstaged by even more intense hurricanes over the next century as the earth's climate is warmed by increasing levels of greenhouse gases in the atmosphere. Although we cannot say at present whether more or fewer hurricanes will occur in the future due to global warming, the hurricanes that do occur near the end of the 21st century are expected to be stronger, more intense, significantly more intense rainfall than the present day climate." No one is saying this is caused by global warming, but some have suggested before you spend $20 billion in rebuilding New Orleans and Mississippi, we should do so with a national policy that pays attention to science and that your administration, says the hurricanes will become more intense in the next several decades, and you are encouraging a policy which occurs in the development to reduce the barrier island protection, you cut the budget dealing with levees, we have now had the most horrendous hurricane in American history. Do you think the administration should rethink its refusal to consider global warming?

Mr. GARMAN. I think the president has said in February 2003, with regards to global warming, it is a serious issue. I think the President repeated that at the G-8 meetings in Gleneagles most recently this year. We believe that we have a response. And I think if you actually look at actual carbon emissions performance of the U.S. Versus the EU and other nations, you will see that we have a very, very good record. Our rhetoric may not be as forward leading as some of our EU partners, but the performance in actually avoiding greenhouse gas emissions is actually among the very best among the signatories to the framework convention on climate change.

Mr. ROSS. Let me suggest a different viewpoint. If the administration policy on global warming is similar to the policy on levees in New Orleans, which is not listening to science that specifically refused to adopt a elimination of this issue which occurred, which many of us here want to propose, you refuse to block efforts to increase and reduce oil consumption in the energy bill just passed. You refused.

Mr. Chairman, I think I have 8 minutes. I waived my opening statements. I probably have a couple more minutes. Am I right on that?

Mr. HALL. I don't recollect that, but if that is your recollection, I will honor it.

Mr. ROSS. You refuse to embrace something meaningful and that has to do with price with global warming. We have a CAFE standards that are so abysmal. China has better corporate standards than we. In the past years they had improved their output. If you want to talk about the way to reduce prices, let me ask you this, don't we have to find a way to reduce demand in the most effective way? We have advanced in the last few decades what we demonstrated in 1975 and 1973 when we almost doubled a 60 percent increase at least in our fuel economy standards which this administration refuses to take action.

Mr. GARMAN. Sir, again, this administration has raised or proposed increases in corporate average fuel economy standards twice—the first increases since the 1996 model year. I can understand that we can have disagreements about the scope of that increase and whether that increase should be more or less, and that is what we propose.

Mr. ROSS. Perhaps there will be a reconsideration, I don't know, but we hope you will reconsider. And I will tell one would think after the hurricane, after we have seen these prices, outrageous prices at the pump, after we have seen the destruction or diminishing of our refineries on the southern coast, which has always been vulnerable to hurricanes, one would think that we would have administration policy to increase CAFE standards enough to at least reduce our dependence on foreign oil.

Now, according to the energy information office, I am told that the policy that the administration has proposed, what they want to do with CAFE, which was almost nothing, would result in our foreign fuel energy policy actually rising in next decade.

Is that true that under the policy that your administration proposed, our dependence on SaudiArabian and Mid East oil—in part because of your refusal to adopt fuel efficiency standards—will actually go up?

Mr. GARMAN. Again, if one were to look at that single policy in isolation, I wouldn't dispute the contention that that single policy in isolation will not reverse our petroleum dependence. But I would submit to you that were we to do that and a good deal more, including opening the Arctic National Wildlife refuge to production and increasing CAFE standards, even beyond that, we would still looking at a situation of increasing dependence on foreign petroleum.

Looking at these policies in isolation will not give us the result that we need. There is no silver bullet. We need comprehensive policies of a variety of different things.

Mr. ROSS. We just feel that we are not doomed if we start to embrace the new technologies. Thank you, Mr. Chairman.

Mr. HALL. Thank you, and the gentleman's time has expired. Let's see. Chair recognizes Mr. Bass, gentleman from New Hampshire, for 5 minutes.

Mr. BASS. Thank you, Mr. Chairman. Mr. Caruso, I was wondering if you could help me understand why retail gasoline prices moved up so rapidly in my neck of the woods in New England, since New England sources a significant amount of its gasoline, finished gasoline supplies from Canada, I believe some of it from the Virgin Islands, but virtually none of it from the gulf. Shouldn't there have been some more protection in our region than there have been in others?

Mr. CARUSO. Well, not all States' prices did rise. The main reason prices rose is that the market's fungible, and, therefore, when gasoline goes up at the NYMEX futures market, often wholesale prices and contracts that retails are indexed to that, so that is part of the reason. I can't really speak specifically to New Hampshire.

Mr. BASS. I guess the fact is that prices for all fuel went up because of the gulf crisis, even though the actual cost to the manufacturer didn't go up, because there was no impact, the refineries, all the fuel, most of the fuels in New England came from—had nothing to do with the gulf.

Mr. CARUSO. That is correct. As I mentioned earlier, it is a global market and prices went up in Europe as well for example.

Mr. BASS. Different subject. You testified earlier that home heating oil would be 30 percent more expensive than last year as a result of Katrina and the market disruptions caused by it. So for low income Americans, do you believe the government's low, major fuel assistance program, LIHEAP, would be need to be funded at a level 30 percent or so higher than last year's just to keep the purchasing power the same. Forgetting the weather issues and how this winter may come, if fuel prices are up 30 percent, should low income energy be up 30 percent?

Mr. CARUSO. Just a clarification, the 30 percent increase includes the fact that prices had already risen before Katrina, so that they would have—it would have been up year on year regardless——

Mr. BASS. So it might be more than 30 percent.

Mr. CARUSO. No. 30 percent includes our latest assessment, including the impact——

Mr. BASS. Fair enough. So you think it is going to take, is it fair to say you think it will take 30 percent more money to fund LIHEAP?

Mr. CARUSO. I am not really familiar with the relationship between the LIHEAP budget and the current price. I would have to——

Mr. BASS. But if any panelist—it is intuitive if price of heating oil goes from 150 to 250 a gallon, that is a third increase, it will take more money to fund LIHEAP?

Mr. CARUSO. Absolutely: whether it is 30 percent or not, I have no idea.

Mr. BASS. I have one further question because I am running out of time.

Mr. Garman, it would be fair to describe some elements of the energy infrastructure as being a total loss. I am not talking about production capacity, but on the consumption side. And are going to go through a period spending a lot of money on rebuilding that infrastructure.

Is there not an opportunity here to implement other kinds of outer energy consumption patterns, distributed energy, distributed—other ideas, if you will, that would lead to a somewhat different infrastructure, a 21st century infrastructure, rather than an early 20th century, or is it going to be the Agency's position, if it has a position, that we are just going to try to duplicate what was there before?

Mr. GARMAN. I would hope that any rebuilding effort that results as a consequence of Katrina will encourage folks to look at new technologies, distributed generation, micro grids, solar, very highly energy efficient housing. I would like to think that as houses are rebuilt, they are rebuilt at a much higher level of energy efficiency. And I would hope that consumers, who are in a position to make those choices, would ask their builders for these new technologies and a higher energy efficiency than the house that they lost.

Mr. BASS. One last question for Mr. Caruso.

Are you aware that there might have been any gasoline stocks that were diverted from, originally designated for the Port of Portsmouth, which is in New Hampshire, or elsewhere in New England, were they diverted to any other region of the country after Katrina?

Are you aware of any? And I don't have a follow-up. If you don't know, would you be willing to look into the possibility that one of the problems of fuel prices was that supplies were diverted away from the northeast for one reason or another?

Mr. CARUSO. I am not aware of any such diversions, but I can certainly check our sources.

Mr. BASS. Thank you. I yield back, Mr. Chairman.

Mr. HALL. Thank the gentleman. The Chair recognizes the gentlelady from Wisconsin, Ms. Baldwin, 5 minutes.

Ms. BALDWIN. Thank you, Mr. Chairman. I wanted to follow a little along the lines of earlier questions by Mr. Upton and Ms. Capps. Mr. Moran, you are the director of the Federal Communications Commission's Office of Homeland Security?

Mr. MORAN. Yes I am.

Ms. BALDWIN. And in addition to the disaster response that you have been describing, what the commission has been up to over the last week, does your office also engage in planning ahead for threats and other disasters that could impact our homeland security?

Mr. MORAN. We—I have responsibilities in that area.

Ms. BALDWIN. And does your office also work with other agencies across the Federal Government in order to come up with such plans and recommendations for——

Mr. MORAN. Yes. We work very closely, especially with national communication systems and the Department of Homeland Security.

Ms. BALDWIN. So, in that role of planning ahead and interagency planning, have plans been developed for emergency communications in the event of a hurricane or other disaster that can be expected to topple power lines, phone lines and wireless towers as we have seen in the past week?

Mr. MORAN. Well, the Federal Government's role primarily in the communication, emergency communications you are talking about, are really to make sure that the carriers' processes and systems can work. And the carriers themselves have—all the major carriers have detailed emergency plans. There is a lot of dialog between the carriers and the FCC and the NCS on these plans, and many——

Ms. BALDWIN. So is what you are saying is you try to prompt the private sector and the regulated sector to do the right thing, but the Government itself does not have its own set of detailed plans on how to have communication occur in the event of a catastrophe like this one?

Mr. MORAN. I would say the biggest role we have is perhaps as a catalyst. We have a number of advisory committees. We have two major advisory committees that look at these things, and we set the direction for the advisory committee. They run for 2 years. One of them is called the National Reliability and Interoperability Council. It is all major carriers, manufacturers participated in this advisory panel, committee. And the Commission establishes the agenda basically for it. Right now, it is primarily focused on actually public safety communications.

Ms. BALDWIN. I am pleased to hear that emphasis. It seems in your testimony you point to, I guess, sort of four issues of communication challenges, problems, things that were hampered during this last week. One was the ability to communicate between first responders within an entity or department. A second was that communication between first responders in various levels and jurisdictions. A third was problems with communication between first responders, government officials and hurricane survivors and telling them about the availability of and where the relief effort was going. And then last, you pointed out too the challenges in assisting survivors to figure out how to communicate with and where are their loved ones, et cetera.

Do Federal plans or recommendations to localities and States exist with regard to all four of these critical areas so that in the future, we have backups and redundancies, perhaps not relying totally on the private and regulated sector, but that we can step in and make sure that this type of lack of communication never happens again to Americans.

Mr. MORAN. Well, approximately 90 percent of all the communications assets that we are talking about that are relevant here, that infrastructure is really a privately owned infrastructure. Our primary focus is to make sure that that is as robust as it can be and that we know that the major—we know that they all have emergency plans. We discussed their plans with them.

I would say that the primary parties who are responsible for the Federal Communications assets that we put to bear on these functions are really not with the FCC.

Ms. BALDWIN. But it does sound like you are saying that this could happen again.

Mr. HALL. Gentlelady's time has expired.

Chair recognizes Mrs. Myrick, the gentlelady from North Carolina, 5 minutes.

Mrs. MYRICK. Thank you, Mr. Chairman. Thank you all for your patience. Next time you need to bring your lunch.

I had a question for Mr. Caruso. Can you tell me if there was any pressure taken off the gasoline prices by the easing of the restrictions on the Clean Air Act probiotic fuels?

Mr. CARUSO. We saw a fairly rapid response to the waiver by the EPA and the other regulatory relief that was granted last week. Within 24 hours of that, NYMEX gasoline prices started to fall.

Mrs. MYRICK. So it was helpful?

Mr. CARUSO. Yes.

Mrs. MYRICK. I appreciate that. We talked a lot about not having a refining capacity cushion in this country. What would you say does our country need? What type or what would you need to do to create that? How much do we need? That kind of thing.

Mr. CARUSO. I think the two most important things are regulatory certainty,, which Mr. Garman mentioned in his comments, and the other one is really not something that we think can be legislated. The return on investment had been so poor in the 1980's and 1990's that that inhibited investment during those 2 decades. And now we have had 3 years or so of pretty high refinery margins.

Whether that is sufficient to attract upstream investment is unclear; so far we haven't seen much. There is one case in Arizona, a project that has continued to languish, I think, partly through regulatory problems, permitting, as well as financing. So I think at least from a Government perspective, the most important thing would be regulatory certainty.

Mrs. MYRICK. I know in our area we have a couple of companies that are looking at nuclear power again very seriously because of what was done in the energy bill, something I feel strongly about, too.

I appreciate your time. Again, I yield back, Mr. Chairman.

Mr. HALL. I thank you for yielding back, and the Chair at this time recognizes a fine member of this committee, Mr. Albert Wynn. Gentleman has 5 minutes.

Mr. WYNN. Thank you, Mr. Chairman. Thank you for your flattery as well.

Mr. Garman and Mr. Seesel today cited the need for increased refining capacity. There is a very significant criticism being leveled at the FTC, Mr. Seesel, that the FTC is not taking a hard enough look at this issue as it reviews acquisitions, according to an article in the Washington Post entitled Refiners Mergers Good For Business Not Consumers. It indicates that the Commission last week approved a purchase of Premcor by Valero making the latter the Nation's largest oil refinery.

Are you familiar with the article?

Mr. SEESEL. Yes, Congressman, I am.

Mr. WYNN. The article States that the FTC and its staff never seem to make the link between industry consolidation, rising energy prices and record profits and suggested there is a gentlemen's agreement against investing too heavily in new capacity that the FTC's analytical approach does not seem to take into account. The article goes on to cite the fact that the rate of return on shareholder equity is 23.9 percent last year and 16 percent over the past decade. I think this is a pretty serious criticism in light of what everyone seems to be saying is the need for more refining capacity. It appears there is no incentive for expanding refinery capacity because of profits being made through mergers and basically maintaining the status quo.

How do you respond to this criticism?

Mr. SEESEL. As Mr. Caruso has pointed out, I think one of the primary reasons refining capacity has really been sort of stalled in recent years is that until the last couple of years, it has fairly low return on investment. So the idea of investing in new refining capacity has been quite unattractive.

Mr. WYNN. 16 percent over the last decade.

Mr. SEESEL. Well, I think it probably reflects the uptick of the last several years because over the last 10 years or so, and Mr. Caruso probably has better figures on that, I think, for example, several years ago I think the return on investment was abysmal. And I don't know exactly what the number was——

Mr. WYNN. Can you get us that information about the return of investment over the past few years because it seems to me that the return actually has been pretty good.

Mr. SEESEL. I will be glad to, Congressman.

Mr. WYNN. So your bottom line response is you don't accept the criticism.

Mr. SEESEL. Well, the author of that article is entitled to have any thoughts he wants about gentlemen's agreements and so forth. We have looked at many, many, many mergers and millions of pages of documents in this industry over the 25 years we have been looking at this and have come up with virtually no evidence of anything like that.

Mr. WYNN. Did you find it interesting that the article says that when President Bush sited the availability of inactive military basis as possible locations for expanded refinery capacity, that the spokesman for the Valero said they weren't interested in that?

Mr. SEESEL. I hadn't really focused that much on that part of the article. I am aware, Congressman, that some refining executives found the proposal about military bases to be interesting, although some had some concerns about how close they were or not to crude oil supplies. So that the idea of siting a refinery on certain military bases didn't have much appeal to them.

Mr. WYNN. Let me move on to another question that has been talked about at some length today, and has to do with price gouging. And I think you, in fact, testified it was very complicated and depended upon circumstances, et cetera. Has FTC ever studied this issue?

Mr. SEESEL. The Commission has looked at pricing issues in the context of claims and allegations that there is collusive activity going on.

Mr. WYNN. Is there a report?

Mr. SEESEL. The Commission did some investigations of petroleum and gasoline prices.

Mr. WYNN. Is there a report on price gouging?

Mr. SEESEL. On price gouging per se, no, sir.

Mr. WYNN. In view of the complaints that you received, don't you think that is an appropriate role for the Agency?

Mr. SEESEL. I certainly think, Congressman, that to the extent we get complaints that are phrased in terms of price gouging, it is appropriate for the FTC to look at whether or not there is any violation of any law.

Mr. WYNN. So can we expect that you will, in fact, conduct a study because you have already testified that you got the complaints and that you forwarded to them to the States' attorneys general. So presumably you think they have some credibility.

Mr. SEESEL. As you know, under the new Energy Act, we are under section 1809, the Commission is going to conduct a study starting right now of manipulation of gasoline prices in this country.

Mr. WYNN. So you are telling us, the committee, that you will be studying and reporting back on price gouging?

Mr. SEESEL. Really, all aspects of possible manipulation of gasoline supply and prices.

Mr. WYNN. I appreciate that. Mr. Garman, you talked about hydrogen. You said 2020 would be the year we would have hydrogen cars. What is the administration doing to speed that up?

Mr. GARMAN. We think that the—frankly because we are dependent on certain technological breakthroughs that aren't necessarily mindful of a timeframe, we have to have some success in the lab. As you and I have talked about before, hydrogen storage is a technical barrier that we are confronting.

We don't know what the answer is. So we are putting more money into that effort to find, you know, to research different and new compounds, halides and chemical and metal hydrides that might be good storage media, but you know what we need in addition to the funding is time—it is a learning process. So I am not certain that there is a lot that can be done to speed that up.

It has been suggested in the past we put more money in it and that would be an approach because it might enable us to study two pathways at once or three pathways or multiple pathways. But what we really need is time.

Right now, we have several—we have a number of hydrogen vehicles on the road. We are collecting data, performance data. And what we need to do is take that data back to the laboratory and then come up with the next iteration. So it is not—there is not a whole lot that we can do to speed that up if the ultimate goal is a product that consumers will choose and consumers will buy.

We can produce a car tomorrow that has performance characteristics that we want but not at a price that consumers can afford. That is going to take some time.

Mr. WYNN. Thank you. I yield back the balance of my time.

Mr. HALL. Thank the gentleman. Chair recognizes Mr. Sullivan, gentleman from Oklahoma, 5 minutes.

Mr. SULLIVAN. Thank you, Mr. Chairman. And you know, this was a horrible tragedy, Katrina.

Mr. HALL. Sir, you get 8 minutes in all.

Mr. SULLIVAN. 8 minutes. It is horrific the things that happened and it disrupted our energy sector, it has affected this country in many ways. I was in the State legislature when the Oklahoma city bombing occurred, too, and that was horrific as well. And one thing that, you know, does come out of these horrific tragedies is sometimes something good. And I think that we do need to look at our—examine our energy infrastructure, energy needs in general, even better than we did on the last energy bill. I don't think that went far enough. And we need to look at the long-term overall strategy of energy looking at nuclear power, all the alternative energy sources, because I do believe that some day, and it won't be in our lifetimes or our kids' lifetimes or even our grandkids' lifetimes, but we will run out of oil and gas or it will become too expensive to produce. And one thing I do want to bring to peoples' attention is, I guess, Mr. Garman, I will focus this to you.

Right now in this country if we were to find, let's say you and I found a billion barrel reserve today somewhere here in this country, would we be able to refine it?

Mr. GARMAN. We would be competing with many other—with others.

Mr. SULLIVAN. Outside the country?

Mr. GARMAN. Probably what would happen is we would displace foreign oil into that—domestically produced crude would displace foreign oil that was coming into those refineries is what most likely would happen, is my estimation.

Mr. SULLIVAN. But we are at maximum capacity with our existing refineries; now, would you agree?

Mr. GARMAN. That is correct. We actually do not have refining capacity to refine all of our needs today.

Mr. SULLIVAN. And there is a place, I don't know if you are familiar with Cushing, Oklahoma, it is just outside my district, but there are 23-some odd pipelines go through that area, and there have been refineries there in the 1920's and I think up to the 1970's. Kerr-McGee had a refinery there. They had like Citgo, Conoco Philips refinery, Embridge, Shell Sunoco, Texas Eastern Pipeline Partners, Magellan, Plains All American, just to name a few. Many pipelines intersect in that area in Oklahoma. And also it is the delivery point for NYMEX crude oil futures contracts.

Would this be a good idea? And I have always said since I got elected, wouldn't it be a great idea to build a super mega refinery in that area? We have the supply coming in. We are not next to an ocean. We do have a great infrastructure of pipelines there, as you know.

But could you see that becoming a reality if maybe we lessened some of the burdensome government regulations and the permitting, kind of like what we did in the House version of the energy bill, maybe we can go back and revisit that and make it even better, but to build a megs refinery there, maybe making let's say 2 million barrels a day. I don't know. But would that be a reality?

Mr. GARMAN. The Nation needs more refineries. And if Oklahoma is willing to host a refinery, then I hope investors are listening, and if there is anything we can do to help make that come to pass, we would be happy to do that.

Mr. SULLIVAN. Well, if Oklahoma decides that is something we want to do, I think that is something Oklahomans want to do, would you be willing to help us along the EPA and all of that as well?

Mr. GARMAN. Yes, sir.

Mr. SULLIVAN. Are you committed to doing that? Thank you very much.

Mr. HALL. Gentleman yield back.

Mr. SULLIVAN. I will yield back, yes.

Mr. HALL. Chair recognizes Mr. Markey, 5 minutes.

Mr. MARKEY. Thank you, Mr. Chairman. Over the course of the last few years, the oil companies have earned record profits as this chart indicates. Exxon-Mobil's profits have risen from $11 billion to a projected $31 billion this year. Chevron-Texaco's profits have risen from $1 billion to $13 billion; BP's from $8 billion to $21 billion; Shell's profits have risen from $10 billion to $20 billion.

These are huge numbers. And they are the direct result of soaring oil and natural gas costs. Gasoline has risen from $1.85 to $3.04 on the average.

In the last year, heating oil has risen from $1.36 a gallon in 2003 to a projected $2.22 a gallon this winter. Natural gas has risen from $9.85 per thousand cubic feet a few years ago to a projected $12.81 this winter. We need to know why. This is all before Katrina. This is just what has been going on in the market.

Mr. Garman, in the last 10 years, at least 30 refineries have closed.In the last 10 years, at least 30 refineries have closed. These refineries were all fully permitted and were producing gasoline for the American public.

Do you know of a single refinery the oil industry is seeking to reopen to produce gasoline in the U.S. Market?

Mr. GARMAN. I am not aware of one, no, sir.

Mr. MARKEY. No. Mr. Caruso, last year, Business Week reported that refineries are running near capacity because they have little incentive to build more. For starters, they make more money when supplies are tight, says Business Week.

Do you agree that reduced refining capacity means higher profits for oil companies?

Mr. CARUSO. Well, profitability really has more to do with the demand, the competitiveness in the world market than that single data point that you just mentioned. But that is one of the components.

Mr. MARKEY. Haven't refinery margins increased, that is, refinery profits increased?

Mr. CARUSO. They have increased.

Mr. MARKEY. The less refining capacity, that is, the more refining capacity that American oil companies have closed is the more money refiners are making. Their profits are going up. Is that correct?

Mr. CARUSO. That is accurate and it is also accurate on a global basis, not just in the United States.

Mr. MARKEY. You are saying the whole world is shutting down refineries?

Mr. CARUSO. The refining capacity on a global basis is tight. Yes.

Mr. MARKEY. Is tight. So this is a global pattern where the largest oil companies, not only here, but across the world, have been shutting down refining capacity, without a government mandate to do so over the last 10 years.

Now Mr. Garman, in the past, oil industry has suggested that somehow environmental permitting requirements were to blame for the industries failure to build new refineries. However, let me read to you from an internal Chevron document in 1995, "If the U.S. Petroleum industry doesn't reduce its refining capacity, it will never see any substantial increase in refining margins."

So Mr. Garman, by closing 30 refineries since 1995, not building new ones, but closing 30 already existing operating refineries, the oil industry seems to have achieved their goal of 1995 of higher refining profits, have they not?

Mr. GARMAN. My understanding is that smaller, less efficient refineries have shut while existing refineries have expanded capacity. And the strategy—and I am not the right person to ask. You should ask—you will have a witness in the next panel to ask specifically what is their motivation, but my observation has been there is a component of the environmental standards to comply with environmental standards and maintain an update refineries to——

Mr. MARKEY. But this was not building—these are not building new ones. These were the old ones. And instead of continuing to maintain them, they just decided to shut them down. But they could have, with their profits, maintained them and kept them going.

Mr. GARMAN. Actually many times—and Mr. Caruso has better data—but many times, refinery margins have been quite small and not been conducive to new invest and expansion.

Mr. MARKEY. Exactly. Chevron said that in its document in 1995, they said, we will never see any substantial increase in refining margins if we don't reduce—if we don't reduce, that is, Chevron doesn't reduce its refining capacity.

And Mr. Seesel, the FTC is supposed to be in charge of monitoring the oil and gas industry for evidence of anti competitive or manipulative activities.

Has the FTC examined whether current situations that now have, with respect to refining capacity, may be the result of a deliberate strategy by the oil industry to reduce capacity in order to drive up the profit margins and prices to consumers? Have you ever had an investigation?

Chairman BARTON. If you can answer that please answer that and that will be his last question.

Mr. SEESEL. Congressman, I don't think the Commission is aware there is any evidence, there is any collusive or anti competitive scheme among oil companies to reduce refining capacity.

Mr. MARKEY. Have you ever investigated it?

Mr. SEESEL. The Commission looked at the Shell Bakersfield refinery situation in California. The situations we are aware of are individual unilateral decisions by refineries.

Mr. MARKEY. Did you ever look at what—30 refineries all shut down?

Chairman BARTON. Gentleman's time has expired. The gentleman is obviously entitled to provide written questions to this panel in addition to the questions he has already asked.

Mr. Murphy of Pennsylvania has 8 minutes, if you chose to use them.

Mr. MURPHY. Thank you, Mr. Chairman. I am trying to summarize what I have learned so far in the last 5 hours.

It comes to this, that I think what you are saying is when it comes to defining these price jumps, price gouging, it is much like the Supreme Court's definition of obscenity. You can't tell us what it is, but you know it when you see it.

Which doesn't leave us, our constituents, or Americans in general, with a lot of comfort, although I say that tongue in cheek.

I just want to review a few things, and whoever is best to answer this, I appreciate that. We do not have enough oil to meet the needs of our citizens. Therefore, we have to import. When we do produce more oil to meet our needs, other areas like OPEC reduce their production in order to keep prices high. When we have catastrophes such as what we just experienced in the Gulf Coast, we have to raise prices to pay for future costs, increased costs of gasoline and also anticipated costs. If we are importing more, other countries can also raise that price. Am I correct so far, anybody, Mr. Caruso?

Mr. CARUSO. I think that is generally accurate.

Mr. MURPHY. Now on this, I have a question. I want to know what the Department of Energy has done on this particular issue. I want to read a couple of quotes from something, and I would like to ask unanimous consent an article from the October 2004 National Geographic be entered into the record. This article made a chilling prediction of this whole event. I will read a couple quotes from this.

It says the "Federal Emergency Management Agency lists a hurricane strike on New Orleans as one of the most dire threats to the Nation, up there with a large earthquake in California or terrorist attack in New York City. Even the Red Cross no longer opens hurricane shelters in the city claiming the risk to its workers is too great." It goes on to say, "the most startling impact has only recently come to light. From concerns about tidal surges, the effect of oil and gas and petroleum subsidence rates—there is another aspect there. For decades geologists believed that the petroleum deposits were too deep and the geology of the coast too complex for drilling to have any impact on the surface. But 2 years ago, petroleum geologist Bob Morton, now with the U.S. Geological Survey, noticed the highest rates of wetland loss occurred during or just after the periods of peak oil and gas production in the 1970's and 1980's and concluded that that had an impact on reducing some of the areas of the wetlands."

We knew before then that this area was prime for huge devastation from tidal surges, and we knew we had huge loss of wetlands, and some of this might have been gue to oil exploration. Was there something the Department of Energy did or should have done with regard to alerting the oil companies and saying we can't have 25 percent of our oil production or refinery production situated in an area which is considered by FEMA to be at extremely high risk for devastation.

Did we know it was coming? And did we do anything about it?

Mr. GARMAN. No, sir. I cannot say that we made a connection between oil production and subsidence. I would note that I haven't seen the scientific literature behind that National Geographic article. My observation would be the old warning of every statistics professor that correlation does not necessarily mean causation. But that is a very interesting possibility——

Mr. MURPHY. But still we knew there was a large loss of the marshlands which were the natural buffer for storm surges, and we did know that with all the oil refineries clustered around there that there would be trouble for category 3, 4, especially 5 hurricanes. I am curious if anybody from the Department of Energy began to raise questions and say we need to put some pressure on oil companies to change this and not wait 30 years.

Mr. GARMAN. My understanding is that refinery siting—and we had a discussion I think about some of this while you were out of the room, but refineries are sited where they can be based on market conditions and the willingness of local population to accept them.

Yes. In a perfect world, it would be better to have refineries distributed geographically around the country. And I think that recognition is well understood. I am not sure that we have the tools or the capability to force anyone to do that distribution. I dare say that the market and the insurance market and the reinsurance market might as a consequence of these losses. It would probably be more difficult in the future to site a refinery or some of this infrastructure that close to the coast. And I think the market will respond, and folks looking to site a new refinery will site their new refinery elsewhere.

Mr. MURPHY. So what you are saying is perhaps our minds will change, at least the minds' of those who are otherwise opposed to siting refineries and distributing them around the country. Otherwise we could remain extremely vulnerable to a natural disaster or terrorist attack.

Mr. GARMAN. I think it would be a mistake for us to ignore a lesson that has been so devastatingly made clear to us in this instance, yes.

Mr. MURPHY. Does anybody else on the panel have a comment on those issues? Mr. Chairman, it comes down to this: A lot of our constituents are enraged about fuel prices. And seems sometimes the best we can offer them is what people have said, is either there is some intentional price gouging, or it is the marketplace and a shortage and we don't have the refinery capacity.

What the American people look upon in times like this is that we have to show them that we are working together in a bipartisan way to come up with some solid solutions on this. And that is why

I am really hoping we can move vigorously forward in a couple of areas, and that is that we have to explore for more oil in this country, we have to move more vigorously toward clean coal technology and nuclear energy, and we have to build more refineries because to wait longer is going to have a more devastating and far-reaching effects on our economy. And with that, I will yield back the balance of my time.

Chairman BARTON. Gentleman yields back. The gentlelady from California, long, patient Mrs. Solis is recognized.

Ms. SOLIS. Thank you, Mr. Chairman. I appreciate that.

My question is for Mr. Caruso. I wanted to bring to your attention report that I came across. The investment firm, Friedman, Billings, Ramsey & Co. noted that in early August 2005, refinery margins rose 54 percent, and that these profit margins were responsible for 60 percent of increased cost of fuel at the pump. Other estimates say as much as two thirds of the increased cost of gas at the pump is a direct result of profit margins of refiners.

Murphy Oil, a company with refineries impacted by the gulf coast, yesterday lamented the fact that it has refineries offline and is missing out on record margins. It seems from these reports that the refinery business is quite profitable, more profitable than any other sectors of our economy.

Do you agree with this assessment of these reports which identify a link between the increase of refinery profits and the cost of gasoline at the pump?

Mr. CARUSO. I haven't seen those specific reports, but clearly, even before Katrina, particularly in July and August, there was a significant run-up in refinery margins as a result of the very tight gasoline supply situation during the time of peak driving. So while I can't subscribe to those numbers because I haven't seen them, they are consistent with the general trend in prices and margins. But I would also caution that this was a very short term situation. Over a long-term period, the refinery sector has not had the kind of margins that you have just referred to.

Ms. SOLIS. It is unusually high, though? Do you agree? And just a comment, of the 95 percent of the Bush administration energy plan which has been implemented, what specific parts, in your opinion, would address the costs to consumers at the pump from the high profit margin of refiners?

Mr. GARMAN. I would point out that, you know, this legislation that has just passed, while we are proud of it, and the President signed it, was a product of this Congress and this committee and other committees of the Congress, developed as a compromise. And I think the Secretary has said, clearly, that the bill is not expected—we cannot reasonably expect the bill in the short term to do much to deliver relief at the pump. It is a long-term bill. It is primarily a research and development bill focused on the opportunity to move us toward new alternatives. But that will not happen overnight.

Ms. SOLIS. One of my other questions that I wanted to raise was with respect to where some of these refineries are sited. And this is for Mr. Garman.

It appears that the Chevron Texaco refinery in Pascagoula—excuse my pronunciation—Mississippi and Conoco Phillips refinery in

Belle Chasse may have suffered the most significant damage from Hurricane Katrina. No. 1, does the Department of Energy or any other Federal agency have regulations which require refineries, such as these, which are constructed in areas of high risk, such as a "hurricane," is there any standards that would prevent a refinery from being placed in an area that we know could possibly be affected by a disaster of this magnitude?

Mr. GARMAN. The two refineries you mentioned are very important refineries, with a combined capacity exceeding 500,000 barrels a day between those two. They represent a substantial investment by the private sector investors.

It is also our understanding that both of those refineries have suffered major damage, and that they will take some time bringing back.

Ms. SOLIS. We currently know there are standards in place to protect nuclear power plants. I am wondering is there any discussion in the administration to look at potential safety standards for refineries.

Mr. GARMAN. There are safety standards in place for refineries, to be sure, to protect public health and safety but——

Ms. SOLIS. But to anticipate a hurricane at the force of category 4, is that something that the administration may look at in the future if, given what you just said, that these are two very important refineries?

Mr. GARMAN. We are willing to look at any variety of ideas and to work with the Congress on any variety of ideas and thoughts that you all may have. But my threshold observation would be someone spent a tremendous amount of money building this refinery.

It is a potential hundreds of millions or billion-dollar investments. And I think that, you know, perhaps they bet wrong, putting such a high investment, high value investment right there at the coast. And they weighed that when they made that investment. I am not sure this is something that is right for some kind of Federal intrusion into the market. But as I said, we are willing to take and consider any ideas and discuss them with this Congress and this committee that you might deem appropriate.

Chairman BARTON. The gentlelady's time has expired. Dr. Burgess has 8 minutes.

Mr. BURGESS. Thank you, Mr. Chairman. Mr. Garman, continuing on that same line, I can't help but observe, we are just a day or two away from the 105th anniversary of the big storm, the Galveston. Galveston, at the time, was the largest city in Texas and after that storm, they never totally recovered. In fact, it was 50 years before they got back to the population they enjoyed in 1900. I don't think there is any question that we will see the location of things change as a result of this storm, regardless of our intention here in Congress. I think insurance companies—just people's behavior would have to question whether or not it is reasonable to live or develop infrastructure in an area that has been proven to be unsafe.

Let's talk—Mr. Seesel, let's talk about price caps because it seems like that is what is on everyone' mind. Now, Hawaii did price caps about a month ago. What has been the experience with Hawaiian price cap? Do they work?

Mr. SEESEL. Actually, Congressman, I believe the price caps in Hawaii went in effect September 1. So it has been a little bit early to tell. The Hawaii price caps, as you know, are geared toward prices on the west coast, the east coast and the gulf coast.

So probably contrary to the expectations in Hawaii, some of the price there may have gone up, along with what has happened to the prices on the gulf coast, which is obviously lower than other parts of the country. But I think it is something that time will tell what will happen with Hawaii's situation. It is hard to tell. As you know, there is cap on the wholesale level not retail level.

Mr. BURGESS. Do you think it is good policy, one that should be practiced in other areas of the country?

Mr. SEESEL. I don't think—in fact, the FTC has testified against price caps, including Hawaii price caps a couple of years ago. And I think efforts to cap prices like that are probably going to result in—a reasonable prediction is that they could result in shortages and decisions by businesses in the market to leave the market and other unintended consequences of that.

Mr. BURGESS. Is there any thought to perhaps allowing States to have price caps if they have a refinery within their borders? I will withdraw the question. I was just wondering about Massachusetts not having any refineries? I am shocked that they do not. Let's talk about the——

Chairman BARTON. You are easily shocked.

Mr. BURGESS. I understand. Round up the usual suspects. I had my staff, a couple of weeks ago when I was getting bombarded with questions by constituents about why not do something about gas prices, and I asked them to just break down for me, when gas was $2.29 a gallon for regular, give me a breakdown on what the—what were the components, what made up that $2.29.

And I was given these figures. Tell me if they are correct or not: $1.25 for crude; 43.9 cents for taxes, State and Federal, I am in Texas; 40 cents for refining; 18.3 cents for distribution and marketing; and total profit of about 17 cents. Is that—would you agree with that breakdown? Is that an accurate representation?

Mr. SEESEL. Congressman, those numbers are fairly consistent with what I am familiar with, but I might defer to my more expert colleagues on that, too, and see what they say.

Mr. CARUSO. I believe those are accurate.

Mr. BURGESS. So just going back to a couple of weeks ago, in prehurricane terms, 17 cents a gallon profit, that is okay, but that is not exorbitant. So the high profits that Mr. Markey showed us on his graph, which was before the hurricane, it seems to me those high profits would indicate that companies are selling a lot of gas. Is that right? If they are only making 17.4 cents a gallon, when it is selling for $2.29, the profit is because there is a lot of those 17-cent gallons that are sold. Is that correct?

Mr. SEESEL. I presume the 17-cent profit is at the retail level?

Mr. BURGESS. Well, even if it was 19 or 22 cents, it is a smidgen of what the total cost of a gallon of gasoline is. The profits are not coming from the $2.29; they are coming from that very narrow bit that is the gasoline or the oil company's profit.

I guess I would be interested to know, and if I could ask one of you to follow up with my office, what would that breakdown be now with gas at $3.10 or whatever it is, again, remembering that I am in Texas, and our State taxes are about a quarter a gallon? I would be very interested to know what that breakdown is now, and perhaps then we could make a judgment if that 17 cents has jumped and is now 34 cents or 50 cents profit per gallon, then perhaps people have a case to be made for excess profits. Otherwise, it is an argument that we should probably abandon.

I could not help but think the day—the Wednesday when you realized that all of the water was in New Orleans—and with all due respect to my colleague here, the wetlands would not have stopped that, the hurricane remember, the puff of dry air that somehow Bush managed to push the hurricane over a little bit so he could do maximum damage to New Orleans, the hurricane did not come across the wetlands, it went in in Biloxi.

But the day that all of the water came into New Orleans, I found myself asking, where is the contingency plan that we have for this type of disaster? Mr. Garman, is there a contingency plan for an energy emergency that you can pull off the shelf in the Department of Energy? And if so, what is the plan, and why wasn't it enacted?

Mr. GARMAN. Well, we do have—our major contingency plan and our major asset for a supply disruption is the Strategic Petroleum Reserve. And as I indicated in the testimony, within 48 hours of the time that we had a request for a loan or a diversion of oil from the Strategic Petroleum Reserve, we approved that loan, and that oil was flowing very, very quickly.

So I would submit that that is our primary method of responding to a severe petroleum supply disruption. And in this instance, it was used, and it was used quickly.

Mr. BURGESS. Are there any other levers that we can pull to manage an emergency? Is that the only tool in our tool box?

Mr. GARMAN. With respect to crude oil and product, the Strategic Petroleum Reserve is the primary tool that we have got. We do not have, as some our Nations have, for example, a refined product or requiring refiners to keep a certain amount of refined product in stock as a reserve. We do not do that.

Mr. BURGESS. Do you think that is policy worth pursuing?

Mr. GARMAN. It is something that I think that—within a full range of things that we ought to be thinking about it. It is hard to dismiss anything out of hand.

Mr. BURGESS. My time is drawing short. There are four locations for the Strategic Petroleum Reserve, two in Louisiana, two in my State of Texas. Do we need to think about locating other areas in the country for the Strategic Petroleum Reserve, since both of these States share the gulf coast and the inherent vulnerability of this type of storm.

Mr. GARMAN. Well, I think it is again instructive, and we will continue to learn from—but we are offering strategic—we are offering oil from terminals that were hit directly by the storm. So it shows that the reserve is quite robust, and the infrastructure that we have around the reserve is robust, and we have the capability to respond, even in this seemly worst-case scenario. So I think that speaks well of the planning that went into the reserve itself.

Mr. BURGESS. Very well. I yield back.

Chairman BARTON. I thank the gentleman. The gentlelady from Tennessee, Mrs. Blackburn.

Mrs. BLACKBURN. Thank you, Mr. Chairman. Thank you all for your patience. We appreciate that very much. It has been so interesting sitting here listening to this today. I think this is an industry when you talk about the petroleum industry, it is an industry we have all got a love-hate relationship with.

And I think you all have shown that today, and you have probably heard it from the questions. I also sit here, and I realize that much of what we are asking and saying today probably to many of our constituents appears to be Monday morning quarter-backing.

And to our friends in Mississippi and Louisiana, we extend our condolences and hopefully understanding hearts that this is a really rough, rough time for you all. I have been in Mississippi, as I said, when I waived my opening statement. And I have been there where there was no cell service; nothing was working. I have stood in gas lines with people that have driven 150 miles to get to an open tank so that they could fill up their drums, 55-gallon drums to go run a generator.

I have been at a shelter trying to run down somebody from a Federal agency who could help somebody with something else. And there are plenty of lessons learned. And there is plenty of education and character-building that has probably taken place through this for everyone involved.

And I thank you all for taking the time to come here and spend a good part of your day with us. And I hope that those who have watched this hearing today understand that we do this in the spirit of trying to be certain that everybody functions well, and that we learn how we responded, we learn what problems were with communication, that we learned a lot about leadership and different leadership styles, and how we handled a team effort from the local, State and Federal agencies that are to be involved with this.

I hope also that we are going to see some changes come out of this. I hope that we will see some changes when it comes to looking at burdensome regulation that makes it very difficult for the petroleum industry to operate in this country.

I have a father who is 80-years old who sells oil field production equipment and goes to work every single day, every single day, and has for many years. I hope that we will see the need to address taxes. I hope that we will see the need to address rules. And I hope we will apply some common sense to this, that we do use it as an opportunity to learn and that we as Members of Congress accept our part of the responsibility in making the appropriate changes.

Transportation permitting, environment, how those regulations affect every bit of this is going to be important. But rather than spending a lot of time on questions, and I have already used a good bit of my time, I want to pose some questions for you all to answer, not now, but in the next 30 days.

And for those of you who are on the next panel, these questions also go to you. I am not going to keep you here that long. Let us look at this, looking at it long range. I want to know what your people in the field say. I want to know what your people that are out there are going to tell you that they learned from this. I would like to know how many States removed or reduced their tax on gas and diesel?

How many are going to move forward and do that? In Tennessee, we have got 21.4 cents on a gallon of gas and 18.4 on diesel. How many are going to step up to the plate and work with us on this? How many are going to report daily price fluctuations in their district? You know, seeing a dollar change, a 50-cent change in a day, that is something that infuriates my constituents. We saw that in Tennessee. We saw that in Tennessee. And it is something that people are not happy about. Our Governor is working on that issue now, so are some of our State legislators.

Mr. Moran, this one would really come through your plan, and the folks in the telecommunication agencies that are going to speak next. How many local governments have a communications plan when everything fails? When you do not have cell phones, when you do not have hardwired phone service, how many have a back-up plan with satellite or with radios or some other frequency?

How many companies have emergency disaster communications plans? How many local governments have a plan for getting those first-responder vehicles filled so that the tanks are full, so that they are able to carry on with the work that they have to do? How many of them were just planning on people having gas in the pick-up to get fuel out to the areas where it was needed? I would also like to know how many of our State governments use all of The Homeland Security funds that are allocated to them? How much are they drawing down, and then how much are they sending on to those local governments? And are those State governments working with those local governments on these energy distribution plans, on these communication plans?

Are they working with their major employers to be certain that there is some kind of back-up system there? And also from the communication and who is taking the responsibility? Is it going to be Red Cross, is it someone else, to be certain that there is a way for individuals to communicate?

It is so difficult standing in one of those shelters when you have got people who desperately, desperately want to find their relatives, and there is no gas within 150 miles, and they drove out on Sunday to come to a shelter, and they ended up there with about an eighth of a tank of gas left. That is a pretty tough spot, pretty tough spot to be in.

In my minute that is left, Mr. Caruso, I did have some questions, I think are most appropriately directed to you, in having listened to your testimony. When we talk about refining capacity here in the U.S., we know we have pretty much been at capacity, we have been at about 94 percent of capacity for refining, and people wonder why we do not have refineries all over the country.

One reason is transporting the fuel. You know, we have got a refinery in Memphis right at the edge of my district, and sometimes you have to go dredge the river in order to get enough depth to be able to unload those barges. So you get a whole other set of problems when you move away from the coast. But we talked about refinery capacity.

Are we higher or lower than the worldwide average on refinery capacity when we talk about other nations and their capacity? How are we measuring up there? And I am going to run out of time, so I will just let you answer at a later time.

And then if we had opened ANWR in 2001, when there was a debate about opening it in 2001, and oil was currently being produced, what effect would that have on the cost of crude today? I would like answers to those, too. And, Mr. Chairman, I am over, so I will yield back.

Chairman BARTON. We thank the gentlelady. The gentlelady from Illinois, Ms. Schakowsky, is recognized for 5 minutes.

Ms. SCHAKOWSKY. Thank you, Mr. Chairman. I just wanted to say that before we start seeing as the main solution to the high gasoline cost the eliminating of gasoline taxes, money that right now is desperately needed by States and communities, not only those that are affected by Katrina but many others who have been affected by budget cuts, health care needs, education, environmental protection, housing, many of which are also being called upon to address other problems, I think we ought to look first at the record profits of the oil companies and start asking our companies to share sacrifice.

You know, we have seen now a million people displaced, and we have seen lots of families, now that we have sort of lifted the veil on poverty in this country, who are barely making ends meet and who are suffering in cities and rural areas all around our country, that maybe those taxes, we have seen tax cuts at the Federal level, for those who have the most, maybe we ought to first look at some of the companies who are profiting most, and maybe we ought to even consider rolling back some of the tax cuts that have already been given.

But, what I am concerned about now and wanted to look ahead a little, in addition to the cost of gasoline, in Chicago before Katrina was paying the highest prices for gasoline, what about the winter heating season? And what about natural gas, and heating oil, and what can we expect in the way of price increases? I just feel so strongly that we need to be planning for that potential eventuality, and I would like to hear what you think the odds are in order of magnitude if we are going to see price increases? Anyone can answer for whom it is appropriate.

Mr. CARUSO. We released our latest short term energy outlook this morning. It indicates that heating oil will be up about 30 percent this winter compared with last winter. Unless there is a significant improvement in the natural gas situation, we think the natural gas prices for heating this winter will have an even higher percentage increase than that. The details are in our report released this morning. I would be happy to make that available directly to you.

But, the bottom line is, heating oil, natural gas and propane will all have significant year-on-year increases this winter compared with last winter.

Ms. SCHAKOWSKY. Well, and I think that that ought to be, sound the alarm for this committee and for this Congress to, you know, we do not want anybody saying we did not expect the levees to break. In many ways, for many families, these kinds of increases in heating bills are—I do not want to say, get equivalency, but is a serious crisis that could put poor families particularly over the edge, but not just, you know, my constituents really cannot afford to pay, middle-class constituents, $1,000 a month to heat their homes in Chicago where we rely mostly on natural gas and have already seen major increases in the price of natural gas.

Small businesses that are, you know, struggling right now and could potentially go under and farmers who rely on natural gas, and you know, so we need to start planning now about what we are going to do. And it is not just about LIHEAP, I want that to go on the record as well. It is not just about LIHEAP. That is needed to expand the funding for LIHEAP, but it goes way beyond that, and we need to have better planning.

I want to—Mr. Caruso, your agency had predicted that as a result of the energy bill that was passed, that at least potentially, gas prices could go up. As I understood, not just understood, we had quotes from the report that gas prices—this is before Katrina—that you know there was a lot of talk about how great the energy bill was, but when it comes to prices at the pump, my understanding was that you thought that that bill could actually raise prices?

Mr. CARUSO. There was an analysis done of the House version of the bill that indicated that there might be some—I think there was a little bit of mischaracterization of that, in that one component of the gasoline mix could actually increase in order to meet the requirements in the bill. Overall, we did not expect the bill to increase gasoline prices. But, I will provide that specifically.

Chairman BARTON. The gentlewoman's time has expired.

We are going to recognize Mr. Walden. I believe he is the last questioner for this panel, and I will announce to the audience our next panel at the conclusion of this panel. We are going to take a very short 5-minute break, just to give people a chance to do personal conveniences and things like that. But we will reconvene very quickly.

So Mr. Walden is recognized for 8 minutes.

Mr. WALDEN. Thank you very much, Mr. Chairman.

First of all, along with my other colleagues on both sides of the aisle, we express our deep sorrow for those who have suffered so much in the South, and we will do everything we can to help them.

In fact, my own State of Oregon is opening its door to 1,000 evacuees, coming all of the way up to Oregon. We are sending about as many as 1,700 National Guard troops down to the Gulf States to lend a hand and do what we can.

Mr. Caruso, I want to follow up real briefly on the tail end of Ms. Schakowsky's question. So your analysis never showed that the congressionally passed energy bill was going to drive up the price of gasoline overall?

Mr. CARUSO. As I understand it, there were certain types of gasoline for which the price would go up; I believe the reformulated component. But, as I mentioned to——

Mr. WALDEN. Is that like the ethanol version?

Mr. CARUSO. I believe it was either the ethanol or the MTBE, the combination of the MTBE ban being replaced by——

Mr. WALDEN. So the MTBE ban and replacement fuel might drive up the cost of gasoline?

Mr. CARUSO. Yes. But I would like to provide the actual——

Mr. WALDEN. That would be good. I want to move onto a couple of other issues, because these are certainly ones that I am concerned about. They relate to the markets, both CFTC and NYMEX and I want to—I do not know who is best to address this, maybe Mr. Seesel.

But in the September 2 issue of Dow Jones Newswire, a Mr. Addison Armstrong, manager of the exchange traded markets TPS Energy Futures LLC in Stanford, Connecticut, said, and I quote, there are, and in parens, oil commodities, quote, traders who made so much money this week following Hurricane Katrina, they will not have to punch a ticket for the rest of the year.

Is anybody here concerned about this whole trading issue? I, along with Ms. Baldwin and about 18 other Members of the House have initiated a letter to the Government Accountability Office asking for a full investigation. We did that back in May of the trading market. Is this something you all have looked at at FTC, the volatility? Does the hedging affect the volatility of the spot market?

Mr. SEESEL. Congressman, that is really not an area that the Federal Trade Commission has looked at very much. I know, obviously, that the CFTC has the great bulk of expertise on that. And perhaps some of my colleagues here do, too. But we have really not focused on the NYMEX markets.

Mr. WALDEN. Is that something that you have the authority to focus on?

Mr. SEESEL. I think the regulatory authority is in the CFTC commission.

Mr. WALDEN. All right. Mr. Garman, Mr. Caruso. Mr. Moran.

Mr. Moran, I want to follow up on the question my colleague from California asked about the emergency alert system and the national notification system. By way of record, I am a broadcaster, so I am intimately familiar, even wired them in and run the test.

It would be highly unusual for the President to trigger a national emergency alert notification on a regional problem, wouldn't it?

Mr. MORAN. Well, that is up to the President. That has not happened. I guess that would be unusual. It has not happened.

Mr. WALDEN. It would be very unusual, wouldn't it, announcing a hurricane off the gulf coast, in Oregon or in New Hampshire, there wouldn't be much relevance to trigger a national EAS, would there?

Mr. MORAN. It would be totally up to the President. However——

Mr. WALDEN. But no president has never done that on a regional event, have they?

Mr. MORAN. That is correct.

Mr. WALDEN. Isn't there a hierarchy of who does notify? Aren't there emergency plans in every community, and generally, they are triggered by whoever the emergency coordinator is in that community?

Mr. MORAN. There are State—local and State plans. The State plans, most of them are filed with the FCC. We are aware of the plans. There is a whole hierarchy of how the various broadcast stations——

Mr. WALDEN. It is built from the grass roots up. I have to have one in each of my studios. You all—you require that. So you know that that has to be the case, that they are triggered from bottom up, unless there is some national emergency.

Mr. MORAN. But if there is a national emergency, the President could—it is automatic.

Mr. WALDEN. What was the status of the broadcast facilities at the time the levees broke? Were any of them on the air? If power was out and towers were down, didn't you testify there may be two AM stations?

Mr. MORAN. There were two on the air. They had—they were on emergency power. And we got word—the FCC mobilized. We had people there 24/7. We had our watch center there. We were working with the NCS and the FEMA. We were notified early on that one of the AM stations, actually I believe several AM—several stations got together and were operating off the same tower.

And they said early on that they were nearly running out of fuel. Getting fuel in there was very, very difficult. It wasn't a matter of getting pickup trucks. It was a matter of getting tankers in there. And it was extremely difficult. I believe that—that that last station in downtown New Orleans, it stayed up the whole time because within I believe hours of when it was going to run out of fuel, it got a tanker in there.

Mr. WALDEN. They were broadcasting full time?

Mr. MORAN. Yes, they were.

Mr. WALDEN. It was all focused on the hurricane? They weren't playing music?

Mr. MORAN. Absolutely. As I recall, ultimately, they had to relocate their studio, I believe up to Baton Rouge, I believe. And they actually were provided special housing in the dorm up there, I believe.

Mr. WALDEN. Were they given notification that the levee might fail, and did they broadcast that, do you know?

Mr. MORAN. I really don't know the answer to that.

Mr. WALDEN. I assume if they would have been given notification——

Mr. MORAN. We have actually—one of the things we did when this happened was, we didn't know who was up and who was down. And we actually have some equipment at the FCC where we can actually sort of scan the air to figure out what is up and what is down.

And after that, by the way, we made calls to every single station in the area. When I last checked here this morning, we hadn't actually contacted all of them. We believe in some cases the phone systems are out, so we could not get to them. But we have contacted most of them, and we have a pretty good idea of the status. And actually if you—we would perhaps be able to ask what it was they knew.

Mr. WALDEN. My experience—and in Oregon, we do not have many hurricanes thankfully, but we do get ice storms, and we get some floods and things like that—is that most stations just go immediately 24/7 doing whatever the emergency report is.

And, I mean, we went through a flood, and we did, you know, trigger an EAS occasionally. But that is generally triggered by the local sheriff or the State police.

Mr. MORAN. Right. So a lot of those sorts of things, the message is getting—if the message is getting out, it does not necessarily have to come from the EAS.

Mr. WALDEN. I remember in our post-9/11 hearing here that the then-chairman of the Commission, Mr. Powell, suggested that in New York, that they actually told the broadcasters to stop using the EAS, because it was scaring people. They actually shut down, asked broadcasters not to do that, not to use EAS, because they were all reporting everything anyway.

So I just wanted to clarify, and you have helped me clarify in terms of how the emergency alert system works. We have to do tests every week. We have to record certain monthly tests. It is—and you rigorously enforced all of that, don't you?

Mr. MORAN. That is absolutely right. And we work with FEMA on that.

Mr. WALDEN. Mr. Chairman, my time has expired. Thank you. And thank you, Mr. Moran.

Chairman BARTON. Thank you. I think that is all for the first panel. I have to commend you, gentlemen. I did not see any of you take a bathroom break in almost 6 hours. That has got to be a record. So go to it.

We are going to take a recess until 5 p.m. So we are going to reconvene with our second panel at 5 p.m.

[Brief recess.]

Chairman BARTON. The committee will come to order. We are now going to begin our second panel. I think we have nine distinguished witnesses, which is not a record, we have had 10 distinguished witnesses on one panel. So you are one away from the record, but you may be the record for distinguished-ness.

We are going to start with Mr. Angelle, who is representing the Louisiana Department of Natural Resources. We will give each of you 7 minutes. And we will just go right down the aisle. There are going to be a series of votes beginning in the next 10 to 20 minutes, but we will attempt to keep the hearing going.

So we thank you folks for your patience and recognize Mr. Angelle for 7 minutes.

STATEMENTS OF SCOTT A. ANGELLE, SECRETARY, LOUISIANA DEPARTMENT OF NATURAL RESOURCES; RED CAVANEY, PRESIDENT, AMERICAN PETROLEUM INSTITUTE; BOB SLAUGHTER, PRESIDENT, NATIONAL PETROCHEMICAL AND REFINERS ASSOCIATION; JAMES NEWSOME, PRESIDENT, NEW YORK MERCANTILE EXCHANGE, WORLD FINANCIAL CENTER; BENJAMIN S. COOPER, EXECUTIVE DIRECTOR, ASSOCIATION OF OIL PIPELINES; BILL DOUGLASS, CEO, DOUGLASS DISTRIBUTING COMPANY, ON BEHALF OF THE NATIONAL ASSOCIATION OF CONVENIENCE STORES AND THE SOCIETY OF INDEPENDENT GASOLINE MARKETERS OF AMERICA; WILLIAM L. SMITH, CHIEF TECHNOLOGY OFFICER, BELLSOUTH CORPORATION; DANIEL A. LASHOF, SCIENCE DIRECTOR, CLIMATE CENTER, NATIONAL RESOURCES DEFENSE COUNCIL; AND MARK N. COOPER, RESEARCH DIRECTOR, CONSUMER FEDERATION OF AMERICA

Mr. ANGELLE. Thank you, Mr. Chairman. It is with a heavy heart that I come to our Nation's capital today. Although we are here to discuss the effects of Hurricane Katrina on our national energy supply, let us all be reminded of the human tragedy on the gulf coast.

A special thanks to you, Mr. Chairman, and to the ranking member for your fight to help coastal producing States in the recent energy bill. Both of you were stand-up guys for Louisiana. Over strong objections of the administration, you gave us hope by providing resources for coastal restoration. And it is only fitting that we return and thank you and now ask your assistance for what is now our very survival.

The citizens of my State are still in the eye of Hurricane Katrina's wake, and many are experiencing the tragedy that is still unfolding; 899,000 people were without power and, currently, 503,000 now. On behalf of our great people, I thank you for your assistance in our rescue and recovery operations. Together, we know we can rebuild a strong and great Louisiana. So I come here today seeking help, bipartisan help, not assessing blame.

It was a wise Thomas Jefferson some 200 years ago who sought what would become the most valuable acquisition in the history of this country, the Louisiana Purchase, including the Orleans Territory. He understood the strategic importance of New Orleans and the Mississippi River for navigation interests and economic prosperity, but he had no way of knowing then the additional resources this Nation would acquire from Louisiana's rich delta land and the bounty off its shore. When it comes to energy production, energy refining, energy distribution and, indeed, America's energy security, this is the most important piece of real estate from sea to shining sea, and every American is connected to it through the gas pump and family energy costs.

We must do everything we can to protect it because most of America has resisted energy development. In fact, it has been over 25 years since America has built a new refinery in the continental United States. On the other hand, Louisiana has a strong and distinguished history of oil and gas production.

Let me tell you a little bit about my Louisiana. We host more than 80 percent of America's offshore oil and gas production and distribution, 34 percent, of the Nation's natural gas supply, and almost 30 percent of the Nation's crude oil supply is either produced in Louisiana, produced offshore Louisiana, or moves through the State and its coastal wetlands.

This production is connected to nearly 50 percent of the country's refining capacity, and Louisiana alone hosts more than 16 percent of the total U.S. refining capacity, second only to the great State of Texas. We host the Strategic Petroleum Reserve. Port Fourchon alone services 16 percent of the Nation's oil supply.

The Louisiana offshore oil port is the only port in the Nation that can handle the large super tankers from the Persian Gulf. This port alone is responsible for some 13 percent of America's foreign oil supply. We are home to America's most recently permitted LNG facility, as well as America's largest LNG facility, and we do all of this at the same time we produce 30 percent of our Nation's fisheries; catch and drain 41 percent of the Continental United States.

We have embraced the concept that we can improve the quality of life for all Americans with the responsible management of our natural resources, and we do all of this when most coastal States continue to say no, and not in my back yard.

We all know good relationships are like bank accounts, and it takes a few deposits to make a few withdrawals. When it comes to energy production, the 18th State of this Great Union has made its share of deposits, and it is in desperate need of a major withdrawal.

Louisiana Governors and Congressmen and DNR secretaries before me, along with Federal agency heads, scientists, economists, business and industry leaders, environmental representatives, have together sounded the alarm for years and respectfully, Mr. Chairman, neither Congress nor the White House, past or present, have answered the call.

We have continuously asked for the Federal commitment to restore our wetlands that protect this Nation's strategic energy infrastructure off the coast of Louisiana, that protect its No. 1 port system, the great city of New Orleans, and our coastal residents from storm surge.

But we have been told that we should scale back our plans and be satisfied with business as usual, that our Nation simply cannot afford it right now. Yet Louisiana State University research indicates that every 2.7 miles of healthy marsh can reduce storm surge by a critical 12 inches.

We have the science and technology to make a difference. We simply need the financial resources. We have asked for the Federal commitment it would take to raise our levees and build and upgrade flood and hurricane protection for our citizens and for the most strategic of American real estate, but so far, we have been shortchanged.

We have continuously asked, pleaded and begged for a true sharing of OCS revenues for the coastal producing States. We were on a course to adopt a constitutional amendment next fall in Louisiana that would dedicate any future OCS funds the State receives to rebuilding our wetlands.

Simply put, unless we invest in protecting the huge concentration of energy assets in Louisiana by restoring our wetlands and building levees, America's energy supply remains exposed.

Gratefully, because of your help, Mr. Chairman, we received the first step in that sharing through coastal impact assistance for 4 years in the recently passed energy bill.

But even that is woefully inadequate for such a challenge. We need true permanent revenue sharing like that with States that produce oil and gas on Federal lands on shore so that we may have the resources to protect our infrastructure.

You can imagine how amazed we were in July when our Nation's Energy Secretary wrote a letter to the House and Senate leaders opposing the sharing of OCS revenues through direct spending and authorized appropriations for coastal States.

What more must Louisiana do when it comes to energy leadership and development to get a share of these resources so that they can be used to help protect the energy infrastructure of our Nation? I think every American would agree that it just makes good common sense to take a portion of the OCS revenues to protect the infrastructure that makes this production possible.

In his letter dated July 15, 2005, the Energy Secretary said, "We can't afford to share revenues with the coastal producing States that host our Nation's energy production." It is right here in writing.

Well, let me share with you what we can't afford: A 50 cent increase in the average cost per gallon of gasoline because infrastructure was exposed. That equates to nearly $1.3 billion a week in increased fuel costs based on the daily consumption of America.

That says nothing of the increased cost of plastics, building materials, home energy costs, and transportation of products. When the Department of Energy doesn't think it is important to share OCS revenues to allow Louisiana to protect a high concentration of energy assets, Washington, we have a problem.

I hear a lot about SPR. That will do nothing to reduce the price of natural gas, and old man winter is just around the corner. Concerted voices, both Republicans and Democrats, have sounded the alarm: If the commitment wasn't made, the Nation would pay a far greater price. But the Office of Management and Budget continued to demand we justify the cost of our project through years of feasibility studies. We have had studies to study studies.

We do not have the luxury of time, especially now, and we ask OMB: Is the cost now justified? We branded Louisiana's coast America's wetland, and sounded the alarm that it is of great significance to the world ecology and that it impacts the Nation's economy and economic security. Restoration must be treated as a special circumstance because there is no comparison with how this coastline benefits the Nation or how it impacts the Nation if it is lost.

We sounded the alarm that what would happen if the big one ever hit New Orleans both in human cost and in energy infrastructure cost. And we are seeing those results firsthand. Our wounds are still gaping, and if these words sound strident, I'm sure you agree that this is not rhetoric.

It is just amazing just how accurate the October 2004 edition of the National Geographic was in laying out the tragic predictions that actually played out this week. Yet the opposition for revenue sharing for coastal producing States continues in Washington. It is no wonder many other States won't allow drilling offshore.

The worst case scenario the experts have long predicted is now reality. But yet in the midst of an ongoing crisis, Louisiana remains committed to the fueling of this great Nation as a world energy leader. Energy companies are working to reestablish families, so that the work to rebuild may begin.

I hear a lot of things about ExxonMobil on the screen up here, but keep in mind that 91 percent of the wells that were drilled last year in Louisiana were by independents who, along with the majors, will need Federal assistance to repair infrastructure, low- or no-interest loans, permit streamlining and immunity from outside litigation during this rebuilding process.

What sits off Louisiana's coast cannot be compromised. Estimated depreciated investment in offshore production facilities is more than $85 billion; pipeline infrastructure, more than $10 billion; and public coastal port facilities, $2 billion.

Production off Louisiana's shore alone contributes an average of $5 billion a year to the Federal Treasury, and that was when oil was less than the $68-a-barrel-plus today. A week after Katrina's landfall, a whopping 58 percent of oil production and 42 percent of natural gas from the OCSs remains shut in.

As of yesterday, we still have six refineries in Louisiana shut down from storm damage or lack of electric power. And a huge unknown in all of this is the condition of the pipeline infrastructure. When Hurricane Ivan made landfall two States away last year, pipeline infrastructure took months and months to rebuild.

As more of the protection from Louisiana's barrier islands and coastal wetlands wash away, more onshore and offshore production will be damaged or destroyed by storms. And according to scientists, the increase in frequency and strength of gulf hurricanes will be with us for years to come.

Louisiana needs America more than any State has ever needed her mother country. And yet, America needs Louisiana more than ever. It is vital to the Nation's security and economic future that Louisiana is not only restored, both its infrastructure and its wetlands, but that it is strengthened in the process.

Thank you for inviting me here to be with you. And to the American people for the outpouring of your generosity, we say thank you in this time of need.

May God continue to bless America and may God restore Louisiana. Thank you.

[The prepared statement of Scott A. Angelle follows:]

PREPARED STATEMENT OF SCOTT A. ANGELLE, SECRETARY, LOUISIANA DEPARTMENT OF NATURAL RESOURCES

INTRODUCTION

Mr. Chairman, Mr. Ranking Member, and distinguished members of the House Committee on Energy and Commerce, thank you for your gracious invitation to address your Committee. Unfortunately, as you know, I come to you today with a somber heart from the frontlines of the worst natural disaster in our nation's history.

My home state will never be the same again, nor will America. Almost no enemy of this nation, or terrorist of any kind, could have wrought the terror and devastation to my state and to this nation as the fury of nature with the name of Katrina did on August 29 and the ensuing days. Overnight, upwards of a hundred thousand citizens of my state and our neighbors in Mississippi and Alabama lost everything they had—homes, jobs, businesses, cars, and for some, their very lives. Hundreds of thousands of others were dramatically affected to a lesser, but significant degree. Suddenly, we find ourselves in the midst of an ongoing crisis, faced with restoring the basic elements of civilization—food, safe drinking water, shelter, clothing, fuel, and sanitation.

I want you to know that the people of Louisiana are deeply touched by the outpouring of concern, prayers, help, and generosity from Americans from every walk of life from all over the country. To all of you, we give you our heartfelt thanks.

Now, I will focus on the subject of this hearing—the impact of Hurricane Katrina on gasoline and petroleum supplies.

SUPPLYING THE NATION: LOUISIANA—AMERICA'S ENERGY CORRIDOR

Louisiana has a long and distinguished history of oil and gas production, both onshore and offshore. Currently, approximately 34% of the nation's natural gas supply and almost 30% of the nation's crude oil supply is either produced in Louisiana, produced offshore Louisiana, or moves through the state and its coastal wetlands. Together with the infrastructure in the rest of the state, this production is connected to nearly 50% of the total refining capacity in the United States. Based on its energy producing value to the nation, acre for acre, Louisiana is the most valuable real-estate in the nation.

Louisiana has 17 petroleum refineries, most of them large, world scale facilities, with a combined crude oil distillation capacity of approximately 2.77 million barrels per calendar day, which is 16.2% of total U.S. refinery capacity of 17.1 million barrels per day, the second highest in the nation after our sister Gulf Coast state, Texas. Louisiana produces approximately 42.1 million gallons of gasoline per day and 29.9 million gallons of distillate fuel (that is, jet fuel and diesel fuel) per day. Two of the four Strategic Petroleum Reserve storage facilities are also in Louisiana. The other two are in Texas.

Louisiana is not some far off energy producing colony. Louisiana and its citizens are fundamental elements from which this great nation was forged. Dating back to Thomas Jefferson's signing of the Louisiana Purchase in 1803, Louisiana has indelibly stamped its mark on this country, becoming the 18th state in the Union in 1812. Even today, Louisiana has provided more national guardsmen to the war against terrorism in Afghanistan and Iraq than any other state, though we rank only 22nd in population. Approximately 41% of the continental land mass of the U.S. drains through Louisiana via the Mississippi River. The Port of greater New Orleans is the largest port in total tonnage the U.S., and the port of Baton Rouge is 10th.

When it comes to developing the nation's offshore petroleum resources, there simply would not be much if it were not for Louisiana's leadership and participation. The offshore territory off Louisiana's coast is the most extensively developed offshore territory in the entire world. As most of you know, the offshore area beyond 3 miles from Louisiana's coast is federal territory called the Outer Continental Shelf, or OCS. Other than in a 3-mile transition zone, the federal government receives ALL of the mineral revenue from production in the OCS. Based on 2004 data, OCS production off Louisiana's coast constitutes 91% of oil and 75% of natural gas production from all U.S. OCS areas combined. Additionally, Louisiana OCS territory has produced 88.8% of the 14.9 billion barrels of crude oil and condensate and 82.3% of the 150 trillion cubic feet of natural gas ever extracted from all federal OCS territories since the beginning of time.

Offshore Energy Development and Economic Prosperity

This service that Louisiana provides to the nation is one of the largest contributing factors to America's strategic security and economic prosperity, which make possible the high standard of living that we all enjoy in this country. Let's look at just one example of how this translates to you. Prior to Hurricane Katrina, the pump price of gasoline was already hitting the $2.50 per gallon range in many parts of the country. If it were not for Louisiana's role in the petroleum supply of the nation, you and your constituents would likely have been paying in the range of $4.00 per gallon for gasoline pre-Katrina, and more than that post-Katrina. And, that does not address how sky-high prices would be for electricity, food, and all of the other things fueled by, or made from, oil and natural gas.

Offshore petroleum production is not only good for the country, but it is essential to the well-being of the USA. Offshore production is also good for coastal producing states, and there are not many of us—coastal states, that is, that allow new production off our coasts. The list currently consists of only Alabama, Alaska, Mississippi, Louisiana, and Texas. Even without being able to share in the mineral revenue produced for the federal treasury off our coasts, offshore production produces economic prosperity for coastal states in the form of jobs for the service industries providing the logistics support for the offshore industry. This includes, among others: equipment and materials suppliers; food service; helicopter and boat transportation; communications services; engineers, geologists, boat and rig crews; other industry staff and employees; and many others. The offshore industry also supports many jobs far removed from the coastal states, including a multitude of employees who, because of the week on, week off type of schedules, commute up to 500 miles or more from places like Arkansas, Tennessee, and Georgia to work offshore in the Gulf.

Offshore Development Includes LNG

Stepping up to the plate to help the nation obtain new supplies of energy including LNG (liquefied natural gas), Louisiana is the home of the largest throughput facility (Southern Union in Lake Charles) of the four existing LNG import terminals in the U.S., and it is undergoing more than a doubling of capacity from 1 billion cubic feet per day to 2.5 billion cubic feet per day. While almost every state in the nation is trying to prevent the siting of any new LNG facilities, Louisiana is the site of the largest permitted LNG import terminal in the nation (Cheniere Energy's 2.6 billion cubic feet per day facility in Sabine Parish).

Offshore Development and Preserving the Environment Are Compatible

I am also here to tell you, that oil and gas production is compatible with protecting and preserving the environment. Louisiana can look at experience and footnote that offshore development and the associated onshore infrastructure construction and operations are done in an environmentally responsible way today and are done so under the oversight of several state and federal regulatory agencies.

Louisiana has suffered some negative impacts in the past from offshore production. And, yes, we still have to deal with some of those legacies of the past, but that is because Louisiana pioneered offshore production in the days before modern technology, before the awakening of America's environmental consciousness, and before the advent of environmental regulatory agencies and regulations.

Louisiana's first well (a dry hole) was drilled in 1868. Our first oil well was drilled in 1901. The first oil well over water in the world was in Louisiana in 1910 in Caddo Lake. The first well drilled off the coast of Louisiana was in 1938 near Creole, Louisiana. Louisiana was the site of the first well drilled out of sight of land in 1947. Things have changed dramatically since 1910, 1938, 1947, or even 1960, 1970, or 1980. Simply put, it was like the old Wild West out there. Just as in other industries in other parts of the country in other times, there was once a time, long ago, when almost anything in the name of progress was accepted. Everything is different now. That era and those practices have nothing more in common with modern exploration, production, and environmental techniques than transportation by horse and buggy in 1800's has in common with jet airliners flying overhead today.

THE CONSEQUENCES OF CONCENTRATING OIL & GAS DEVELOPMENT IN ONE AREA

This country now faces an energy disaster of both short-term and long-term causes, implications, and solutions. Our present energy crisis is caused by the immediate effects of Hurricane Katrina, compounded by the long term consequence of decades of having had no meaningful energy policy, concentrating energy production and processing in the Gulf Coast area, the aversion to energy development in most other areas of the country, and this country's insatiable appetite for energy. The Energy Policy Act of 2005 (EPAct 2005) that was just enacted is a good step in the right direction, but it is not soon enough and not enough. For the foreseeable future, EPAct 2005 will not meaningfully reduce this country's increasing energy appetite. It will not reduce this country's increasing dependence on unreliable foreign sources of crude oil AND, NOW, liquefied natural gas. It will not significantly increase domestic energy supply or diversity. And it will not protect, much less rebuild, the Louisiana energy production infrastructure and the eroding and decimated coastal wetlands that protected and made the offshore production possible off Louisiana.

We are all familiar with the old adage, "Don't put all of your eggs in one basket." We all also know the reason for that: If you drop that basket, what are you going to do? Well, ladies and gentlemen, this nation's oil and gas offshore production, foreign import capability, refining, and basic petrochemical eggs have been placed in one basket called the Louisiana and the Gulf Coast, and that basket has not only been dropped, it has been run over by Hurricane Katrina.

I am not here to chastise anyone from those states that will not allow drilling off their coasts, or drilling rigs, petroleum refineries, or petrochemical plants in their states. What I am here to say is that since Louisiana has welcomed those facilities and operations and has become America's Energy Corridor, help us. And, by helping us, you are helping yourselves and all Americans.

Energy is the lifeblood of an industrialized nation and a prosperous society, and none is more of both than this country. The mainline artery supplying that sustaining life blood of oil, natural gas, petroleum products such as gasoline, jet fuel, and diesel fuel, is Louisiana. Louisiana has over 40,000 miles of pipelines just within our state as part of the infrastructure that receives offshore and foreign oil and gas, and feeds it through processing facilities, refineries, and petrochemical plants that then distribute it to the rest of the nation.

A PLAN NEEDED TO REBUILD LOUISIANA

Most of this offshore and onshore production is shutdown, and much of the onshore infrastructure is either shutdown, damaged, destroyed, or underwater. We will not know the full extent of either the short-term or long-term damage for some time. Until that information is available, a reasonable assessment of the cost and time to repair or replace it and to restore energy flow to the pre-storm level will not be known.

Here are just a few of the challenges we face in even determining the damage:

The communications infrastructure is in ruins.

Telephone lines, cell phone towers, radio towers, repeaters and remote data telemetry are either destroyed or have no power.

Advance rescue and assessment teams have to resort to carrying in satellite phones just to communicate from sites they are able to reach.

Accessibility to wells, pipeline pumping stations, and processing facilities is limited by flood waters, downed trees, washed out roads, lack of vehicle fuel and other impediments.

Complicating this even further, hundreds of thousands of people have been dislocated to other cities throughout Louisiana and other states.

The people who are most familiar with the damaged areas and who operate the affected oil and gas facilities are among the hundreds of thousands of displaced citizens.

Untold tens of thousands, or even hundreds of thousands of these evacuees cannot return to homes for months, if they still have homes to return to. Even the facilities that can be restarted and operated soon, need the people who operate them, and those people need food, water, and a place to live. The people and their needs cannot be separated from the infrastructure.

Refineries are shut down, wells are shut in, and bodies are floating in the streets. As the floodwaters recede, fires are burning uncontrolled in New Orleans because there are no firefighters to put them out. Businesses have been destroyed. Most of the oil and gas exploration and development onshore in Louisiana, and a large portion in the shallow waters offshore are done by independent companies. These are small operations, many with only a half dozen to a couple of dozen employees. These people would be your typical neighbors, not large corporations with extensive resources. Without help, many of them will never drill another well, because their employees are dislocated, their equipment ruined, their offices and workshops destroyed, and their financial resources gone.

It is expected that unemployment in Louisiana has almost overnight, jumped to about 25%. Tens of thousands of people who once had jobs, many in the oil and gas industry, have now lost homes, jobs, or both.

These are extraordinary times, and extraordinary times call for extraordinary measures. Louisiana needs the rest of America more now than ever before, and America needs Louisiana and its lifeblood energy supply more now than ever before. The U.S. had a Marshall Plan to rebuild Germany, the defeated enemy, after World War II; the U.S. now needs to institute a massive rebuilding plan for its own people in Louisiana, Mississippi, and Alabama.

A Rebuild Program from the Past to Inspire Us Today

In 1932, there was a cry for help from a desperate people near panic. The nation turned to its leaders searching for an end to the rampant unemployment and economic chaos that gripped the country. They were not disappointed. A plan was needed to fight soil erosion and declining timber resources, utilizing the unemployed of large urban areas. Congress and the President initiated several actions, one of which was the Emergency Conservation Work (ECW) Act, more commonly known as the Civilian Conservation Corps. With this action, two wasted resources were brought to bear, the young men and the land, in an effort to save both.

President Roosevelt called the 73rd Congress into Emergency Session on March 9, 1933, to hear and authorize the program. It included recruiting thousands of unemployed young men, enrolling them in a peacetime army, and sending them into battle against destruction and erosion of the nation's natural resources. Before it was over, over three million young men engaged in a massive salvage and public works operation. We are all familiar with the public works facilities these hard working men built throughout the country. These facilities—post offices, other public buildings, roads, parks, fire towers, telephone lines and many other facilities that Americans still use today.

A massive rebuilding program is needed to replace and restore all that Katrina destroyed. This includes the whole infrastructure of a modern civilization such as housing, public buildings, communications, energy production facilities, offices, etc. As the infrastructure is rebuilt and financial assistance is provided, more businesses can be reopened, creating more jobs, reducing unemployment, and restarting the decimated economy of the area. Today, skilled, hard-working men and women of Louisiana, who until a few days ago, were going to their jobs and returning home each day, need America's help, not charity, to restore those jobs, homes, and lives.

Maybe the legacies of the Marshall Plan and the Civilian Conservation Corps can serve as an inspiration for developing the rebuild program direly needed today for Louisiana, Mississippi, and Alabama.

LOUISIANA'S ROLE AS A PRODUCING AND CONSUMING STATE

A reliable and affordable supply of energy is necessary for economic development, prosperity, and expansion. Although technological improvements and investments in energy efficiency have reduced this country's energy consumption per unit of Gross Domestic Product over the past 20 years, increased economic prosperity is still dependent on increased energy consumption. In the U.S., the availability of energy has generally been taken for granted, but recent blackouts in California and other parts of the country, the emergence of 70 plus dollar per barrel oil and $11 to $12 per million BTU natural gas, and the drive to build terminals to import foreign natural gas in the form of a cryogenic liquid, have highlighted the need for addressing energy supply.

I come to you representing a state to which energy is its middle name. The words Louisiana and energy are almost synonymous. Among the 50 states, Louisiana ranks (2004 Energy Information Administration—EIA data):

1st in crude oil production
2nd in natural gas production
2nd in total energy production from all sources

The importance of energy to Louisiana is further highlighted in the following rankings in which Louisiana is (2003 EIA data latest available):

2nd in petroleum refining capacity
2nd in primary petrochemical production
3rd in industrial energy consumption
3rd in natural gas consumption
5th in petroleum consumption
8th in total energy consumption
But, only 22nd in residential energy consumption

Usually, when national energy issues are discussed, Louisiana is cast in the image of a rich producing state floating in a sea of oil and gas that is being inequitably shared with the consuming states. Often misunderstood or overlooked, is the fact that about two thirds of the production from the state is in the Louisiana federal OCS territory and, hence, produces no revenue for the state, while at the same time incurring significant infrastructure support costs to the state, which I will discuss in more detail later.

Also often overlooked or not explained, is the fact that, though Louisiana is the 2nd highest energy producing state in the nation, Louisiana is also 8th highest in total energy consumption. Therefore, Louisiana is more of a consuming state than 42 other states! This story is never told, nor are Louisiana's difficulties as a key consuming state given much concern at the federal energy policy level. Thus, when Louisiana, the energy producing state speaks, it is also Louisiana, the energy consuming state speaking. Louisiana is inexorably tied into the issues of all states in the nation, whether considered producing states or consuming states. However goes the energy situation in Louisiana, so goes the energy situation in the United States of America, and things are not going well for Louisiana today.

Louisiana's Role as a Through-Processor of Hydrocarbons for the Nation

All of the preceding represents only the direct supply line of oil and natural gas. Additionally, Louisiana's 8th highest ranking among the states in energy consumption is attributable to the fact that Louisiana is consuming most of this energy as a through-processor of energy supplies for the rest of the nation, consuming colossal amounts of energy for their benefit.

An example of how Louisiana is consuming energy resources for the primary benefit of other states is petroleum refining. The energy equivalent of 10% of Louisiana's entire petroleum product consumption is required just to fuel the processes that refine crude oil into gasoline, diesel fuel, jet fuel, heating oil and other products consumed out of state. The oil refining industry employs only about 10,400 workers in the state; whereas tens of millions of jobs throughout the country are dependent on the affordability and availability of the products from the continued operation of these refineries and associated petrochemical facilities in Louisiana.

Many other examples could be cited of the numerous energy intensive natural gas and oil derived chemical products Louisiana (and also Texas, Oklahoma, and California) through-processes for the rest of the U.S. Per unit of output, these industrial processes in Louisiana are characterized as capital (equipment), energy, raw material, and pollution discharge intensive, and low in labor requirements and dollar value added, essentially the opposite of the downstream industries in other states that upgrade these chemicals into ultimate end products. Much of the energy Louisiana technically consumes is really the transformation of oil and gas into primary chemical building blocks that are shipped to other states where the final products are made, whether it be plastic toys, pharmaceuticals, automobile dash boards, bumpers and upholstery, electronic components and cabinets, synthetic fibers, or thousands of other products dependent on this flow of energy and high energy content materials out of Louisiana.

OCS INFRASTRUCTURE AND ITS IMPACTS AND NEEDS

It is important to understand that there is no free lunch. Louisiana, like other coastal producing states, sustains impacts on coastal communities and bears the costs of onshore infrastructure required to support this production activity.

Saving Louisiana's Wetlands that Protect Offshore and Onshore Production Infrastructure

Louisiana's unique and fragile coastal wetlands introduce yet an additional issue: land loss. Prior to Hurricane Katrina, Louisiana was losing more than 24 square miles of our coastal land each year. In fact, if what is happening today in coastal Louisiana were happening in our nation's capital, the Potomac River would be washing away the steps of the Capitol today, the White House next year, and the Pentagon soon after that. In fact, during the course of this morning alone, Louisiana will lose a football field wide area from the Capitol Building to the Washington Monument. It is feared that the ferocity of Hurricane Katrina may have accelerated the land loss by several years.

There are many causes of this coastal erosion in Louisiana, including what may be the most significant factor: building levees and channeling the Mississippi River. Whatever the cause of its demise, the health and restoration of Louisiana's coastal wetlands are vital to protecting the offshore and onshore infrastructure that is essential for the continuation, as well as the expansion, of offshore energy production in the Gulf of Mexico.

Once the State realized the magnitude of the coastal erosion problem, we got serious about doing something about it. In 1980, the coastal restoration permitting program was moved to the Department of Natural Resources (DNR). In 1981, $40 million of state oil and gas revenue was set aside in a legislative trust fund for coastal restoration projects. The State has a dedicated revenue stream of up to $25 million per year, depending on the level of revenue collections from oil and gas production within the state, to replenish the fund. In the past few years, that replenishment stream has been at the $25 million level. In 1989, the Office of Coastal Restoration and Management was created in DNR, and the magnitude of the program was greatly expanded.

The War against the Elements

Let me emphasize something extremely important to this nation's energy supply. Here along the coast, WE ARE AT WAR. It is a war in which the enemy is nature. It is an enemy with names like Andrew, Ivan, Dennis, and Katrina—hurricanes. It is an enemy with names like wave erosion, storm surges, sedimentary subsidence, soil consolidation, salt water intrusion, and leveeing of the Mississippi River. As Hurricane Katrina demonstrated last week, it is a war we are losing in Louisiana.

Prior to Hurricane Katrina, Louisiana needed a minimum of $14 billion (in today's dollars) over the next 20 to 30 years for coastal restoration projects. Louisiana has quite a unique geology relative to the rest of the country. The Louisiana coast is geologically the youngest part of the U.S. and, prior to manmade interference from leveeing and channeling the Mississippi River and other activities, was still accreting land mass faster than it was losing it to subsidence, erosion, salt water intrusion, sea level rise from global warming, and other causes. The science of coastal geology and the expertise of coastal engineering to counter these forces is in its infancy, as it has never in the history of civilization, been attempted on the scale it must be implemented in South Louisiana. Also, we are dealing with a situation that is continuously subject to changing dynamics, such as more frequent and more powerful hurricanes, the apparently increasing effects of global warming, etc.

Extent of Louisiana Infrastructure Supporting OCS Production

The total value of the Louisiana OCS infrastructure and the onshore infrastructure supporting it is difficult to ascertain. The estimated depreciated investment in offshore production facilities is over $85 billion, depreciated offshore pipeline infrastructure is over $10 billion, and public coastal port facilities is $2 billion, for a total of approximately $100 billion, depreciated, and not counting highways, sewer, water, fire and police protection, schools, and other public works structures that also have ongoing operation and maintenance costs. The replacement of all of this would be several times the $100 billion depreciated figure. It also does not count the onshore coastal infrastructure of pipelines, storage facilities, pumping stations, processing facilities, etc.

This infrastructure is vulnerable if not protected by the State's barrier islands and marshes. As these erode and disappear, infrastructure is exposed to the open sea and all of its fury. As the coast recedes, near shore facilities become further offshore and subject to greater forces of nature, including subsidence, currents, and mudslides. Erosion in the coastal zone is already beginning to expose pipelines that were once buried.

A Wake-up Call from Hurricane Ivan

To bring home the point of infrastructure vulnerability, we need only look back to this past Summer. Hurricane Ivan was not even a direct hit on Louisiana's offshore and coastal oil and gas infrastructure, striking two states away; yet, its effects on the nation's supply of oil and gas were significant, even many months after it hit. Most of the damage occurred along pipeline routes rather than actual structural damage to the producing platforms. As of February 14, 2005, when the Minerals Management Service (MMS) released its final impact report on Ivan, 7.42% of daily oil production and 1.19% of daily gas production in the Gulf of Mexico was still shut-in. The cumulative shut-in production through February 14 was 43.8 million barrels or 7.25% of annual Gulf of Mexico OCS production and 172.3 billion cubic feet of natural gas or 3.9% of annual Gulf of Mexico OCS gas production.

With Katrina, that infrastructure has sustained a direct hit. As of Saturday, September 3, the Minerals Management Service (MMS) reported that 70% of manned platforms and 71% of the drilling rigs in the Gulf were not operating. Saturday's shut-in oil production was 1.2 million barrels per day, or 79% of Gulf production. Shut-in gas production in the Gulf was 5.8 billion cubic feet per day, or 58% of daily gas production in the Gulf.

Also, as of noon Sunday, 7 refineries in Louisiana and 1 in Mississippi were still shutdown from storm damage and/or lack of electric power. An additional 4 refineries in Louisiana were operating at reduced rates due to storm damage or lack of crude supply.

As more of the protection from Louisiana's barrier islands and coastal wetlands wash away, increasingly more onshore and offshore production will be damaged or destroyed by even less powerful storms than Ivan and Katrina, and particularly by storms whose paths more directly pass through the producing areas off of Louisiana's coast, as did Katrina. Direct hits to the prime production area by hurricanes and tropical storms will cause incalculable damage to this production infrastructure, as well as to the onshore support infrastructure, as Katrina is proving.

HOW TO INCREASE OFFSHORE ENERGY PRODUCTION

Share Offshore Revenue with the States that Allow Offshore Production

The most effective way to help is to assist those states that make offshore energy production possible off their coasts. This can be accomplished by sharing with those coastal producing states some of the offshore revenues generated off their coasts. This would encourage those states to pursue more development, and it would help offset infrastructure costs those states incur that is associated with that development. Louisiana, like other coastal producing states, sustains impacts on coastal communities and bears the costs of onshore infrastructure to support this production activity.

When states like Wyoming, New Mexico, Colorado, and others host drilling on federal lands onshore, they receive 50% of those revenues in direct payments, and consequently have the financial resources to support that infrastructure. In Fiscal Year 2004, Wyoming and New Mexico together received about $928 million from those revenues, which IS an appropriate revenue sharing procedure.

In contrast, for example in 2001, of the $7.5 BILLION in revenues produced in the federal OCS area, only a fraction of one percent came back to those coastal states. The inequity is truly profound.

We are pleased this committee is investigating gasoline supply and pricing. The need to sustain the existing supply that Louisiana provides must simultaneously be addressed. The most effective answer to both issues is to share offshore revenues with the coastal producing states that make that production possible. It is critical that coastal producing states receive a fair share of revenues to build and maintain onshore infrastructure and, in Louisiana's case, to help stem our dramatic land loss, which is occurring at a rate believed to be the fastest on the planet.

Production off Louisiana shores alone contributes an average of $5 BILLION dollars a year to the federal treasury, its second largest source of revenue. And, that was when oil was less than half of the $60 plus per barrel price it is selling for today.

Does it not make sense to encourage the coastal producing states which provide that revenue for the benefit of the rest of the nation? Does it not make sense, that when so many, like the U.S. Ocean Commission, are targeting offshore OCS revenues to pay for worthwhile preservation of natural resources, that this nation first protect those who make these resources possible?

Prior to Katrina, in Louisiana's coastal zone, many of the pipelines and other infrastructure that our wetlands have historically protected had become exposed to open Gulf of Mexico conditions. I shudder to think of the extent of production infrastructure damage that we will learn that Katrina caused once we are able to get a full damage assessment.

To maintain, much less increase, production from off our coasts, we must reinvest in the infrastructure that makes all of the activity possible, whether it be port facilities, roads to transport equipment and supplies, erosion control, or barrier island and wetlands storm protection.

Assistance from the Energy Policy Act of 2005

The Coastal Impact Assistance Money provided in the Energy Policy Act of 2005 that you just helped pass is tremendously good news for the state's coastal restoration efforts. Yet, the $540 million provided over four years for coastal restoration is only a drop in the bucket compared to the total of $14 billion needed, prior to Katrina, over 20 to 30 years for Louisiana's unique coastal restoration needs.

Enact Legislation to Extend Section 29 Tax Credits to Deep and Ultra-Deep Production in States Allowing Offshore Production

Section 29 of the Internal Revenue Service (IRS) Code granted a tax credit for the production of natural gas from unconventional resources (coal bed methane and tight sands gas). The effect of the application to coal bed methane gas production was astounding in those areas of the country that have significant deposits of this kind, which is not along the Gulf Coast. Natural gas reserves from coal bed methane rose from 6.3% of U. S. reserves at the end of 1993 to 9.9% at the end of 2003. Annual natural gas production from coal bed methane rose from 4.2% of U. S. dry gas production in 1993 to 8.2% by the end of 2003.

Deep natural gas reserves (15,000-24,999 feet sub-surface) and ultra-deep gas reserves (greater than 25,000 feet sub-surface) are the most immediately available resources capable of providing a substantial increase in domestic production of natural gas. Substantial deep gas reserves are known to exist, and a deep gas well can have the productive capacity many fold over that of coal seam wells and as much as five to ten times that of conventional shallower wells. For example, a typical coal seam gas well may produce 100,000 cubic feet (CF) per day, a good conventional 15,000 foot well could produce 1 to 2 million CF per day, and a deep gas well could produce in excess of 50 million CF per day. The richest deep gas domain known in the U.S. underlies the onshore area and adjacent offshore shallow water shelf of the Gulf of Mexico. A 1998 study of the Potential Gas Committee put estimates of the U.S. deep gas resource base at possibly 170 Trillion Cubic Feet. The deep gas domain along the Gulf Coast underlies the existing surface infrastructure of pipelines, gas processing plants, and other drilling/production support infrastructure to move this gas into the U.S. gas supply immediately.

One problem is that, while productivity increases with depth in elevated reservoir pressure wells, drilling costs rise exponentially with well depth, and the drilling of one deep well takes a year or more. For example, conventional wells less than 15,000 feet normally cost between $100,000 and $2 million to drill. The deeper 15,000, plus foot range wells average around $6 million, 20,000 foot wells about $16 million, and 25,000 to 30,000 foot wells are in the range of $25 million, plus. Hence, the capital at risk for a dry hole is substantial, which makes the ability to fund such ventures difficult. Additionally, deep wells require leading edge drilling technology. Due to the limited amount of deep drilling done, few companies have the experience, technological capabilities, and financial resources to undertake this high return, but high risk activity. Of the few companies that have the ability to drill in this domain, most are the major oil companies, who have focused their financial resources on the more lucrative oil reserves of the deep water Gulf and drilling in foreign countries. Substantial new financial incentives could significantly reduce the entry hurtle, increase the reward to risk ratio, and reduce barriers to capital access, particularly for the independent companies who now do most of the onshore drilling in this country.

Immediately Share with the States A Percentage of Royalties from Deep Drilling in the Shallow Waters of the Gulf:

Another thing that is needed immediately, is to share with coastal producing states 50% of the royalties from new deep drilling in the shallow federal waters on the shelf. The MMS royalty deep shelf suspension program is a good program, but it is draining investment from our parishes by shifting drilling across the boundary line into federal waters, causing loss of investment and tax revenue from lost drilling in state territory. Louisiana should receive a substantive percentage of royalties from deep drilling on the shelf immediately.

Encourage New Energy Sources and Technology

Recent studies show that the Gulf of Mexico has a significant wind energy potential. Although wind power does not have the energy density of petroleum, it is an inexhaustible, renewable source of clean energy. Again, much to my consternation, it appears that there are many parts of the country that use a lot of energy and want it at low prices, but do not want production of any kind, anywhere near them, including wind energy. Again, Louisiana is stepping up to help encourage this clean energy source. The State of Louisiana is currently working with private sector investors who are interested in developing wind farms in state and federal waters off Louisiana's coasts. My office submitted wind power legislation which the Louisiana Legislature passed earlier this year to facilitate offshore wind power development in Louisiana's State offshore waters.

Natural gas hydrates probably offer the greatest untapped energy resource the nation has. *The Oil and Gas Journal* recently reported that the U.S. Geological Survey estimates that methane hydrate deposits are greater than all other forms of fossil fuels combined. Large deposits of gas hydrates are believed to lie below the offshore waters of the U.S. Unfortunately, technology to tap these resources needs to be developed. Once the technology is available, the first areas to be developed will be the areas adjacent to the existing offshore producing areas where the infrastructure is in place to get it to shore and into the nation's pipeline distribution system. The federal government needs to fund meaningful research into developing the technology to produce gas hydrates, assessing the resource base, and delivering it.

IN CONCLUSION

It is vital to the nation's security and prosperity that new energy sources be developed. The federal government has proven that it has the ability to steer investment, as in the case of deep water drilling in the Gulf and coal seam gas. In addition to its significance in producing 30% of oil and 23% of natural gas produced domestically, which is mostly off Louisiana, the OCS is probably the single most promising area for the U.S. to obtain significant new energy supplies. These supplies, whether conventional oil and gas, imported oil, imported LNG, wind and ocean energy, or gas hydrates, need the support of coastal states to cooperate and to supply and maintain critical production and support infrastructure.

LNG facilities are being built where the existing U.S. pipeline infrastructure exists (essentially Louisiana and Texas) in order to get the gas from the coast into the delivery system to supply the nation. The same will be true when the technology is developed to commercialize methane hydrate production off the coasts. This Louisiana and Texas infrastructure will also be used when deep and ultra-deep shelf production comes on stream. This is another reason why offshore revenue should be shared with the coastal producing states and why the extension of Section 29 tax credits should be extended to deep gas exploration at least in the states that are allowing onshore and offshore drilling and allowing the siting of LNG facilities to make energy available to the rest of the country.

With effective policies and incentives, the federal government can steer investment into the offshore areas, and by receiving an equitable share of revenue generated offshore, the coastal producing states can be in a position to ensure that this production will be made available to the rest of the nation. Louisiana desperately needs immediate revenue sharing financial assistance from a source not subject to annual appropriations, to continue to maintain existing, and to develop future energy supplies for the nation.

Although the Congress enacted national energy legislation that included direct payments to the coastal producing states for four years for coastal impact assistance, it did not enact true sharing of OCS revenues on a permanent basis that would be similar to the automatic payments for drilling on federal lands onshore. This must be addressed.

Now that Hurricane Katrina has laid waste to Louisiana's largest city, the entire southeastern portion of the state, the state's coastal oil and gas infrastructure and its protective wetlands, a massive national rescue and rebuilding program is imperative to bring the state back from this crisis and to enable us to continue to supply a critically needed portion of this nations energy needs.

Thank you for this opportunity to appear before you.

Mr. HALL [presiding]. We thank you, Mr. Angelle. And thank you for your patience today, and thank you for the things you have seen and witnessed and suffered through the last several days.

All right. The Chair recognizes Mr. Red Cavaney. Thank you, too. And thank you for the courtesy you extended to the President of the United States out in New Mexico 2 weeks ago. Appreciate that very much.

I recognize you for—we are not going to blow the whistle on you. You have been so patient. You are really valuable people. You have expended a lot of time and money to get here, and you still got to go home sometime tonight, maybe. The Chair is willing to recognize you for as long as it takes, but roll around, if you can, pretty quick.

STATEMENT OF RED CAVANEY

Mr. CAVANEY. Mr. Chairman, we will give you some time back.

The gulf coast is the very heartland of our industry, and our prayers and support go out to each and everyone there. We are not just responding to this disaster; we are living it.

Thousands of our husband and wives, sons and daughters and friends and neighbors are suffering the hardships of others living in this devastated region. They are the ones restoring the production, bringing the refineries back on line and restarting the pipelines. Facilities are coming back on line. And we are grateful to the administration for access to the Strategic Petroleum Reserve and for waivers to expedite the flow of fuels, particularly to emergency responders.

The gulf coast region includes some 4,000 offshore platforms in Federal waters, a dozen refineries, and hundreds of production, transportation and marketing facilities. There is a reason for this geographic concentration in the high-risk hurricane area. Government policies have largely limited offshore exploration and production to the central and western gulf, and our on-shore facilities have been welcomed in the communities in the region.

Unfortunately, offshore oil and natural gas development has been barred elsewhere, including the eastern half of the gulf and the entire Atlantic and Pacific coasts. On-shore construction has been held back by Government restrictions, permitting delays and the not-in-my-back-yard or NIMBY sentiments.

It is ironic that we talk so much about diversifying the sources of our energy supplies from abroad; yet we have done so little to geographically diversify our oil and natural gas presence here at home.

In an area of much recent concern has been the need to bring additional clean-burning natural gas to industries and consumers nationwide. Yet, efforts to increase domestic natural gas production, both in the Rocky Mountain west and offshore, have largely been stymied. And efforts to build more terminals outside the gulf region to permit increased imports of liquified natural gas or LNG have also been largely blocked.

Oil companies recognize the urgent need to expand refining capacity. But they cannot do it alone. Chairman Barton, and the rest of you, are particularly appreciated for your leadership in this area. Government policies are needed to create a climate conducive to investments to expand refining capacity.

For example, the Federal Government should take steps to streamline the permitting process, to expand capacity at both existing refineries and possibly even to build a new refinery or two. We know that Hurricane Katrina's effects on our industry are having a nationwide impact through skyrocketing prices for gasoline and other fuels.

Our fuels are sold at more than 168,000 retail outlets nationwide, and less than 10 percent of those outlets are actually owned by the large oil companies. The remaining 150,000 outlets are owned by independent small businessmen and women. They are making business judgments each and every day, as is their right.

However, for any of us that break the law, prosecution must follow. Let me be very clear. API and its member companies condemn price gouging. History provides an important guide here. Our industry has repeatedly been investigated over many decades by the Federal Trade Commission, other Federal enforcement agencies and States attorneys general, among a few.

In each and every instance, our companies have been exonerated of price gouging or other anticompetitive behavior. In conclusion, let us all not be diverted from the serious work needed to ensure Americans continue to get the fuel they deserve. We look forward to working with the committee in that regard. Thank you, Mr. Chairman.

[The prepared statement of Red Cavaney follows:]

PREPARED STATEMENT OF RED CAVANEY, PRESIDENT AND CEO, AMERICAN PETROLEUM INSTITUTE

I am Red Cavaney, President and CEO of the American Petroleum Institute—the national trade association for the U.S. oil and natural gas industry, representing all sectors of the industry, including companies that make and market gasoline.

The Gulf Coast is the very heartland of our industry. We are not just responding to this disaster. We are living it. Thousands of our husbands and wives, sons and daughters, and friends and neighbors are suffering the hardships of those living in this devastated region. Fitch Ratings, a leading global ratings agency, reports that

Hurricane Katrina has caused the largest insured loss in U.S. history—more than 9/11 and more than any previous natural disaster.

Facilities are starting to come back online, and we are grateful to the Administration for access to the Strategic Petroleum Reserve and for waivers to expedite the flow of fuels, particularly to emergency responders.

The Gulf Coast region includes some 4,000 offshore platforms in federal waters, major refineries, and hundreds of production, transportation and marketing facilities. There is a reason for this geographic concentration in a high-risk weather area. Government policies have largely limited offshore exploration and production to the Central and Western Gulf—and our onshore facilities, including refineries, have been welcomed in communities in the region. Unfortunately, offshore oil and natural gas development has been barred elsewhere—including the eastern half of the Gulf and the entire Atlantic and Pacific Coasts. Onshore construction has been held back by government restrictions, permitting delays, and not-in-my-backyard NIMBY sentiments.

It is ironic that we talk so much about diversifying the sources of our energy supplies from abroad, yet we have done so little to geographically diversify our oil and natural gas industry here at home.

An area of much recent concern has been the need to bring additional clean-burning natural gas to industries and consumers nationwide. Yet, efforts to increase domestic natural gas production, both in the Rocky Mountain West and offshore, have been stymied—and efforts to build more terminals outside the Gulf region to permit increased imports of LNG have also been largely blocked.

Impact of Hurricane Katrina

While it is still too soon to know the full effects of Hurricane Katrina on production and refinery facilities in and along the Gulf of Mexico, it is clear that the impact of this devastating storm on oil and natural gas operations will be significant and protracted.

I know that I speak for every one of our member companies when I say that our first concern—from the moment it becomes evident that a hurricane is approaching the Gulf—is for the wellbeing and safety of the thousands of men and women from across the country who work on offshore facilities, on the vessels that serve them, in the refineries, distribution networks, and retail outlets around the Gulf coast.

Equally as important is the welfare and recovery of the communities in the Gulf region. Millions of people in the area are experiencing firsthand the physical and emotional hardship of the death and devastation caused by Katrina, and our hearts and our prayers are with them.

API is working with the American Red Cross to facilitate U.S. oil and natural gas industry efforts to help people throughout the Gulf region. We have informed our companies that the Red Cross has described how they can help relief efforts through corporate contributions and by encouraging customer and employee contributions.

Effects of Hurricane Katrina on Industry Facilities

We are concerned, also, about our facilities in the area. While they are designed to withstand the forces of the most severe storms, extraordinary circumstances do occur. Therefore, one of our industry's top goals is always to ensure minimal impact on the Gulf of Mexico and coastal environments. The industry takes pride in its outstanding record for safety and environmental protection in the Gulf region, and we intend to live up to that record. Let me review the latest information (as of September 4) we have from the Department of Energy (DOE) and the Minerals Management Service (MMS) on the status of our facilities:

Offshore Production Facilities. According to the latest MMS reports, 30 percent of the 819 manned platforms and 29 percent of the 137 rigs are currently operating in the Gulf of Mexico. Shut-in oil production is at 1,184,747 barrels of oil per day, which is equivalent to 78.9 percent of the daily oil production in the Gulf. Shut-in gas production is 5.779 billion cubic feet per day, which is equivalent to 57.8 percent of the daily gas production in the Gulf.

Refineries. A significant volume of refining capacity in the Gulf Coast and Midwest remains impacted by Katrina. According to DOE, 11 percent of U.S. refinery capacity is shut-in, and refineries representing another 14 percent of U.S. capacity are operating at reduced levels because of a lack of crude supplies. Lack of electricity has also been an issue in restarting refineries. Much progress has been made and Entergy reports that it has restored electricity to all but three refineries in the New Orleans area.

Pipelines. DOE reports that the Colonial and Plantation pipelines, critical for distributing petroleum products from the Gulf Coast to the Southeast and Mid-Atlantic regions, have resumed operations, albeit at reduced rates. Colonial is operating at

66 percent of normal operating capacity. Both gasoline and distillates are currently being transported and delivered. Colonial's capacity is about 2.4 million barrels per day. Plantation announced it would be 100 percent operational by late on September 2. Plantation moves about 620,000 barrels of gasoline, diesel, and jet fuel per day. The Capline pipeline is also now operational at reduced rates, according to DOE. Capline will operate at reduced rates until the Louisiana Offshore Oil Port (LOOP) is fully operational. Capline runs roughly 1.2 million barrels a day of crude oil to the Midwest.

LOOP. LOOP is operational at the Clovelly terminal. Entergy energized a line to Clovelly and the terminal is now capable of operating at approximately 75 percent of capacity. The Fourchon terminal remains shut down.

Katrina Impact on Jet Fuel Supply

The Committee has expressed interest in the impact of Hurricane Katrina on jet fuel supply. It is too soon to assess that impact, but we are hopeful that restoration of refineries and pipelines to at least partial operation will increasingly alleviate whatever supply shortfalls are caused by the hurricane.

The Louisiana Gulf Coast District, the region hit by Katrina, accounts for about 23 percent of U.S. jet fuel production. In 2004, the region's refineries produced 355,000 barrels per day of the national output of 1.547 million barrels per day. The Gulf Coast region as a whole accounts for about half of U.S. jet fuel production, or 779,000 barrels per day in 2004.

The Gulf Coast region ships about two-thirds of what it produces to the East Coast (about 500,000 barrels per day), and more than 80 percent of those shipments are by pipeline. Some jet fuel is also shipped by tanker and barge to the East Coast, mainly to the South Atlantic states. The Gulf Coast region ships approximately another 135,000 barrels per day to the Midwest, mostly by pipeline. The United States also imports about 125,000 barrels per day of jet fuel.

Responding to Hurricane Katrina

In the coming days and weeks, we are committed doing our best to minimize the impact of Hurricane Katrina on the flow of fuels to consumers.

Even before the hurricane's devastating impact, American consumers were concerned over the rising cost of gasoline, diesel and other fuels. Katrina's aftermath, however, underscores the need for all drivers to take seriously common-sense energy conservation recommendations—found on API's website and elsewhere—for reducing the amount of fuel they consume.

We also want to thank President Bush for making available crude oil from the Strategic Petroleum Reserve to address circumstances for which it was intended and appreciate the IEA member nations' contributions as well. We are also grateful that EPA and the Department of Transportation have granted waivers to expedite the flow of fuels, particularly to emergency responders—an action that is very helpful at a time when logistics and distribution of fuels are extremely difficult and critical. The Departments of Energy and Homeland Security have also been helpful in many ways.

We believe Congress can take action to help alleviate the hardships Americans are suffering from Hurricane Katrina. One action involves LNG. I earlier mentioned the importance of siting LNG receiving terminals in areas beyond the Gulf region. This diversification is helpful, and your support in facilitating it would be much appreciated.

These and other positive steps by government can be most helpful in dealing with this catastrophe. We believe it is particularly important for government officials at the federal, state and local levels to urge citizens nationwide to use energy wisely, particularly in terms of not hoarding gasoline and not "topping off" their vehicle tanks. Effective conservation measures are vital in helping meet the fuel needs of U.S. consumers in this difficult situation.

In attempting to meet the challenges we face, it is also most important to do no harm. The worst thing Congress could do in these challenging times would be to repeat the mistakes of some past energy policies by trampling the structures of the free marketplace. Imposing new controls, allocation schemes, or other obstacles will only serve to make a bad situation much worse. (See the attachment, "Hurricane Katrina and U.S. Energy Policy: Do No Harm.")

Why Have Gasoline Prices Risen?

We know that Hurricane Katrina's effects on our industry are having a nationwide impact. We understand how Americans throughout the country are facing skyrocketing prices for gasoline and other fuels. What follows is background on two key components of the price of gasoline: crude oil price and taxes.

Crude Oil Price. Before Hurricane Katrina struck, the price of gasoline was rising primarily because U.S. refiners are paying more for crude oil, the principal cost component of a gallon of gasoline. In fact, the Federal Trade Commission noted this exact point in a report this July: "To understand U.S. gasoline prices over the past three decades, including why gasoline prices rose so high and sharply in 2004 and 2005, we must begin with crude oil. The world price of crude oil is the most important factor in the price of gasoline. Over the last 20 years, changes in crude oil prices have explained 85 percent of the changes in the price of gasoline in the U.S." The crude oil price is set in the international oil marketplace by the forces of supply and demand for oil worldwide.

Tax Component. While more than half the cost of gasoline is for crude oil, every time a motorist pulls up at the pump, he or she pays 46 cents in federal and state taxes per gallon of gasoline. The remainder is the cost to refine and market the gasoline. The average price of a gallon of regular gasoline reached $2.85 on September 2, according to AAA. When the price of a barrel of crude oil is $67, as it was at the end of last week, a refiner paid about $1.61 per gallon for the crude oil in order to make a single gallon of gasoline. As noted above, taxes average 46 cents per gallon nationwide. The remaining 78 cents per gallon includes the cost of running refineries, transporting the finished gasoline to markets via pipelines and tank trucks, and operating retail outlets. The cost to refine, market and distribute gasoline has been trending downward for many years. The recent price spikes are a direct consequence of disruptions in crude oil and gasoline supplies. (Attached is a chart showing combined federal, state and local gasoline taxes for each state.)

How Fuels Are Marketed. It is important to recognize that our fuels are sold at more than 168,000 retail outlets nationwide—and less than 10 percent of those outlets are actually owned by refiners. The remaining 150,000 outlets are owned by independent small businessmen and women, who are your neighbors. They are making business judgments every day, as is their right. However, if any of us breaks the law, prosecution should follow.

History provides an important guide here. Our industry has been repeatedly investigated over many decades by the Federal Trade Commission, other federal agencies, and state attorneys-general. None has ever found evidence that our companies have engaged in price gouging or other anti-competitive behavior to drive up fuel prices.

The gasoline marketing system has the complexity and flexibility required to meet the varying needs of both companies and consumers. Companies have three basic types of outlet options and may employ any and all in their marketing strategies to maximize efficiencies and compete in the marketplace. First, they can own and operate the retail outlets themselves (company owned and operated outlets). The second option is to franchise the outlet to an independent dealer and directly supply it with gasoline. This option may have three different forms of property ownership: The operator can lease from the refiner, lease from a third party, or own the outlet outright. The third option is to utilize a "jobber," who gains the right to franchise the brand in a particular area. Jobbers can choose to operate some of their outlets with their own employees and franchise other outlets to dealers. The mix of distribution methods varies widely across firms. Different refiners, depending on which type is perceived as most efficient, use different types of outlets.

Supply and Demand in the World Market. Prices are rising because of the forces of supply and demand in the global crude oil market. Supply and demand is in a razor-thin balance in the global market. Small changes in this market have a big impact.

World oil demand reached unprecedented levels in 2004 and continues to grow. Strong economic growth, particularly in China and the United States, is fueling a surge in oil demand. The U.S. Energy Information Administration (EIA) reports that global oil demand in 2004 grew by 3.2 percent—the strongest growth since 1978—and projects growth to increase by about 2.1 percent this year and next. By comparison, world demand between 1993 and 2003 grew at an average rate of 1.6 percent.

At the same time, world oil spare production capacity—crude that can be brought online quickly during a supply emergency or during surges in demand—is at its lowest level in 30 years. Current spare capacity is equal to about 1 percent of world demand. EIA projects spare capacity for 2005 at just over 1.0 million barrels a day. Thus, the world's oil production has lagged, forcing suppliers to struggle to keep up with the strong growth in demand.

The delicate supply/demand balance in the global crude oil market makes this market extremely sensitive to political and economic uncertainty, unusual weather conditions, and other factors. Over the past year, we have seen how the market has reacted to such diverse developments as dollar depreciation, an unusually cold winter, the post-war insurgency in Iraq, hurricanes in the Gulf of Mexico, the continued impact on the Venezuelan sector from the oil workers' strike in 2002-03, uncertainty in the Russian oil patch, ongoing ethnic and civil strife in Nigeria's key oil producing region, recent mass protests targeting Ecuador's oil infrastructure, and decisions by OPEC.

Gasoline Prices Mirror Crude Oil Prices

While consumer concern about high gasoline prices is very understandable, we must recognize that gasoline prices mirror crude oil prices. Crude oil costs make up more than 50 percent of the cost of gasoline. Retail gasoline prices and crude oil prices have historically tracked, rising and falling together. We import more than 60 percent of the crude oil and petroleum products we consume. American refiners pay the world price for crude and distributors pay the world price for imported petroleum products. U.S. oil companies don't set crude oil prices. The world market does. Whether a barrel is produced in Texas or Saudi Arabia, it is sold on the world market, which is comprised of hundreds of thousands of buyers and sellers of crude oil from around the world.

Earnings

There is considerable misunderstanding about the oil and natural gas industry's earnings and how they compare with other industries. The oil and natural gas industry is among the world's largest industries. Its revenues are large, but so are its costs of providing consumers with the energy they need. Included are the costs of finding and producing oil and natural gas and the costs of refining, distributing and marketing it. The energy Americans consume today is brought to them by investments made years or even decades ago. Today's oil and natural gas industry earnings are invested in new technology, new production, and environmental and product quality improvements to meet tomorrow's energy needs.

The industry's earnings are very much in line with other industries and often they are lower. This fact is not well understood, in part, because the reports typically focus on only half the story—the total earnings reported. Earnings reflect the size of an industry, but they're not necessarily a good reflection of financial performance. Earnings per dollar of sales (measured as net income divided by sales) provide a more relevant and accurate measure of a company's or an industry's health, and also provide a useful way of comparing financial performance between industries, large and small.

For the second quarter of 2005, the oil and natural gas industry earned 7.6 cents for every dollar of sales compared to an average of 7.9 cents for all U.S. industry.[1] Many industries earned better returns in the second quarter than the oil and natural gas industry. For example, banks realized earnings of 19.6 cents on the dollar. Pharmaceuticals reached 18.6 cents, software and services averaged 17 cents, consumer services earned 10.9 cents and insurance saw 10.7 cents for every dollar of sales. Last year, the oil and natural gas industry realized earnings of 7 percent compared to an average of 7.2 percent for all U.S. industry. Over the last five years, the oil and natural gas industry's earnings averaged 5.7 cents compared to an average for all U.S. industry of 5.5 cents for every dollar of sales.

Some are calling for reinstatement of a windfall profits tax as a response to the nation's energy challenges. As the figures I just cited demonstrate, our industry's earnings are hardly a "windfall." Strong earnings enable our industry to remain competitive globally, benefit millions of shareholders—your constituents—and enable the industry to invest in innovative technologies that improve our environment and increase energy production to provide for America's future energy needs. Levying new taxes would likely end up harming consumers. As *The Wall Street Journal* editorialized recently, ("China Does Carternomics," August 19), "A windfall profits tax only discourages increases in supply by disincentivizing further production." According to the Congressional Research Service, the windfall profits tax drained $79 billion in industry revenues during the 1980s that could have been used to invest in new oil and natural gas production. In fact, 1.6 billion fewer barrels of oil were produced domestically due to the windfall profits tax—barrels that instead had to be secured from foreign sources.

Perspective: The Role of Oil and Natural Gas

High gasoline prices have caused some to call for us to decrease, if not eliminate, our nation's reliance on oil and natural gas. However, if we are to understand and

---

[1] Earnings equal profits divided by sales calculated from "Corporate Scorecard," Business Week, August 22/29, 2005; and from company financial reports for oil and natural gas figures.

address the causes of the high prices, we need to recognize the energy realities that our nation faces.

These realities could not be clearer: We live in a global economy, and there is a strong link between energy and economic growth. If we are to continue to grow economically, we must be cost-competitive in our use of energy. We need all sources of energy. We do not have the luxury of limiting ourselves to one source to the exclusion of others. Nor can we afford to write off our leading source of energy before we have found a cost-competitive and readily available alternative.

Consider how oil and natural gas enhance our quality of life—fueling growth and jobs in industry and commerce, cooling and warming our homes, and getting us where we need to go. Oil provides about 97 percent of U.S. transportation fuels, which power nearly all of the cars and trucks traveling on our nation's highways. More than 60 million American households are heated and/or cooled by natural gas. And plastics, medicines, fertilizers, and countless other products that extend and enhance our quality of life are derived from oil and natural gas.

The U.S. Energy Information Administration has projected that fossil fuels will continue to dominate U.S. energy consumption, with oil and natural gas providing nearly two-thirds of that consumption in the year 2025, even though energy efficiency and renewables will grow faster than their historical rates. However, renewables, in particular, start from a very small base; and the major shares provided by oil, natural gas, and coal in 2025 are projected to be nearly identical to those in 2003.

Our nation cannot afford to leave the Age of Oil before a realistic substitute is fully in place. We will leave the Age of Oil, not because we will run out of oil. Yes, someday oil will be replaced, but clearly not until a substitute is found—a substitute that is proven more reliable, more versatile, and more cost-competitive than oil.

There is a misperception by some about the time and costs involved in any such transition. Consider what would be involved in replacing the dominant role of oil with a substitute like ethanol, hydrogen, or solar power. Most experts agree that finding and transitioning to a substitute for oil will require dramatic advances in technology and massive capital investments—and that such a displacement will take many years to accomplish.

In the early 1970s, many energy policymakers were "sure" that oil and natural gas would soon be exhausted, and government policy was explicitly aimed at "guiding" the market in a smooth transition away from these fuels to new, more sustainable alternatives. Price controls, allocation schemes, limitations on natural gas, massive subsidies to synthetic fuels, and other measures were funded heavily and implemented.

Unfortunately, the key premises on which these programs were based, namely that oil and gas were nearing exhaustion, and that government "guidance" was desirable to safely transition to new energy sources, are now recognized as having been clearly wrong—and to have resulted in enormously expensive mistakes.

The leading role that oil and natural gas will continue to play makes it all the more important for our government to adopt policies that do not prevent or delay oil and gas development before substitutes are ready to satisfy consumer needs and to meet the economic investment demands.

In considering future U.S. energy needs, we need also to understand that gasoline-powered automobiles have been the dominant mode of transport for the past century. Regardless of fuel, the automobile—likely to be configured far differently from today—will remain the consumer's choice for personal transport for decades to come. The freedom of mobility and the independence it affords consumers are highly valued.

Moreover, we expect that the dominant transport fuels will remain gasoline and diesel for decades—the minimum amount of time required to fully retire any existing and still growing fleet of automobiles and trucks powered by these fuels and to deploy any replacement fuel source throughout our nation. We cannot afford to prematurely retire a century-old champion. And, sulfur-free diesel and sulfur-free gasoline could well live on as the preferred sources for fuel cells well into the future.

Gasoline Prices: What Can Be Done?

The solution to high gasoline prices is more supply of crude oil and gasoline and less demand, but there is no simple strategy to make that happen. Now that the long Congressional debate over energy legislation has come to an end, the United States is at a critical turning point in shaping its future energy policy. The legislation signed by the President signals a first step in a much-needed effort to enhance energy security and ensure the reliable delivery of affordable energy to consumers. But much remains to be done.

The problems we face are very real: growing world demand for energy at a time when many oil-producing countries around the world are increasingly limiting or restricting our industry's access to new resources; a lack of national commitment to develop our abundant domestic energy resources and critical infrastructure; and scant attention to energy efficiency. These factors have resulted in a tight supply/demand balance for U.S. consumers, causing recurring price spikes, greater market volatility, and overall strain on the nation's energy production and delivery systems.

Energy demand continues to grow. The Energy Information Administration (EIA) forecast that by 2025, U.S. energy consumption will increase by 35 percent, with petroleum demand up by 39 percent and natural gas up by 34 percent. These demand increases occur despite expected energy efficiency improvements of 33 percent and renewable energy supply increases of 41 percent.

Additional EIA forecasts point out our basic problem: Domestic energy supplies are not keeping up with increased demand; and we are relying more and more heavily on imports to meet our energy needs. EIA projects that U.S. crude oil production will fall by 17 percent by 2025 (assuming no production from ANWR), while crude oil imports will increase by 67 percent, and net petroleum product imports increase by 90 percent. Given these trends, it comes as no surprise that EIA forecasts that our nation's dependency on foreign sources of petroleum will rise from 59 percent today to 68 percent in 2025.

This increase, to the extent that it reflects import costs lower than domestic supply costs, would represent a gain from trade which should be encouraged. However, when we have resources that can be developed at prices competitive to imports, and we choose not to do so, we place a wasteful and unnecessary burden on our own consumers,

In fact, we do have an abundance of competitive domestic oil and gas resources in the U.S. According to the latest published estimates, there are more than 131 billion barrels of oil and more than 1000 TCF of natural gas remaining to be discovered in the US.

However, 78 percent of this oil and 62 percent of this gas are expected to be found beneath federal lands and coastal waters.

Federal restrictions on leasing put significant volumes of these resources off limits, while post-lease restrictions on operations effectively preclude development of both federal and non-federal resources. The most comprehensive study of the effects of such constraints was the 2003 National Petroleum Council study of natural gas, which included an analysis of federal constraints on U.S. gas supply in two key areas—the Outer Continental Shelf (OCS) and the Rockies. The study found that in key areas of greatest supply potential, federal policy precludes or seriously constrains development. For instance, of the 209 TCF of estimated undiscovered gas in the Rockies, 69 TCF is completely off limits, while another 56 TCF is seriously constrained by federal policy. On the OCS, the entire Atlantic, Pacific, and most of the Eastern Gulf of Mexico are off limits to development. Furthermore, the study found that sustaining these constraints over the next 20 years would cost U.S. consumers more than $300 billion in increased energy costs.

We are aware that opponents of oil and natural gas development still raise environmental concerns. However, we would point out that history provides overwhelming evidence that our industry can find and develop oil and natural gas resources safely and with full protection of the environment, both on land and offshore. For example, according to the U.S. Coast Guard, for the 1980-1999 period, 7.4 billion barrels of oil were produced in federal offshore waters, with less than 0.001 percent spilled. That's a 99.999 percent record for clean operations—a statistic few others can likely match or best, and far less than the volumes of natural seeps that occur on ocean and gulf floors.

Using advanced technology and sound operational practices, our industry has steadily reduced the impact of oil and gas development, both onshore and offshore. The surface presence for exploration and development wells has shrunk significantly. For example, a drilling pad the size of Capitol Hill is all that is needed to access any oil reserves that might exist in the entire 68.2 square mile District of Columbia. Horizontal and directional drilling now enables our industry to drill multiple underground wells from a single pad, sometimes reaching sites as far away as 10 miles from the drilling pad.

Additionally, the U.S. oil and natural gas industry is among the most heavily regulated industries in our country. Every lease contains a standard stipulation to protect air, water, wildlife and historic and cultural resources, but leases may also include any number of a additional stipulations to further protect resources.

The recently enacted energy legislation takes a positive step by requiring an inventory of OCS oil and natural gas resources. It will not, by itself, result in new energy supplies.

We need to build on the energy legislation by encouraging the flow of more natural gas and oil to the marketplace. And, while we must focus on producing more energy here at home, we do not have the luxury of ignoring the global energy situation. In the world of energy, the U.S. operates in a global marketplace. What others do in that market matters greatly.

For the U.S. to secure energy for our economy, government policies must create a level playing field for U.S. companies to ensure international supply competitiveness. With the net effect of current U.S. policy serving to decrease U.S. oil and gas production and to increase our reliance on imports, this international competitiveness point is vital. In fact, it is a matter of national security.

We can no longer wait 15 years, as we have, to address our nation's energy policy. The energy legislation is a foundation, but it must be built upon. More needs to be done and more quickly, particularly increasing access to offshore resources. We have the ingenuity, the technology, and environmental protections. If enactment of the energy legislation means we have a commitment to continued action, then it will truly be a turning point in reshaping U.S. energy policy.

Refineries

We cannot understand or deal with high gasoline prices if we do not consider the state of refineries in the United States. During the 1990s, the oil industry earned relatively poor rates of return on their investments. This was especially true in the refining sector, which was hard hit with the need for new investment in technology and equipment to produce cleaner burning fuels to meet clean air standards set by the Clean Air Act of 1990. The act had a major impact on the operation of refineries in the U.S. and the return on investment realized at the time.

From 1994 to 2003, the industry spent $47.4 billion to bring refineries into compliance with environmental regulations. That included $15.9 billion in capital costs and $31.4 billion in operations and maintenance costs to comply with regulations covering air, water and waste rules. Moreover, by 2010, the U.S. refining industry will have invested upwards of $20 billion to comply with new clean fuel regulations. This is in addition to the cost of compliance with many dozens of other environmental, health, safety and security regulations. All this investment severely reduces the funds available for discretionary capacity expansion projects.

Technological advancements have helped refineries produce more from existing facilities than they did in the past. In addition, the elimination of subsidies under the government price and allocation controls in 1981 led to the closure of many smaller, less-efficient refineries throughout the 1980s and 1990s. Those refineries left standing did a better job of bringing product to market for less.

This consolidation benefited consumers. We can see this in the decline in the refiner/market margin (measured as the difference between the retail price of gasoline minus taxes and minus the refiner's composite crude oil price). Back in 1980, the cost to refine and market and distribute gasoline averaged about 95 cents per gallon (in inflation-adjusted terms). By 1990, it averaged more than 61 cents per gallon, and, by 2000, it was 52 cents per gallon, which is about where it has averaged over the last five years. Multiplying these reductions by the 330 billion gallons of petroleum products consumed translates into billions of dollars of savings for consumers. We all benefit every day from these improvements and efficiency gains.

The Need to Expand Refinery Capacity

The expansion of refinery capacity must be a national priority. The record-high gasoline prices, while primarily caused by increased crude oil prices and exacerbated by Hurricane Katrina, have underscored the fact that U.S. demand for petroleum products has been growing faster than—and now exceeds—domestic refining capacity. While refiners have increased the efficiency, utilization and capacity of existing refineries, these efforts have not enabled the refining industry to keep up with growing demand. Even with a projected expansion of product imports of 90 percent, the Energy Information Administration (EIA) forecasts a need for 5.5 million barrels a day of additional refining capacity by 2025 beyond today's 16.9 million barrels a day of capacity, even with higher utilization rates.

The fact is that—faced with increasingly more challenging fuels regulations—only major refineries have the resources needed to expand their capacity. Smaller refineries are increasingly unable to afford to expand. Moreover, local opposition and not in my backyard (NIMBY) attitudes persist and prevent new refineries from being constructed. The steady growth in U.S. fuels demand must increasingly be met by foreign product imports. Thus, in addition to blocking or delaying refinery expansion, the extensive federal regulatory burden is contributing to increased reliance on foreign product imports. This is a result that neither serves the best interests of U.S. consumers nor bolsters the U.S. economy and American jobs.

Oil companies recognize the urgent need to expand refining capacity, but they cannot do it alone. Government policies are needed to create a climate conducive to investments to expand refining capacity. The President's innovative proposal earlier this year to build new refineries on closed military bases deserves serious consideration. In addition, many of the steps the federal government could take to help the refinery capacity situation are covered in the December 2004 National Petroleum Council (NPC) study, *Observations on Petroleum Product Supply—A Supplement to the NPC Reports "U.S. Petroleum Product Supply—Inventory Dynamics, 1998" and "U.S. Petroleum Refining—Assuring the Adequacy and Affordability of Cleaner Fuels, 2000."* For example, that NPC study suggested that the federal government should take steps to streamline the permitting process to ensure the timely review of federal, state and local permits to expand capacity at existing refineries and possibly even build a new refinery.

In addition to the myriad of permitting issues deterring new refining capacity investments, there are financial constraints as well. Attracting capital for new refinery capacity has been difficult with refining rates of return historically averaging well below the average for S&P Industrials. Over the 10-year 1994-2003 period, the return on investment for the refining sector was 6.2 percent or less than half as much as the 13.5 percent for S&P Industrials.

U.S. tax policy has also hindered the refining industry's ability to attract new investment capital. New refinery investments are depreciated over 10 years, while comparable assets in other industries are recovered over five or seven years. The recently enacted energy legislation takes a small, but positive, step in addressing this inequity by allowing 50 percent of those investments to be currently expensed through 2011. However, much more needs to be done to make U.S. refinery investments more economically attractive, and, thus, better able to compete for limited available capital.

Conclusion

The U.S. oil and natural gas industry recognizes the catastrophic impact that Hurricane Katrina has had on millions of Americans and our industry is working with government and others in the private sector to do all we can to alleviate their suffering.

If we all do our part—industry providing supplies and repairs as expeditiously as possible, government facilitating needed approvals, and consumers adjusting their driving habits to consume less fuel—Americans can overcome this challenge as we have others in our nation's history.

ATTACHMENT: HURRICANE KATRINA AND U.S. ENERGY POLICY: DO NO HARM

Hurricane Katrina has brought devastation to much of the Gulf Coast, interrupting operation of significant parts of the nation's oil and natural gas production facilities, refineries and pipelines. In addressing this catastrophe, energy policymakers should do no harm. They should avoid repeating past energy policy mistakes which could make a bad situation much worse. The following are examples of actions that should be avoided:

- *Windfall Profits Tax:* This was tried before. Backers of the 1980 tax claimed it would raise revenue and prevent oil companies from benefiting from higher crude oil prices and the removal of price controls. The tax drained $79 billion in industry revenues that could have been used to invest in new oil production—leading to 1.6 billion fewer barrels of oil being produced domestically. The industry uses profits to invest in new technology, new production, and environmental and product quality improvements. The National Petroleum Council projects that producers will have to invest a total of almost $1.2 trillion through 2025 to fund U.S. and Canadian natural gas exploration and production activities. Investments of this magnitude require long-term fiscal stability.
- *Price Controls:* As seen the 1970s, price controls further reduce product availability as suppliers are unable or unwilling to bring product to market if they cannot recover the cost of doing so. The result is less product available, potential outages, and long lines at service stations.
- *Rationing/Product Allocation:* Rationing results in too much product being sent to some areas and too little product being sent to other areas. The reason is that rationing ignores the market price signal that producers use to decide which areas are in greatest need of product. The result would be an inefficient distribution of product with some areas of the country having too much motor fuel while shortages develop in other areas.
- *Moratorium on Mergers:* As noted by the Federal Trade Commission in its August 2004 report, *The Petroleum Industry: Mergers, Structural Change, and Antitrust*

Enforcement, merger activity in the U.S. refining sector over the last several years has not adversely affected competition in the sector, and has resulted in greater operational efficiencies in the refining sector and lower costs to consumers. Government policy prohibiting mergers would slow or reverse this positive trend and ultimately result in higher fuel costs to the motoring public.

- *Regional Strategic Reserves of Refined Products:* While the concept behind the Strategic Petroleum Reserve (SPR) has merit, the same cannot be said of regional strategic reserves of refined products. Holding and managing refined products is much more complex and impractical than holding and managing crude. The large number of boutique fuels (17) would require a diverse number of storage facilities for each chosen location. Additionally, product degradation means that the product in the reserves would have to be continuously rotated. Because of this it is unlikely that there would be sufficient product of the right specification in the right location to be helpful during a supply disruption.[2]
- *Mandatory Minimum Inventory Levels:* Since fuel producers have considerable incentive to maintain sufficient inventories so as to not forfeit sales, a minimum inventory mandate could result in an inefficient level of inventory being held. Inventory is considered working capital and as such is a cost of doing business. Inefficient levels of inventory arising from mandatory minimum inventory levels would unnecessarily raise the cost of providing fuel to consumers.
- *Price Trigger for the SPR:* Industry has long supported government holdings of strategic stocks in the SPR, under one condition: that it be used only to replace volumes of oil lost in an emergency, not as an instrument for government price tinkering. The current mechanism allows the President a wide range of discretion to determine what constitutes an emergency. Some argue that this essentially makes the SPR a political instrument, subject to the President's whim. Setting a price trigger, some argue, would leave the trigger decision to the market. However, setting the price for the trigger is no less arbitrary than the existing trigger, and puts the government directly in the role of manipulating price.
- *Oil Import Tariff:* Oil import tariffs have been proposed, and used, in the past as an instrument of energy policy. The key motive of such an approach stems from a belief that reducing imports is unambiguously beneficial. However, when we look carefully at each of the claimed benefits, we find them all to be dubious at best, not to mention illegal under existing trade agreements with many of our trading partners.

Chairman BARTON. We thank you, Mr. Cavaney.

We now would like to hear from Mr. Bob Slaughter, who is president, and is representing the National Petroleum and Refiners Association.

Welcome, Mr. Slaughter. You are recognized for 7 minutes.

STATEMENT OF BOB SLAUGHTER

Ms. SLAUGHTER. Thank you, Mr. Chairman. I wanted to associate NPRA with the comments that API has just made, particularly, especially mentioning no toleration of profiteering or price gouging. And I think that is a very important matter. Also, I think it is very important to take note of the statement that the industry has been intensively investigated many times, and a monitoring project is ongoing at FTC for gasoline prices in 360 cities, and the industry has never—there has never been any evidence of gouging or any kind of price manipulative behavior on the part of our members.

I particularly want to focus, however, on the refining questions that were raised earlier, and I want to actually show a couple of charts. The first chart just shows the very strong relationship between crude price and the price of gasoline. You will see the curves are essentially the same. And the FTC, in its publication a couple of months ago, it was a very large study on gasoline, said that 85

[2] National Petroleum Council, *Observations on Petroleum Product Supply,* December, 2004 p. II-4

percent of the price movements in gasoline over the last 20 years were attributable to changes in the price of crude.

The second chart, if we can show it, points out—and this is an EIA chart—it shows the extent to which gasoline prices are determined mostly by crude but, second, by taxes.

And if you put your crude price, which is the price of our raw material, together with taxes on gasoline, when the gasoline gets to the pump, 80 percent of it is out of our control. The cost of actually refining and then the cost of distributing and marketing is quite small. And those numbers hold over considerable time.

The third chart, if I could. I would like to hold there for a second. One thing I would like to say and I wouldn't like to forget is, my first summer in Washington was 1971, and that was the summer in which President Nixon imposed wage and price controls. And it was 10 years before the wage and the price controls on energy could be removed, 1981.

Gas price controls were put on in 1952, and they weren't removed until 1983. So I just want to caution the committee in everything that it does that we do not want to take a giant step backward into the world of price control or other government intervention in this market. It takes a great deal to get rid of the shortages, lines and other negatives that result from that policy.

This particular chart shows the many programs that refiners have to comply with over this decade. The red ones are fuel controls. The blue ones are stationary source controls. There is well over $20 billion worth of investment on that chart. And frankly, most of it did not get a very good review for impact on supply.

One of our strong recommendations is supply, particularly oil and gas supply, needs to be job No. 1. Those are the fuels we depend on, and frankly, they always end up being the second priority, behind whatever people wanted to really do at that time.

Environmental regulations should go forward. We spend a great deal of money on them. And the industry has contributed greatly to environmental clean-up. But we should also look seriously at the impact on supply of these regulations. And these have not even been well sequenced, so we get many, many expensive regulations, one on top of another.

I asked a gentleman who has been in the refining industry for many years just a couple of years ago, why he found it difficult. He has been involved in transactions involving refineries. To get why it has been difficult to get new people into the refining business, he said, Well, because—he said—they know that it will take millions, maybe up to a billion dollars to get in the industry. And then they have the fear that Government at some level, whether it be legislative or regulatory, will come along and suddenly impose additional hundreds of millions of investment on them.

And if you will see that chart, that is pretty much what happens. That is one of the deterrents to more investment in the refining industry. A lot of people have jumped to the conclusion that there hasn't been a lot of investment in the industry. Mr. Chairman, there has been a great deal, billions and billions of dollars.

Over the 1990's, the industry invested billions of dollars to comply with the requirements of the Clean Air Act. You see what it is going to have to comply with this decade. On top of this, these are environmental programs, you have got to basically spend billions of dollars to stay in business and hopefully increase capacity. So there has been a lot of money spent on the business.

Someone mentioned mergers earlier, and said that that might be anticompetitive and lead to less capacity. But, I can just mention one case, I know Valero, and this was quoted in the Post this morning. That part of the article previously read wasn't mentioned. Valero has added 380,000 barrels of oil a day capacity to the refineries that it has purchased over the last 8 or 9 years. So often times mergers, you know, someone gets an asset who sees new possibilities in it and will put additional investment in it.

I would like to see that next chart if we could for just a minute. That really shows what has to be done in 2006 and 2007. It includes also the few programs that were imposed by the Energy Act. And there are a couple of things there that the industry has to do, but that is the agenda just for the next couple of years. You can see that refineries have a lot on their plate.

I know it is not in your jurisdiction, Mr. Chairman, but the recent Energy Bill also included some tax treatment for refinery investments. And it basically would allow expensing 50 percent of the cost of increasing refinery capacity by more than 5 percent. It would be very useful also to have a look at the depreciation rate for refining investments where 10-year property now, and all of their businesses like ours are 5-year property.

That would basically allow more investment in the industry. And the other thing would be, you know, there is going to have to be a lot of reconstruction done in these areas that have been affected by the flood and the hurricane. It would certainly be helpful to have assistance in getting the necessary permits to rebuild, and perhaps harden those facilities.

You know, we have heard several times today comments that it is bad that the industry is centered in this area of the country, as if we could pick it up and put it anywhere else in the country. The fact of the matter is that we have got to keep those assets there, hope they keep operating, which I believe they will, and help them harden themselves against hurricanes, because they are either there or they are nowhere is the history that we have seen.

Although, we would hope that other areas of the U.S. would take refineries, we have not found them willing to do that. So thank you, Mr. Chairman.

[The prepared statement of Bob Slaughter follows:]

PREPARED STATEMENT OF BOB SLAUGHTER, PRESIDENT, NATIONAL PETROCHEMICAL & REFINERS ASSOCIATION

Mr. Chairman and members of the Committee, thank you for the opportunity to appear today to discuss the impact of the wide-spread devastation caused by Hurricane Katrina on transportation fuels markets. While I will focus on that urgent matter, I will also discuss the many other factors impacting current transportation fuels markets. My name is Bob Slaughter and I am President of NPRA, the National Petrochemical & Refiners Association. NPRA is a national trade association with 450 members, including those who own or operate virtually all U.S. refining capacity, and most U.S. petrochemical manufacturers.

PART I. RESPONDING TO HURRICANE KATRINA

In the aftermath of Hurricane Katrina our nation confronts death, injuries and devastation of staggering proportions. The images of the tragedy displayed in the last several days on television and other media underscore the human toll and seeming hopelessness in ways more eloquent and compelling than could ever be captured in testimony. We share both the sense of dismay and increased humility felt by all Americans before this latest reminder of nature's power to devastate and confound the best efforts of human beings. NPRA offers our sympathy and prayers to those who have suffered the loss of loved ones among family members, or their neighbors and colleagues, as well as to those who have lost much or all of their personal assets and livelihood in this worst U.S. natural disaster.

Today's hearing has been called to inquire into the impact of Hurricane Katrina on the nation's energy supply. It is appropriate that Congress turn immediately to such questions because of the huge impact of that storm on the Gulf Coast, the energy heartland of the United States. This is a time when national attention is and should be focused on human needs. Many industry employees and their families have been victims as you will hear. Nevertheless, NPRA appreciates the committee's immediate attention to the issue of energy supply, which was the subject of considerable debate and attention even before the hurricane disaster occurred. We also appreciate the opportunity to respond to the committee's questions in person on this matter of critical national importance. Because our expertise lies in the area of refining and petrochemicals, we will focus on those areas, but will try to provide other available information insofar as is possible.

Thus, on behalf of our refining and petrochemical industry members we have attempted to respond to the questions most asked about Hurricane Katrina's impact on the industry and energy supply, as follows:

1. How much of the nation's oil and gas supplies come from this region?

According to the U.S. Energy Information Administration (EIA), the Gulf of Mexico produces 1.582 million barrels per day (mmb/d) of federal crude production, which is **28.5% of the U.S. total crude production** (5.488 million barrels per day).

Again according to EIA, the region contains 8.068 million barrels per day of refining capacity, **47.4% of the nation's total refining capacity** (17 million barrels per day).

The Gulf Coast region receives 6.490 mmb/d of crude oil imports, **60.4% of the nation's total crude oil imports** (10.753 mmb/d). (23.5% of the nation's total comes into ports in Louisiana, Mississippi and Alabama, and 8.5% of the nation's total crude imports come into the LOOP.)

The Gulf Coast region produces 10.4 billion cubic feet (bcf/d) of natural gas per day, **19.2% of the nation's total natural gas production** (54.1 bcf/d).

2. How extensive was the damage?

Crude Oil, Natural Gas Production

According to the U.S. Minerals Management Service (MMS), **as of September 2, 88.53% (1.328 mmb/d) of Gulf crude oil production was shut-in,** and **72.48% (7.248 bcf/d) of Gulf natural gas production was shut-in.** This amounts to 25% of total federal crude production and 14% of the nation's natural gas production.

Crude Oil Import Facilities

The storm resulted in temporary closure of LOOP, the Louisiana Offshore Oil Port. More than 10% (900,000 b/d) of the nation's crude oil imports enter through LOOP. Roughly 500,000 b/d of crude produced offshore is also unloaded at LOOP, which ceased operations on Sunday, August 28 as the storm approached.

Refineries

The following refineries were directly affected by Hurricane Katrina:

Belle Chasse, Louisiana (ConocoPhillips) 247,000 b/d; shut
Chalmette, Louisiana (ExxonMobil/PDVSA) 190,000 b/d; shut
Convent, Louisiana (Motiva) 235,000 b/d; shut
Garyville, Louisiana (Marathon) 245,000 b/d; shut
Meraux, Louisiana (Murphy) 125,000 b/d; shut
Norco, Louisiana (Motiva) 227,000 b/d; shut
Pascagoula, Mississippi (Chevron) 325,000 b/d; shut
Port Allen, Louisiana (Placid) 48,500 b/d; shut
St. Charles, Louisiana (Valero) 260,000 b/d; shut
Vicksburg, Mississippi (Ergon) 23000; shut

Together, these facilities constitute about 2 mm/b/d, 12% of the nation's total refining capacity (17 mmb/d).

In addition, the following refineries were forced to reduce operations because of the impact of Hurricane Kristina:

Baton Rouge, Louisiana (ExxonMobil) 488,000 b/d; reduced runs
Krotz Springs, Louisiana (Valero) 85,000 b/d; reduced runs
Memphis, Tennessee (Valero) 180,000; reduced runs
Port Arthur, Texas (Total) 285,000 b/d; reduced runs
Tuscaloosa, Alabama (Hunt Refining Co.), 35,000 b/d; reduced runs

In addition, several Midwestern refineries were affected by shutdown of the Capline Pipeline, which supplies crude oil from the Gulf region to refineries in the Midwest (16% of the nation's refining capacity is in the Midwest). For example, Marathon's refineries at Catlettsburg, West Virginia (222,000) and Robinson, Illinois (192,000) were affected by Capline's closure, as were other Midwestern facilities.

In total, we believe that **at least 20% of the nation's refining capacity** (3.4 mmb/d) ceased operations or reduced runs at some time due to the direct impact of Hurricane Katrina and the loss of crude supplies from pipelines affected by the storm. This is probably a conservative estimate.

Recent reports indicate that many of these refineries are either up and running or anticipate start-up as early as this week. But, unfortunately, there are some refineries representing a significant amount of capacity that will remain shut for an undetermined period.

The Gulf refineries were first impacted by the need to protect the personal and family safety of employees, as well as the high likelihood of wind and flood damage as a result of the hurricane. After the hurricane passed, many of these facilities remained totally off-line as damages were assessed. In some instances companies could not physically enter the facilities to conduct an assessment for several days, and had to first depend on flyovers to study the plant. Damages included flooding, wind damage, and lack of electricity.

Pipelines

In addition, the widespread damage caused by the storm disrupted the electricity supply, which affected all industry operations. From a refiner's point of view, among the most serious was closure of three pipelines:

The Colonial Pipeline, 5,500 miles of pipeline originating in Houston and ending in New York Harbor, carries a daily average of 100 million gallons of gasoline, diesel and other petroleum products from refineries in the Gulf to customers in the South and Eastern United States.

The Plantation Pipe Line, 3,100 miles of pipeline, performs a similar function along a slightly different route, delivering a total of 620,000 barrels (26 million gallons) of refined petroleum products per day to Birmingham, Alabama; Atlanta, Georgia; Charlotte, North Carolina; and Washington, D.C., among other cities.

The Capline Pipeline (previously mentioned), which carries 1.1 million b/d of crude oil to refineries in the Midwest where it is refined to produce gasoline, diesel and other petroleum products for distribution primarily in the Midwest.

All three of these pipelines were totally or partially out of service due to disruption of electricity supplies as a result of Hurricane Katrina. As a result, **the major supply lines of refined products to the Southern and Eastern states were unavailable for shipment in whole or in part, during the initial period after the storm. Midwestern gasoline and diesel production was affected by lack of supply from the Capline Pipeline. This led to reduced supplies of gasoline, diesel, and other products in parts of the country often far removed from the Gulf area.**

Petrochemical Facilities

The Gulf region is home to many of America's petrochemical plants, which manufacture plastics and other products made from oil and natural gas feedstocks, and which rely on these energy sources for fuel and electricity for power. The impact of Hurricane Katrina on these facilities is not currently known but is potentially quite serious, both in terms of facility damage due to water or wind damage and temporary closure or reduced operations due to feedstock shortages, lack of fuel or electricity and transportation problems.

Petrochemical products serve as the building blocks for many ultimate products such as computers, medicines and other medical products, plastic packaging for food, and also automobile components, to name just a few. Disruption of petrochemical production due to the storm, if it continues, could affect the economy considerably due to the economic importance of petrochemical-based products.

Other Facilities

In addition to the major impacts outlined above, company pipelines and shore facilities and other operations were impacted by the hurricane, but information on these matters is less readily available to us. Company and government statements indicate that many of these facilities were not operating due to lack of electricity or because other related facilities (e.g. refineries) were down. Some natural gas processing plants were affected but NPRA does not have more information on this sector of the industry.

3. What is the current state of repairs?

The many different sectors of the energy industry, working around the clock together with core service providers and with important help from local, state and federal government agencies, have made considerable progress in restoring some of the operations affected by the storm.

The magnitude of the impact outlined above clearly dictates caution in any assessment of when the energy production, refining, distribution and related facilities will be back in service and industry conditions will return to normal. Clearly, our national energy infrastructure has suffered a setback from which it will take some time to emerge completely.

Crude Oil, Natural Gas

According to the MMS as of Saturday, September 3, 78.98% of Gulf of Mexico crude oil offshore production remained shut-in, an improvement of 10% over Friday. Shut-in Gulf natural gas production stood at 57.80% of total Gulf gas marketed production, an improvement of 21% over Friday's figure. The number of manned offshore platforms that are evacuated declined by 25% over the same period. Thus, important but limited progress has been made both in restoring the flow of crude and natural gas necessary for refiners to manufacture gasoline, diesel, jet fuel and other petroleum products and to meet the needs of petrochemical manufacturers. In addition, it is reported that LOOP is operating at 75% of capacity.

These figures still leave significant amounts of offshore Gulf crude oil and natural gas shut-in, and oil and gas volumes not produced in the past several days are large. During the period 8/26-9/3 9.8 million barrels were shut-in, totaling 1.8% of yearly crude oil production in the Gulf. During the same period 53.2 billion cubic feet of natural gas were shut-in, roughly 1.45% of annual gas marketed production from offshore.

There are indications of progress as well regarding refineries. Marathon announced this weekend that, barring unforeseen problems, all seven of its refineries would be operating at capacity on Monday. This includes the Midwestern refineries impacted by the Capline Pipeline closure as well as the Garyville, Louisiana refinery impacted directly by the hurricane. Valero has announced that its St Charles refinery will probably return to operation in the next two weeks. Shell has stated that the Convent refinery may be restarted Sunday and the Norco refinery midweek. Those refineries will be returned to full production gradually and safely as soon as start-ups take place. Assessments of physical damage to the Chalmette and Meraux refineries last week helped ascertain the extent of damage was limited; no start-up date has been set.

The Colonial Pipe Line expected to return to 86% capacity service by the end of the Labor Day weekend. Plantation Pipe Line has returned to 100% operation as has the Capline crude oil pipeline. This means that major pipeline links to the Midwest, South and East have been gradually restored. Serious problems remain, however, due to the significant loss of product and crude volumes which would have been shipped on these lines last week.

In addition, it remains unclear when many, if not most, of the refineries impacted directly by Hurricane Katrina in the Gulf can return to service. Problems with wind and water damage, electricity supply and other infrastructure remain to be addressed despite the best efforts of facility owners and operators. Thus, although some of the affected refineries may restart and return to capacity or near-capacity levels this week, there are indications that several facilities may be out of service for a longer period.

The industry is committed to operation of these facilities as soon as possible, but employee safety and overall safe start-up and operation concerns are paramount. Significant flooding and damage still affects some facilities. However, some refiners with operating facilities have indicated that they will be able to ramp-up production from currently reduced levels at refineries near the affected areas which should have a positive impact on product supplies.

4. What else is industry doing to improve the situation?

As indicated above, the industry has moved with considerable speed to restart the nation's energy infrastructure so severely damaged by Hurricane Katrina. Even more important than assessing and repairing physical damage however, was the need to locate and assist employees, many of whom experienced significant personal losses of family or friends in the tragedy as well as loss of or severe damage to their homes. (All industry companies throughout this region have been deeply involved in locating and providing for the needs of their employees at the same time they were attempting to assess and respond to facility damages and restore energy production).

Many companies are offering varying types of assistance to personnel and their families who were impacted by the hurricane. These include interest free loans; temporary living supplements for housing and food; pay continuation while facilities are closed; transportation assistance; paid time off; medical and prescription drug assistance; temporary housing, including trailers, tents, and other available housing.

The oil, gas and petrochemical industries have already contributed millions of dollars to the American Red Cross and other relief agencies involved in assisting all residents of the affected communities. They are also matching employee contributions. Companies are also supplying in-kind assistance, often including fuel, for relief efforts as well. The industry will doubtless maintain its deep commitment to help end the suffering in the affected communities and to begin planning for the future.

5. What has the federal government done to address these emergency conditions?

Federal authorities have taken several decisive actions to help relieve the many energy-related problems left in the wake of Hurricane Katrina.

SPR Release

The Administration has released 9 million barrels of crude oil from the Strategic Petroleum Reserve (SPR) to assist refiners who are short crude supplies as a result of hurricane damage. The recipients will use this crude to manufacture more gasoline, diesel, jet fuel and home heating oil to be supplied to consumers across the nation. This is a dynamic process, and additional volumes may be needed as more refineries restart.

The current situation is precisely the type of event meant to trigger SPR release. It demonstrates the importance of careful SPR management.

Waivers to Increase Fuel Flexibility

EPA has provided temporary fuel waivers that will make it easier to provide fuels to affected areas. This action pertains to both gasoline and diesel specifications, and will help alleviate some of the supply problems in these areas by increasing the available supply of both domestic production and imports. Affected states participated in the EPA's decision process on this action.

Jones Act Waiver

DOT has temporarily lifted Jones Act requirements to allow non-U.S. flag vessels to transport much needed refined products from one U.S port to another.

IEA (International Energy Agency) Exchange

The Secretary of Energy has announced that the IEA will make available 60 million barrels of petroleum. This will provide relief in the form of refined products (gasoline, diesel, jet fuel, home heating oil) which are much needed due to disrupted supplies from several refineries. These products should begin to reach the U.S in one to two weeks. The agreement with the IEA also requires the U.S. to release an additional 30 million barrels of SPR crude.

Industry appreciates these actions, which were taken by the Administration with bipartisan support from the Congress. They will be very helpful in dealing with the serious supply problems that have resulted from Hurricane Katrina.

6. What is the impact on fuel supply? When will the situation return to normal?

As indicated above, Hurricane Katrina's direct hit on the energy heartland of America resulted in significant damage to offshore energy production in the Gulf, to facilities that are critically important to imported oil supplies, to refineries in the affected states and beyond, and to pipelines that serve as the major providers of refined products and crude to large parts of the East, South and Midwest.

All segments of the industry are working together in an intensive effort to repair as much of the damage as is possible at this time in order to increase the flow of crude oil to refineries and refined products to consumers throughout the country. Safety considerations and the immediate needs of the industry's workforce are of course taken into account at all times.

Industry and government are working together to provide available supplies of product to areas that are experiencing supply concerns. The fuel and Jones Act waivers mentioned above will be of immediate and near-term assistance. Increased product imports through the IEA should also help when they arrive. Refiners who have the ability to do so will attempt to increase production to help meet the needs of the affected areas. The release of oil from the SPR will be helpful in supplying them with some of the crude needed to make these products.

Despite this hopeful news, our nation faces a disruption of the fuel supply system that should not be understated. The hurricane temporarily affected more than 90% of the Gulf's oil production and 80% of its gas production. It effectively removed 10% of the nation's gasoline supply by its impact on U.S. refining capacity located near the Gulf. It also impacted refineries hundreds of miles away that lost access to crude oil supplies. Although important progress has been made through the efforts of government and industry, and with some help from abroad, full recovery will take time. Hard work and cooperation throughout this difficult period will certainly help speed the return to normal conditions. The direct and indirect impact of the hurricane on energy demand, which cannot yet be determined, will also be a major factor during this period.

7. Should we continue to rely on free market forces during this period?

Absolutely. Continued reliance on market forces provides appropriate market signals to help balance supply and demand even during difficult times. President Reagan eliminated price controls on oil products immediately upon taking office in 1981. He was outspoken about the inefficiencies and added costs to consumers as a result of America's ten-year experiment with energy price controls.

The energy price and allocation controls of the 1970s resulted in supply shortages in the form of long gas lines. Studies have shown that, although intended to reduce costs, they actually resulted in increased costs and greater inconvenience for consumers. The benefits of market pricing became clear soon after their elimination. The U.S. Federal Trade Commission stated in an extensive study published this June that "Gasoline supply, demand and competition produced relatively low and stable annual average real U.S gasoline prices from 1984 until 2004, despite substantial increases in U.S. gasoline consumption" and "...For most of the past 20 years, real annual average retail gasoline process in the U.S., including taxes, have been lower than at any time since 1919." Price caps and other forms of price regulation are no more effective in the 21st century than they turned out to be in the 1970s. Interference in market forces always creates inefficiencies in the marketplace and extra costs for consumers.

The same holds true for "windfall profit taxes." The U.S. had a "windfall profit tax" on crude oil from 1980 until 1988. That tax, which was actually an ad valorem tax imposed on crude oil, discouraged crude oil production in the United States and resulted in other market distortions. It was repealed in 1988.

Calls for re-imposition of a windfall profits tax on refiners reflect a misunderstanding of refining industry economics. In the ten-year period 1993-2002, average return on investment in the refining industry was only about 5.5%. This is less than half of the S&P industrials average return of 12.7% for the same period. Refining industry profits as a percentage of operating capital are not excessive. In dollars, they seem large due to the massive scale needed to compete in a large, capital-intensive industry. For example, a new medium scale refinery (100,000 to 200,000 b/d) would cost $2 to $3 billion. In short, company revenues can be in the billions, but so, too are the costs of operations.

The FTC June 2005 study cited above had the following comments on industry profits: "Profits play necessary and important roles in a well-functioning market economy. Recent oil company profits are high but have varied widely over time, over industry segments and among firms... Profits also compensate firms for taking risks, such as the risks in the oil industry that war or terrorism may destroy crude production assets or, that new environmental requirements may require substantial new refinery capital investments."

Many other industries enjoy higher earnings than the oil industry. Among these are telecommunication services, software, semiconductors, banking, pharmaceuticals, coal and real estate, to name just a few. Imposition of a windfall profits tax on the industry would discourage investment at a time when significant capital commitments to all parts of the industry, including refining, will be needed.

Tight gasoline market conditions have often led to calls for industry investigations. More than two dozen federal and state investigations over the last several decades have found no evidence of wrongdoing or illegal activity on our industry's part. For example, after a 9-month FTC investigation into the causes of price spikes in local markets in the Midwest during the spring and summer of 2000, former FTC Chairman Robert Pitofsky stated, "There were many causes for the extraordinary price spikes in Midwest markets. Importantly, there is no evidence that the price increases were a result of conspiracy or any other antitrust violation. Indeed, most of the causes were beyond the immediate control of the oil companies." Similar investigations before and since have reached the same conclusion.

There have been, however, reports of price gouging by unscrupulous individuals who seek to profit during this time of national emergency and crisis. Federal and state laws prohibit actions of this kind in emergency situations like the present. Each alleged situation should be thoroughly investigated by the appropriate state and federal authorities and prosecuted when the law has been broken.

PART II. A SHORT DISCUSSION OF OIL AND OIL PRODUCT SUPPLY DRIVERS

1. INTRODUCTION

This hearing was originally intended to inquire into the factors affecting the gasoline market. The natural disaster resulting from Hurricane Katrina required an understandable shift in emphasis to the human needs damages resulting from that storm and only then to supply impacts. But it is important to remember that the effect of Hurricane Katrina is an overlay on a pre-existing condition. That was and is a situation characterized by high crude prices, strong demand for gasoline, diesel and other petroleum products, and a challenged energy infrastructure, especially in refining. In the interest of space and time, NPRA has shortened the following discussion of these conditions and policy recommendations for improving them. We urge members of the committee to consider the need for policy changes to increase the nation's supply of oil, oil products and natural gas as soon as possible.

As the nation moves forward in its resolve to address and overcome the effects of Katrina and the transportation fuels production and distribution systems regain much-needed pre-storm productivity levels, an underlying domestic fuel supply problem remains that requires immediate, bold, and perhaps politically unpopular actions. NPRA believes that policy changes must be put in place to enhance domestically-produced supplies of oil, oil products and natural gas. NPRA has consistently urged policy makers in Congress and the Administration to support environmentally sound, economically justifiable policies that encourage the production of an abundant supply of petroleum and natural gas products for U.S. consumers.

NPRA supports requirements for the orderly production and use of cleaner-burning fuels to address health and environmental concerns, while at the same time maintaining the flow of adequate and affordable gasoline and diesel supplies to the consuming public. Since 1970, clean fuels and clean vehicles have accounted for about 70% of all U.S. emission reductions from all sources, according to EPA. Over the past 10 years, U.S. refiners have invested about $47 billion in environmental improvements, much of that to make cleaner fuels. For example, according to EPA, the new Tier 2 low sulfur gasoline program, initiated in January 2004, will have the same effect as removing 164 million cars from the road when fully implemented.

Unfortunately, however, federal environmental policies have often neglected to consider fully the impact of environmental regulations on fuel supply. Frankly, policy makers have often taken supply for granted, except in times of obvious market instability. This attitude must end. A healthy and growing U.S. economy requires a steady, secure, and predictable supply of petroleum products.

Unfortunately, there are no silver bullet solutions for balancing supply and demand. Indeed most of the problems in today's gasoline market—without factoring the market disruptions caused by Katrina—result from the high price of crude oil due to economic recovery abroad together with strong U.S. demand for gasoline and diesel due to the improving U.S. economy.

2. UNDERSTANDING GASOLINE MARKET FUNDAMENTALS: HIGH CRUDE PRICES; STRONG GASOLINE DEMAND GROWTH

It is important to recognize the overwhelming factor affecting gasoline prices: crude oil. In June of this year the U.S. Federal Trade Commission released a landmark study titled: "Gasoline Price Changes: The Dynamic of Supply, Demand and Competition." To quote from the FTC's findings: "Worldwide supply, demand, and competition for crude oil are the most important factors in the national average price of gasoline in the U.S." and "The world price of crude oil is the most important factor in the price of gasoline. Over the last 20 years, changes in crude oil prices have explained 85 percent of the changes in the price of gasoline in the U.S."

Crude prices have been steadily increasing since 2004, largely because of surprising levels of growth in oil demand in countries such as China and India, and in the United States as well. Actual demand growth for oil and oil products in these countries in 2004 exceeded the experts" predictions and has remained strong this year. As a result, world demand for crude is bumping up against the worldwide ability to produce crude.

Strong demand for crude has dissipated the cushion of excess available worldwide oil supply, just as strong U.S. demand for refined products has eliminated excess refining capacity in the United States. The good news is that producing countries will probably be able to add crude production capacity in the years to come. The bad news is that the United States has thus far shown only limited willingness to face up to its own energy supply problems.

As shown in Attachment I, gasoline costs closely track the cost of crude oil. Before hurricane Katrina, gasoline price increases lagged crude oil price increases on a gallon for gallon basis. This means that refiners did not pass through all of the increased costs in their raw material, crude oil. Crude oil accounts for 55-60% of the price of gasoline seen at the service station.

The cost of federal and state taxes adds another 19% to the cost of a finished gallon of gasoline. Therefore under current conditions, 74-79% of the total cost of a gallon of gasoline is pre-determined before the crude is delivered to the refiner for manufacture into gasoline. (See Attachment 2)

Another contributor to gasoline costs is tightness in our nation's gasoline markets. While U.S. refiners are producing huge volumes of products, strong demand has tightened supply. Gasoline demand currently averages approximately 9 million barrels per day. Domestic refineries produce about 90 percent of U.S. gasoline supply, while about 10 percent is imported.

Thus, strong and increasing demand can only be met by either adding new domestic refinery capacity or by relying on more foreign gasoline imports. Unfortunately, the desire for more domestic gasoline production capacity is often thwarted by other public priorities.

3. U. S. POLICY SHOULD ENCOURAGE ADDITIONAL DOMESTIC REFINING CAPACITY.

Domestic refining capacity is a scarce asset. There are currently 148 U.S. refineries owned by 55 companies in 33 states, with total crude oil processing capacity at roughly 17 million barrels per day. In 1981, there were 325 refineries in the U.S. with a capacity of 18.6 million barrels per day. Thus, while U.S. demand for gasoline has increased over 20% in the last twenty years, U.S. refining capacity has decreased by 10%. No new refinery has been built in the United States since 1976, and it will be difficult to change this situation. This is due to economic, public policy and political considerations, including siting costs, environmental requirements, a history of low refining industry profitability and, significantly, "not in my backyard" (NIMBY) public attitudes.

Nevertheless, existing refineries have been extensively updated to incorporate the technology needed to produce a large and predictable supply of clean fuels with significantly improved environmental performance. Capacity additions have taken place at some facilities as well; several of these projects implemented over several years can actually increase product output as much as a new refinery. But this increase in capacity at existing sites has not kept pace with the growth in U.S. demand for products, meaning that the nation is increasing its reliance on imports of gasoline and other petroleum products each year.

Proposed capacity expansions can often become controversial and contentious at the state and local level, even when necessary to produce cleaner fuels pursuant to regulatory requirements. We hope that policymakers will recognize the importance of domestic refining capacity expansion to the successful implementation of the nation's environmental policies, especially clean fuels programs. The Administration's New Source Review reform program will also provide one tool to help add and update capacity.

NPRA wants to recognize a provision in the recently enacted energy legislation that will help encourage additional refining investment. The provision allows 50% expensing of the costs associated with expanding a refinery's output by more than 5%. The refiner must have a signed contract for the work by 1/1/08, and the equipment must be put in service by 1/1/12.

Common sense dictates that it is in our nation's best interest to manufacture the lion's share of the petroleum products required for U.S. consumption in domestic refineries and petrochemical plants. Nevertheless, we currently import more than 62% of the crude oil and oil products we consume. Reduced U.S. refining capacity clearly affects our supply of refined petroleum products and the flexibility of the supply system, particularly in times of unforeseen disruption or other stress. Unfortunately, EIA currently predicts "substantial growth" in refining capacity only in the Middle East, Central and South America, and the Asia/Pacific region, not in the U.S.

4. THE U.S. REFINING INDUSTRY IS DIVERSE AND COMPETITIVE.

Today's U.S. refining industry is highly competitive. Some suggest past mergers are responsible for higher prices. The data do not support such claims. In fact, companies have become more efficient and continue to compete fiercely. There are 55 refining companies in the U.S., hundreds of wholesale and marketing companies, and more than 165,000 retail outlets. The biggest refiner accounts for only about 13 % of the nation's total refining capacity; and the large integrated companies own and operate only about 10 % of the retail outlets. The Federal Trade Commission (FTC) thoroughly evaluates every merger proposal, holds industry mergers to the highest standards of review, and subjects normal industry operations to a higher level of ongoing scrutiny.

Critics of mergers sometimes suggest that industry is able to affect prices because it has become much more concentrated, with a handful of companies controlling most of the market. This is untrue. According to data compiled by the U.S. Department of Commerce and by Public Citizen, in 2003 the four largest U.S. refining companies controlled a little more than 40 % of the nation's refining capacity. In contrast, the top four companies in the auto manufacturing, brewing, tobacco, floor coverings and breakfast cereals industries controlled between 80% and 90% of the market.

5. INDUSTRY IS WORKING HARD TO KEEP PACE WITH GROWING DEMAND FOR FUEL.

Despite the powerful factors that influence gasoline manufacturing, cost and demand, refiners are addressing current supply challenges and working hard to supply sufficient volumes of gasoline and other petroleum products to the public. Refineries have been running at very high levels, producing gasoline and distillate. Refiners operated at high utilization rates—even before the start of the summer driving season. To put this in perspective, peak utilization rates for other manufacturers average about 82 %. At times during summer, refiners often operate at rates close to 98 %. However, such high rates cannot be sustained for long periods.

In addition to coping with higher fuel costs and growing demand, refiners are implementing significant transitions in major gasoline markets. Nationwide, the amount of sulfur in gasoline will be reduced to an average of 30 parts per million (ppm) effective January 1, 2006, giving refiners an additional challenge in both the manufacture and distribution of fuel. Equally significant, California, New York and Connecticut bans on use of MTBE are in effect. This is a major change affecting one-sixth of the nation's gasoline market. MTBE use as an oxygenate in reformulated gasoline accounted for as much as 11% of RFG supply at its peak; substitution of ethanol for MTBE does not replace all of the volume lost by removing MTBE. (Ethanol's properties generally cause it to replace only about 50% of the volume lost when MTBE is removed.) This lost volume must be supplied by additional gasoline or gasoline blendstocks. Especially during a period of supply concerns it is in the nation's interest to be prudent in taking any action that affects MTBE use. That product still accounts for 1.6% of the nation's gasoline supply on average, but it provides a larger portion of gasoline supplies in areas with RFG requirements that are not subject to an MTBE ban.

Obviously, refiners face a daunting task in completing many changes to deliver the fuels that consumers and the nation's economy require. But they are succeeding. And regardless of recent press stories, we need to remember that American gasoline and other petroleum product prices have long been low when compared to the price consumers in other large industrialized nations pay for those products. The Federal Trade Commission recently found that "Gasoline supply, demand and competition produced relatively low and stable annual average real U.S. gasoline prices from 1984 until 2004, despite substantial increases in U.S. gasoline consumption."

6. REFINERS FACE A BLIZZARD OF REGULATORY REQUIREMENTS AFFECTING BOTH FACILITIES AND PRODUCTS.

Refiners currently face the massive task of complying with fourteen new environmental regulatory programs with significant investment requirements, all in the same 2006-2012 timeframe. (See Attachment 3.) In addition, many programs start soon. (See Attachment 4.) For the most part, these regulations are required by the Clean Air Act. Some will require additional emission reductions at facilities and plants, while others will require further changes in clean fuel specifications. NPRA estimates that refiners are in the process of investing about $20 billion to sharply reduce the sulfur content of gasoline and both highway and off-road diesel. Refiners will face additional investment requirements to deal with limitations on ether use, as well as compliance costs for controls on Mobile Source Air Toxics and other limitations. These costs do not include the significant additional investments needed to comply with stationary source regulations that affect refineries.

Other potential environmental regulations on the horizon could force additional large investment requirements. They are: the challenges posed by increased ethanol use, possible additional changes in diesel fuel content involving cetane, and potential proliferation of new fuel specifications driven by the need for states to comply with the new eight-hour ozone NAAQS standard. The 8-hour standard could also result in more regulations affecting facilities such as refiners and petrochemical plants.

These are just some of the pending and potential air quality challenges that the industry faces. Refineries are also subject to extensive regulations under the Clean Water Act, Toxic Substances Control Act, Safe Drinking Water Act, Oil Pollution Act of 1990, Resource Conservation and Recovery Act, Emergency Planning and Community Right-To-Know (EPCRA), Comprehensive Environmental Response, Compensation, and Liability Act (CERCLA), and other federal statutes. The industry also complies with OSHA standards and many state statutes. A complete list of federal regulations impacting refineries is included with this statement. (See Attachment 5.)

API estimates that, since 1993, about $89 billion (an average of $9 billion per year) has been spent by the oil and gas industry to protect the environment. This amounts to $308 for each person in the United States. More than half of the $89 billion was spent in the refining sector.

Obviously, refiners face a daunting task in completing many changes to deliver the fuels that consumers and the nation's economy require. But they are succeeding. And regardless of recent press stories, we need to remember that American gasoline and other petroleum products have long been low when compared to the price consumers in other large industrialized nations pay for those products. The Federal Trade Commission recently found that "Gasoline supply, demand and competition produced relatively low and stable annual average real U.S. gasoline prices from 1984 until 2004, despite substantial increases in U.S. gasoline consumption."

7. A KEY GOVERNMENT ADVISORY PANEL HAS URGED MORE SENSITIVITY TO SUPPLY CONCERNS.

The National Petroleum Council (NPC) issued a landmark report on the state of the refining industry in 2000. Given the limited return on investment in the industry and the capital requirements of environmental regulations, the NPC urged policymakers to pay special attention to the timing and sequencing of any changes in product specifications. Failing such action, the report cautioned that adverse fuel supply ramifications may result. Unfortunately, this warning has been widely disregarded. On June 22, 2004 Energy Secretary Abraham asked NPC to update and expand its refining study and a report was released last December. NPRA again urges policymakers to take action to implement NPC's study recommendations in order to deal with U.S. refining problems.

8. NPRA RECOMMENDATIONS TO ADD REFINING CAPACITY AND INCREASE FUTURE PRODUCT SUPPLY

- Make increasing the nation's supply of oil, oil products and natural gas a number one public policy priority. Now, and for many years in the past, increasing oil and gas supply has often been a number 2 priority. Thus, oil and gas supply concerns have been secondary and subjugated to whatever policy goal was more politically popular at the time. Enactment of the recent Energy Bill is a first step to making a first priority the supply of energy sources the nation depends upon.
- Remove barriers to increased supplies of domestic oil and gas resources. Recent criticism about the concentration of America's energy infrastructure in the western Gulf is misplaced. Refineries and other important onshore facilities have been welcome in this area but not in many other parts of the country. Policymakers have also restricted access to much-needed offshore oil and natural gas supplies in the eastern Gulf and off the shores of California and the East Coast. These areas must follow the example of Louisiana and many other states in sharing these energy resources with the rest of the nation because they are sorely needed.
- Resist tinkering with market forces when the supply/demand balance is tight. Market interference that may initially be politically popular leads to market inefficiencies and unnecessary costs. Policymakers must resist turning the clock backwards to the failed policies of the past. Experience with price constraints and allocation controls in the 1970s demonstrates the failure of price regulation, which adversely impacted both fuel supply and consumer cost.
- Expand the refining tax incentive provision in the Energy Act. Reduce the depreciation period for refining investments from 10 to seven or five years in order

to remove a current disincentive for refining investment. Allow expensing under the current language to take place as the investment is made rather than when the equipment is actually placed in service. Or the percentage expensed could be increased as per the original legislation introduced by Senator Hatch.

- Review permitting procedures for new refinery construction and refinery capacity additions. Seek ways to encourage state authorities to recognize the national interest in more domestic capacity.
- Keep a close eye on several upcoming regulatory programs that could have significant impacts on gasoline and diesel supply. They are:

—Design and implementation of the credit trading program for the ethanol mandate (RFS) contained in the recent Energy Act. This mechanism is vital to increase the chance that this program can be implemented next year without additional gasoline supply disruption. Additional resources are needed within EPA to accomplish this key task.

—Implementation of the ultra low sulfur diesel highway diesel regulation. The refining industry has made large investments to meet the severe reductions in diesel sulfur that take effect next June. We remain concerned about the distribution system's ability to deliver this material at the required 15 ppm level at retail. If not resolved, these problems could affect America's critical diesel supply. Industry is working with EPA on this issue, but time left to solve this problem is growing short.

—Phase II of the MSAT (mobile source air toxics) rule for gasoline. Many refiners are concerned that this new regulation, which we expect next year, will be overly stringent and impact gasoline supply. We are working with EPA to help develop a rule that protects the environment and avoids a reduction in gasoline supply.

—Implementation of the new 8-hour ozone NAAQS standard. The current implementation schedule determined by EPA has established ozone attainment deadlines for parts of the country that will be impossible to meet. EPA has to date not made changes that would provide realistic attainment dates for the areas. The result is that areas will be required to place sweeping new controls on both stationary and mobile sources, in a vain effort to attain the unattainable. The new lower-sulfur gasoline and ULSD diesel programs will provide significant reductions to emissions within these areas once implemented. But they will not come soon enough to be considered unless the current unrealistic schedule is revised. If not, the result will be additional fuel and stationary source controls which will have an adverse impact on fuel supply and could actually reduce U.S. refining capacity. This issue needs immediate attention.

NPRA's members are dedicated to working cooperatively with government at all levels to resolve the current emergency conditions that result from Hurricane Kristina. But we feel obliged to remind policymakers that action must also be taken to improve energy policy in order to increase supply and strengthen the nation's refining infrastructure. We look forward to answering the Committee's questions.

Attachment 1

Crude Oil and Gasoline Price Outlook

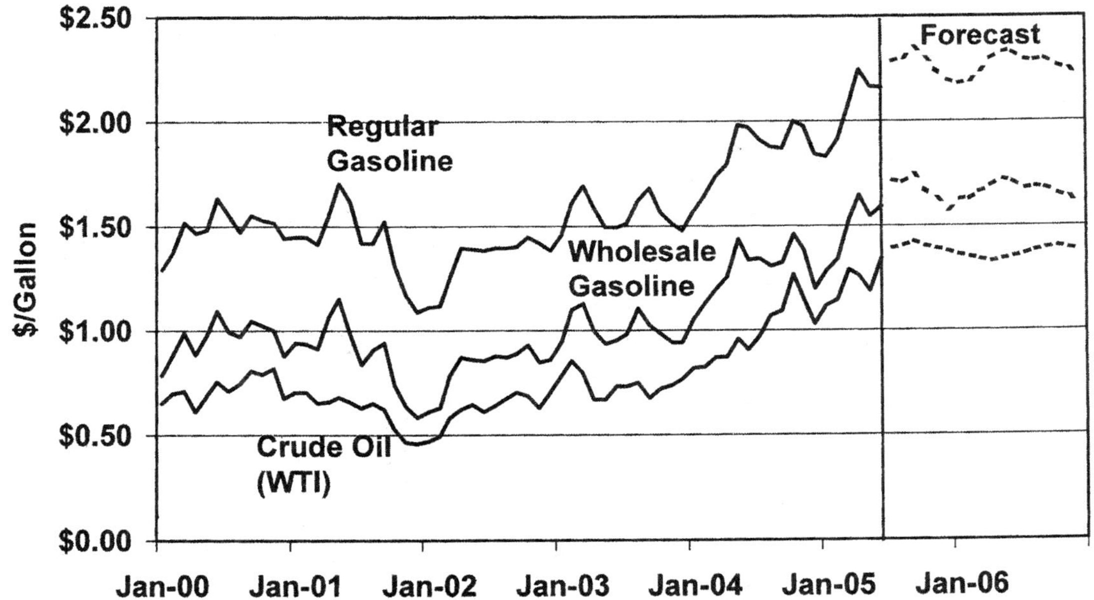

Sources: History: EIA; Projections: Short-Term Energy Outlook, July 2005.

www.eia.doe.gov

Attachment 2

What We Pay for in a Gallon of Regular Gasoline

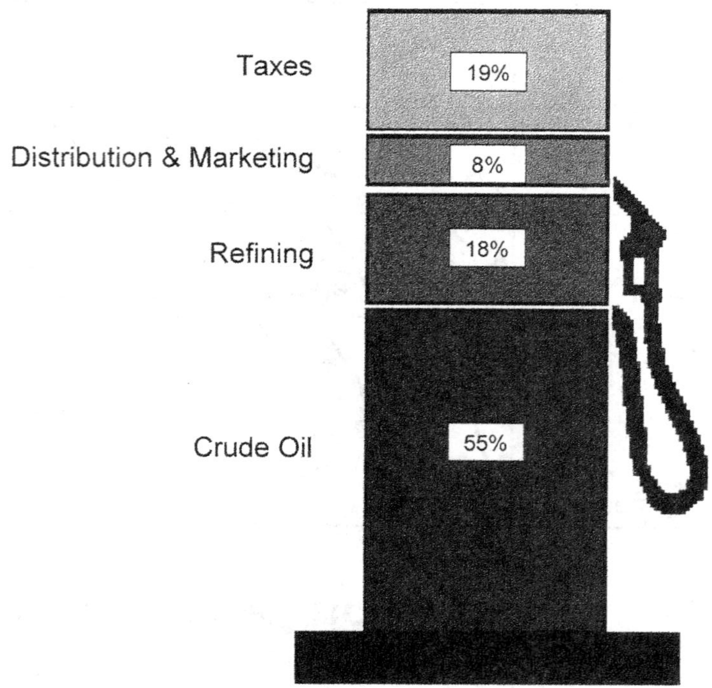

Source: EIA

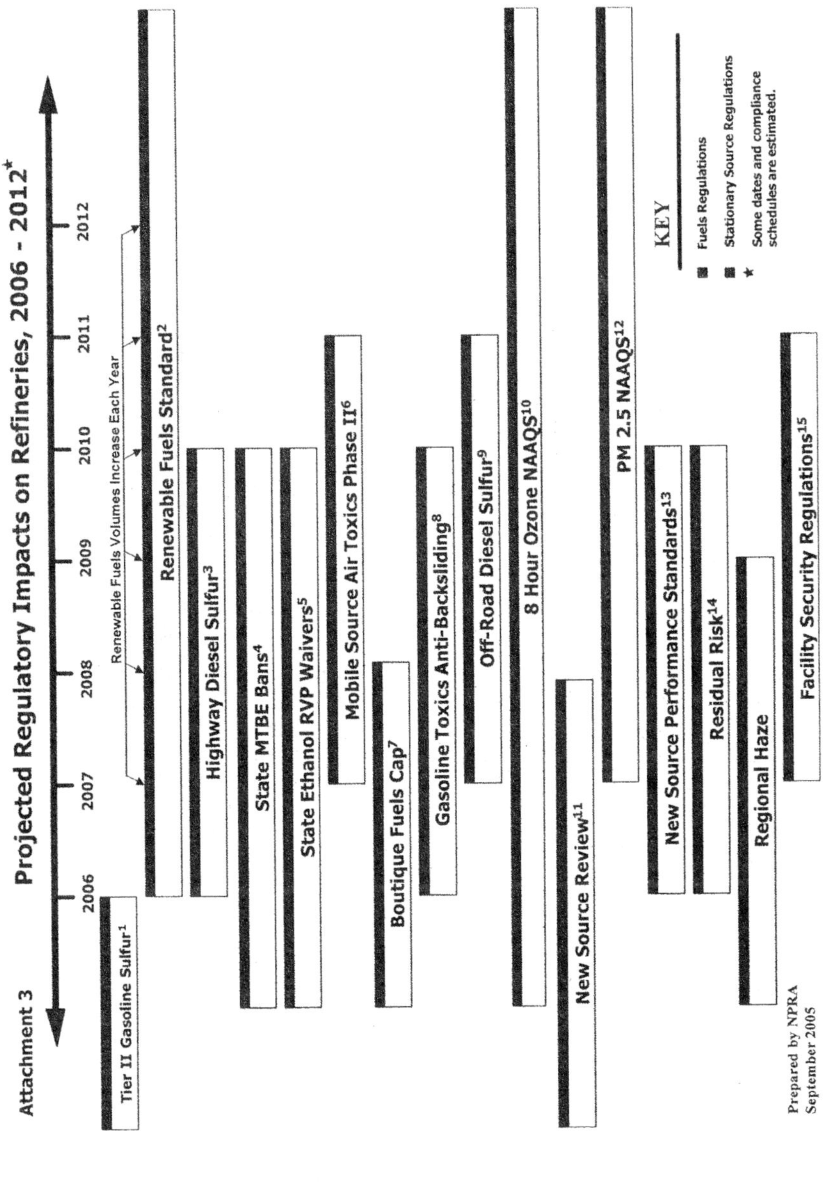
Attachment 3
Projected Regulatory Impacts on Refineries, 2006 - 2012*
2006
2007
2008
2009
2010
2011
2012
Renewable Fuels Volumes Increase Each Year
Tier II Gasoline Sulfur1
Renewable Fuels Standard2
Highway Diesel Sulfur3
State MTBE Bans4
State Ethanol RVP Waivers5
Mobile Source Air Toxics Phase II6
Boutique Fuels Cap7
Gasoline Toxics Anti-Backsliding8
Off-Road Diesel Sulfur9
8 Hour Ozone NAAQS10
New Source Review11
PM 2.5 NAAQS12
New Source Performance Standards13
Residual Risk14
Regional Haze
Facility Security Regulations15
KEY
Fuels Regulations
Stationary Source Regulations
Some dates and compliance schedules are estimated.
Prepared by NPRA
September 2005

Notes:

1. Longer compliance time for refineries in Alaska and Rocky Mountain states as well as small refineries covered by the Small Business Regulatory Enforcement and Flexibility Act (SBREFA). Additional compliance time is available for these refineries if they produce ultra low sulfur highway diesel beginning in 2006.
2. The Energy Policy Act of 2005 includes a renewable fuels standard (RFS) which mandates the use of 4 billion gallons of renewable fuels starting in 2006. The mandate increases to 7.5 billion gallons in 2012. EPA must promulgate regulations by August 2006.
3. Longer compliance time for small refiners covered by SBREFA.
4. Approximately twenty-five states currently have MTBE bans in place and others may pass similar bans in the future.
5. The Energy Policy Act of 2005 allows state governors to petition EPA to eliminate the one pound RVP waiver for summer gasoline blended with ethanol.
6. Phase II Mobile Source Air Toxics Rule to be proposed in February, 2006. Final rule expected in 2007.
7. The Energy Policy Act of 2005 caps the number of motor fuels available for use in State Implementation Plans at the same level as those already in use as of September 1, 2004. EPA must publish a list of approved fuels by state and PADD by November, 2005.
8. Under the Energy Policy Act of 2005 EPA must promulgate a rule to implement RFG anti-backsliding adjustments that will maintain emissions at 2001 and 2002 levels.
9. The first phase of the off-road diesel sulfur program is effective in 2007 and the second phase is effective in 2011.
10. Ozone non-attainment designations made April 2004. State Implementation Plans (SIPs) are due by June 2007. Compliance, depending upon classification, required between 2007 and 2021. EPA promulgated a Phase 1 implementation rule in April 2004, but has not yet promulgated a Phase 2 rule.
11. New Source Review reform (RMRR) is subject to litigation. Refiners face uncertainty in meeting regulatory requirements. The NSR program was upheld in part by the courts however, part of the rule was remanded to EPA. Refiners support the reforms. EPA is continuing enforcement actions under the old rules.
12. EPA set a new PM 2.5 NAAQS in 1997 and designated nonattainment areas in December 2004, but has not yet promulgated implementation standards. EPA is currently conducting a five-year review of the standard.
13. EPA has entered into a consent decree with environmental organizations to review, and possibly revise, the New Source Performance Standards for petroleum refineries.
14. Proposed rule expected mid 2006.
15. The Senate and the Administration support new authority for DHS to regulate chemical security which will impact refiners. Many facilities currently meet Coast Guard regulations under MTSA.

Attachment 4

Fuels Timeline

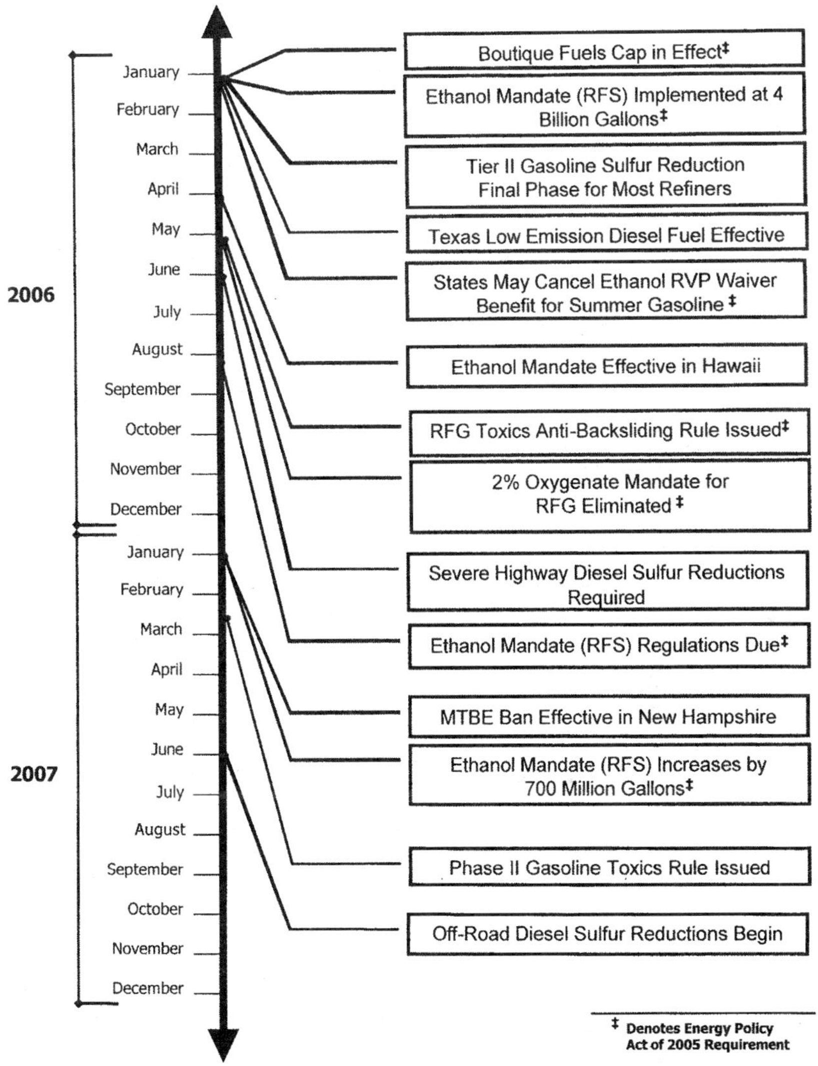

Attachment 5 - Source: API, 1997

Appendix A
PETROLEUM REFINING: APPLICABLE REGULATIONS

Name	Code of Federal Regulation (CFR) Cite	Effective Date
CLEAN AIR ACT (CAA)		
New Source Performance Standards (NSPSs)	40 CFR Part 60	
Subpart A: General Provisions	40 CFR Part 60	mid 1970s
Subpart Cb: Designated Facilities - Existing Sulfuric Acid Units	40 CFR Part 60	1991
Subpart D: Fossil-Fuel Fired Steam Generators Constructed After 8/17/71	40 CFR Part 60	1977
Subpart Da: Electric Utility Steam Generating Units Constructed After 9/18/78	40 CFR Part 60	1978
Subpart Db: Industrial-Commercial-Institutional Steam Generating Units	40 CFR Part 60	1987
Subpart Dc: Small Industrial-Commercial-Institutional Steam Generating Units	40 CFR Part 60	1990
Subpart H: Sulfuric Acid Units	40 CFR Part 60	1977
Subpart J: Petroleum Refineries	40 CFR Part 60	1978
Subpart K: Storage Vessels for Petroleum Liquids Constructed, Reconstructed or Modified between 6/11/73 and 5/19/78	40 CFR Part 60	1977
Subpart Ka: Storage Vessels for Petroleum Liquids Constructed, Reconstructed or Modified between 5/18/78 and 7/23/84	40 CFR Part 60	1980
Subpart Kb: Volatile Organic Liquid Storage	40 CFR Part 60	1987
Subpart GG: Stationary Gas Turbines	40 CFR Part 60	1978
Subpart UU: Asphalt Processing and Roofing Manufacturing	40 CFR Part 60	1982
Subpart VV: Equipment Leaks of VOC in the Synthetic Organic Chemicals Manufacturing Industry (SOCMI)	40 CFR Part 60	1983
Supart XX: Bulk Gasoline Terminals	40 CFR Part 60	1983
Subpart GGG: Equipment Leaks of VOC in Petroleum Refineries	40 CFR Part 60	1984
Subpart III: VOC Emissions for SOCMI Air Oxidation Unit Processes	40 CFR Part 60	1990
Subpart NNN: VOC Emissions for SOCMI Distillation Processes	40 CFR Part 60	1990
Subpart QQQ: VOC Emissions for Petroleum Refinery Wastewater Systems	40 CFR Part 60	1988
Subpart RRR: SOCMI Reactor Processes	40 CFR Part 60	1993
National Emission Standards for Hazardous Air Pollutants (NESHAPs)		
Subpart A: General Provisions	40 CFR Part 61	1973
Subpart J/V: Equipment Leaks (Fugitive Emission Sources) of Benzene	40 CFR Part 61	mid 1980s
Subpart M: Asbestos	40 CFR Part 61	1984
Subpart Y: Benzene Emissions from Benzene Storage Vessels	40 CFR Part 61	mid 1980s
Subpart BB: Benzene Emissions from Benzene Transfer Operations	40 CFR Part 61	mid 1980s
Subpart FF: Benzene Waste Operations	40 CFR Part 61	1993

Name	Code of Federal Regulation (CFR) Cite	Effective Date
NESHAPs for Source Categories		
Subpart A: General Provisions	40 CFR Part 63	1994
Subpart B: Control Technology Determination	40 CFR Part 63	1994
Subpart F: SOCMI	40 CFR Part 63	1994
Subpart G: SOCMI Process Vents, Storage Vessels, Transfer Operations, and Wastewater	40 CFR Part 63	1994
Subpart H: Equipment Leaks	40 CFR Part 63	1994
Subpart I: NESHAP for Organic Hazardous Air Pollutants (HON); Certain Processes Subject to the Negotiated Regulation for Equipment Leaks	40 CFR Part 63	1994
NESHAP for HON (partially under stay pending reconsideration for compressors, surge control vessels, and bottom receivers)	40 CFR Part 63	4/22/94
Subpart Q: Industrial Cooling Towers	40 CFR Part 63	1994
Subpart R: Stage I Gasoline Distribution Facilities	40 CFR Part 63	12/14/94
Subpart T: Halogenated Solvent Cleansing (MACT)	40 CFR Part 63	12/2/94
Subpart Y: NESHAP for Marine Tank Vessel Loading and Unloading Operations (MACT)	40 CFR Parts 9, 63	mid 1995
Subpart CC: NESHAP for Petroleum Refining — Phase I (MACT)	40 CFR Parts 9, 60, 63	mid 1995
Stack Height Provisions	40 CFR Part 51, Subpart G	1986
Control Technology Guidelines (CTGs)		
Petroleum Liquid Storage In External Floating Roof Tanks	40 CFR Part 52	1978
Petroleum Liquid Storage in Fixed Roof Tanks	40 CFR Part 52	1977
Petroleum Refinery Equipment Leaks	40 CFR Part 52	1978
Refinery Vacuum Producing Systems, Wastewater Separators and Process Unit Turnarounds	40 CFR Part 52	1977
SOCMI Air Oxidation Processes	40 CFR Part 52	1984
SOCMI Distillation Operations and Reactor Processes	40 CFR Part 52	1993
Tank Truck Gasoline Loading Terminals	40 CFR Part 52	1977
Fuels		
Fuel and Fuel Additives:		
Registration Requirements	40 CFR Part 79	5/27/94
Interim Requirements for Deposit Control Gasoline Additives	40 CFR Part 80	1/1/95
Reid Vapor Pressure Limitation	40 CFR Part 80	late 1980s
Oxygenated Fuel Requirement	40 CFR Part 80	1992
Lead Phaseout	40 CFR Part 80	12/31/95
Reformulated Gasoline	40 CFR Part 80	1/1/95
Low Sulfur Diesel	40 CFR Part 85	1993
Permits		
State Operating Permit Program - Title V (Revised 8/29/94)	40 CFR Part 70	1992
Prevention of Significant Deterioration (new sources in attainment areas) and New Source Review (new sources in non-attainment areas); LAER requirements (existing source)	40 CFR Part 52	1978
Stratospheric Ozone	40 CFR Part 82	1990-2015

Name	Code of Federal Regulation (CFR) Cite	Effective Date
Acid Rain Provisions	40 CFR Parts 72, 73, 75, 77, 78	ongoing
Nitrogen Oxides Emission Reduction Program	40 CFR Part 76	1994
CLEAN WATER ACT (CWA)		
Discharge of Oil: Notification Requirements	40 CFR Part 110	1987
Designation of Hazardous Substances	40 CFR Part 116	1978
Notice of Discharge of a Reportable Quantity	40 CFR Part 117	late 1970s
Spill Prevention, Control, and Countermeasures (SPCC) Requirements for Oil Storage	40 CFR Part 112	mid 1970s
General Provisions for Effluent Guidelines and Standards	40 CFR Part 401	1974
Toxic Pollutant Effluent Standards	40 CFR Part 129	1977
Effluent Guidelines and Categorical Pretreatment Standards	40 CFR Part 419	late 1970s - mid 1980s
Water Quality Standards for Toxic Pollutants	40 CFR Part 131	2/5/93
General National Pretreatment Standards	40 CFR Part 403	early 1980s
Great Lakes Water Quality Guidance	40 CFR Parts 9, 122, 123, 131, 132	early 1995
NPDES		
Stormwater Application, Permit, and Reporting Requirements Associated with Industrial Activities	40 CFR Part 122	5/4/92
Permit	40 CFR Parts 121-125	early 1980s
OIL POLLUTION ACT (OPA)		
Natural Resource Damage Assessments (NRDA) under National Oceanic and Atmospheric Administration	15 CFR Part 990	early 1996
Response Plans for Marine Transportation-Related Facilities (interim final rule)	33 CFR Parts 150, 154	1/19/93
Oil Pollution Prevention; Non-Transportation-Related Onshore Facilities	40 CFR Parts 9, 112	8/30/94
RESOURCE CONSERVATION AND RECOVERY ACT (RCRA)		
Non-Hazardous Waste Requirements (Subtitle D)	40 CFR Parts 256, 257 (Federal guidelines for state/local requirements)	late 1970s, early 1980s
Subtitle C Requirements		
General Requirements for Hazardous Waste Management	40 CFR Part 260	late 1970s
Identification and Listing of Hazardous Wastes and Toxicity Characteristics	40 CFR Part 261	late 1970s
Standards Applicable to Generators of Hazardous Wastes		
Subpart A: General Provisions	40 CFR Part 262	early 1980s
Subpart B: Shipping Manifest	40 CFR Part 262	early 1980s
Subpart C: Packaging, Labeling, Marking, and Placarding	40 CFR Part 262	early 1980s
Subpart D: Recordkeeping and Reporting	40 CFR Part 262	early 1980s
Subparts E & F: Exports and Imports	40 CFR Part 262	early 1980s
Standards for Owners and Operators of Hazardous Waste Treatment, Storage, and Disposal Facilities (and generally for Interim Status)		
Subparts A & B: General Provisions & Facility Standards	40 CFR Part 264 (265)	early 1980s
Subparts C & D: Preparedness, Prevention, & Emergency Plans	40 CFR Part 264 (265)	early 1980s
Subpart E: Recordkeeping/Reporting Requirements	40 CFR Part 264 (265)	early 1980s

Name	Code of Federal Regulation (CFR) Cite	Effective Date
Subpart F: Releases from Units	40 CFR Part 264	early 1980s
Subpart F: Groundwater Monitoring Requirements (Interim Status only)	40 CFR Part 265	early 1980s
Subpart G: Closure and Post-closure Requirements	40 CFR Part 264 (265)	1986
Subpart H: Financial Responsibility Requirements	40 CFR Part 264 (265)	early 1980s
Subparts I, J, K, & L: Use and Management of Containers, Tank Systems, Surface Impoundments, & Waste Piles	40 CFR Part 264 (265)	early 1980s (except tanks: 1986)
Liners and Leak Detection for Hazardous Waste Land Disposal Units	40 CFR Part 264 (265)	1992
Double Liners and Leachate Collection Systems for Hazardous Waste Disposal Units	40 CFR Parts 144, 264 (265)	1992
Subparts M, N, & O: Land Treatment, Landfills, & Incinerators	40 CFR Part 264 (265)	early 1980s
Subpart S: Corrective Action	40 CFR Part 264 (265)	1985 (1993)
Subparts AA, BB, & CC: Air Emission Standards for Process Vents; Equipment Leaks; & Tanks, Surface Impoundments, and Containers	40 CFR Part 264 (265)	
Phase I	40 CFR Part 264 (265)	1990
Phase II	40 CFR Part 264 (265)	1994
Standards for the Management of Specific Hazardous Wastes	40 CFR Part 266	1985
Land Disposal Restrictions	40 CFR Part 268	1986
Phase I: Contaminated Debris and Newly Identified Wastes, F037 and F038 Petroleum	40 CFR Parts 148, 268	1992, 1993
Phase II: Set Treatment Standards (BDAT) for TC Wastes and Establish Universal Treatment Standards	40 CFR Parts 148, 268	1994
Permits	40 CFR Parts 270, 271, 272	1980s
Standards for the Management of Used Oil: Used Oil Destined for Recycling	40 CFR Part 279	1993
Underground Storage Tanks: Technical Standards and Corrective Action	40 CFR Part 280	1988
SAFE DRINKING WATER ACT (SDWA)		
Underground Injection Control Regulations	40 CFR Parts 144, 146	12/16/93
SUPERFUND (CERCLA)		
Natural Resource Damage Assessments (also under CWA)	43 CFR Part 11	3/17/94
Reportable Quantities Releases (Notification to National Response Center)	40 CFR Part 302	mid 1980s
Extremely Hazardous Substances (EHSs) Emergency Planning	40 CFR Part 355	1987
EHS Release Notification (Notification to State Emergency Response Commission, Local Emergency Response Commission) and Follow-up	60 CFR Part 355	mid 1980s
Community Right-To-Know		
Hazardous Chemicals (Material Safety Data Sheet Chemicals) Inventory Reporting	40 CFR Part 370	late 1980s
Toxic Chemical Release Reporting	40 CFR Part 372	1988
Expansion of TRI List	40 CFR Part 372	11/30/94

Name	Code of Federal Regulation (CFR) Cite	Effective Date
TOXIC SUBSTANCES CONTROL ACT (TSCA)		
General Provisions	40 CFR Part 702	1982
Reporting and Recordkeeping Requirements	40 CFR Parts 704, 710	1988, late 1970s
Chemical Information Rule	40 CFR Part 712	1982
Health & Safety Data Reporting	40 CFR Parts 716	1986
Premanufacture Notification (and Exemptions)	40 CFR Parts 720 (723)	1983 (1995)
Significant New Uses	40 CFR Part 721	1988
Chromium Comfort D Cooling Towers	40 CFR Part 749	1990
Rules for Controlling Polychlorinated Biphenyls	40 CFR Part 761	1979
Asbestos-Containing Products Labelling Requirements	40 CFR Part 763	1979

Update of Appendix A[1]

Name	Code of Federal Regulation (CFR) Cite
CLEAN AIR ACT (CAA)	
New Source Performance Standards	40 CFR Part 60
Subpart CCCC: Commercial and Industrial Solid Waste Incineration Units	40 CFR Part 60
NESHAPS for Source Categories	40 CFR Part 63
Subpart EEE: Hazardous Waste Combustors	40 CFR Part 63
Subpart UUU: Petroleum Refineries: Catalytic Cracking Units, Catalytic Reforming Units, and Sulfur Recovery Units (Refinery MACT II)	40 CFR Part 63
Subpart EEEE: Organiz Liquids Distribution (Non- Gasoline)	40 CFR Part 63
Subpart FFFF: Miscellaneous Organic Chemical Manufacturing	40 CFR Part 63
Subpart YYYY: Stationary Combustion Turbines	40 CFR Part 63
Subpart GGGGG: Site Remidiation	40 CFR Part 63
Subpart LLLLL: Asphalt Roofing and Asphalt Processing	40 CFR Part 63
Subpart DDDDD: Industrial/Commerical/Institutional Boilers and Process Heaters	40 CFR Part 63
Subpart ZZZZZ: Reciprocating Internal Combustion Engines	40 CFR Part 63
Fuels	40 CFR Part 80
Subpart H: Tier II Gasoline Sulfur	40 CFR Part 80
Subpart I: Ultra Low Sulfur Highway Diesel	40 CFR Part 80
Subpart J: Mobile Source Air Toxics	40 CFR Part 80

[1] As of April 2004 Source: NPRA

Chairman BARTON. We thank you. Your time has expired.

We now want to hear from Mr. James Newsome, who is the President of the New York Mercantile Exchange.

You are recognized for 7 minutes, sir.

STATEMENT OF JAMES NEWSOME

Mr. NEWSOME. Thank you, Mr. Chairman. And as a native Mississippian, I want to thank you for holding this committee meeting, for having our Governor Haley Barbour on earlier, and for the leadership that you are providing through what may come from this hearing.

NYMEX is the world's largest forum for trading and clearing physical commodity-based futures contracts, including energy and metals. NYMEX provides an important economic benefit to the public by facilitating competitive price discovery and hedging.

As the benchmark for energy prices around the world, trading on NYMEX is transparent, open, competitive and heavily regulated.

Contrary to some beliefs, NYMEX does not set prices for commodities trading on the exchange. NYMEX does not trade in the market, and, being price neutral, does not influence price movement. NYMEX provides a forum for traders to come together and execute trades at prices which best represent what market participants think prices should be in the future, given today's information.

Periods of market uncertainty and volatility often result from extreme supply disruptions, as we saw with the numerous refineries shut down due to Hurricane Katrina. Price volatility following Hurricane Katrina drove many to the futures markets, as is reflected by the record volumes traded on NYMEX since the hurricane.

Futures markets fulfill two primary functions. They permit hedging, giving market participants the ability to shift risk. And, two, they facilitate price discovery and market transparency.

Transparency involves many factors, including continuous price reporting during the trading session, daily reporting of trading volumes and open interest, and monthly reporting of deliveries against the futures contracts.

NYMEX energy futures markets are highly liquid and transparent, representing the views and expectations of a wide variety of participants from every sector of the energy marketplace. The price agreed upon for sale of any futures contract trade is immediately transmitted to the exchange's electronic price reporting system and to the news wires and information vendors who inform the world of accurate futures prices.

Gasoline is the largest single volume refined product sold in the United States and accounts for almost one-half of national oil consumption. It is a highly diverse market with hundreds of wholesale distributors and thousands of retail outlets, making it subject to intense competition and occasionally price volatility. Average daily volume in these contracts has hit record levels in recent months, and prices have been volatile. These market conditions reflect the basic market fundamentals where there is an imbalance of supply and demand. Tight gasoline supplies due to the lack of refinery capacity compounded by the impact of Hurricane Katrina drove prices upward dramatically in the cash and in the futures markets.

The importance of the gulf coast refineries as a key supply source for the New York Harbor via Colonial Pipeline directly impacts the physical and the futures gasoline markets. During the 1-week period prior to Hurricane Katrina, the cash market price for gulf coast gasoline averaged $1.82 per gallon, which was 8 cents per gallon lower than the weekly average NYMEX futures settlement price. After the supply disruption, the gulf coast gasoline cash market rose more than $1, to $2.84 per gallon for the daily average on August 30, 37 cents higher than the NYMEX futures settlement price on August 30.

A number of refineries in the Gulf of Mexico were damaged beyond immediate repair and critical petroleum supplies were lost. Prior to Hurricane Katrina, the U.S. refineries had already begun running at maximum capacities struggling to keep up with gasoline demand. This disaster in a key refining region only further exasperated an already growing problem.

It is widely theorized, Mr. Chairman, that speculators can drive up prices. Placing blame on speculators may grab the attention of the media but does not accurately reflect the realities of how markets work. With hundreds of commercial participants and instantaneous price dissemination, any speculative price would be met with an equally strong commercial reaction. If markets move in a direction inconsistent with actual market factors, there are a vast number of participants, including energy producers, wholesalers, retailers, and government agencies, that have comparable access to information.

During the August 30 trading session, NYMEX set daily volume records for overall exchange volume and for gasoline and crude oil futures. These volume numbers clearly reflect NYMEX's importance as a transparent trading forum where customers can effectively manage their price risk. It is precisely during such times of market volatility uncertainty that the Exchange's vital role in facilitating price discovery and risk management is most crucial to our customers.

At all times during this period of extreme uncertainty in the market, NYMEX has been the source for transparent prices in the energy markets. Our trading systems and price reporting systems to the world's vendors worked flawlessly and without delay. Even though as consumers we may not necessarily like the result, the NYMEX marketplace performed its responsibility to create open, competitive, and transparent pricing. We can only imagine the market uncertainty and further devastation to consumers if NYMEX were unable to perform its duty and prices were determined behind closed doors.

Thank you, Mr. Chairman, for the chance to be here.

[The prepared statement of James Newsome follows:]

PREPARED STATEMENT OF JAMES NEWSOME, PRESIDENT, NEW YORK MERCANTILE EXCHANGE, INC.

Mr. Chairman and members of the Committee, my name is Jim Newsome and I am the President of the New York Mercantile Exchange (NYMEX or Exchange). NYMEX is the world's largest forum for trading and clearing physical-commodity based futures contracts, including energy and metals products. We have been in the business for 135 years and are a federally chartered marketplace, fully regulated by the Commodity Futures Trading Commission. On behalf of the Exchange, its Board of Directors and shareholders, I thank you and the members of the Committee for the opportunity to participate in today's hearing on Hurricane Katrina's devastating effect on gasoline supply and prices.

First and foremost, we would like to acknowledge that not only has the nation's energy supply been severely affected, but lives have been lost, homes have been destroyed, and entire cities are in ruins. Our thoughts and prayers are with all the families that have suffered from the destruction of Katrina.

INTRODUCTION

NYMEX provides an important economic benefit to the public by facilitating competitive price discovery and hedging. As the benchmark for energy prices around the world, trading on NYMEX is transparent, open and competitive and heavily regulated. Contrary to some beliefs, NYMEX does not set prices for commodities trading on the exchange. NYMEX does not trade in the market and, being price neutral, does not influence price movement. NYMEX provides the forum for traders to come together and execute trades at prices which best represent what market participants think prices should be in the future, given today's information.

Periods of market uncertainty and volatility often result from extreme supply disruptions as we see with the numerous refineries shut down due to Hurricane Katrina, which brings me to the reason I was asked to testify today. There is a strong beneficial and interdependent relationship between the futures and cash markets. The primary motivation for using the futures market is to hedge against price risk in the cash market. Prudent business managers rely on the futures market to protect their business against price swings in the cash market. Price volatility following Hurricane Katrina drove many into the futures markets, as is reflected by the record volumes traded on NYMEX since the hurricane.

Futures markets provide a reference point for use in arranging trades at competitively determined prices. An understanding of the NYMEX market, its pricing mechanism and the relationship between the futures price and the cash price will provide useful instruction and clarity to what is often perceived as an esoteric area of financial dealings.

OVERVIEW

Futures markets fulfill two primary functions: (1) They permit hedging, giving market participants the ability to shift price risk to others who have inverse risk profiles or are willing to assume that risk for profit; and (2) They facilitate price discovery and market transparency. Transparency involves many factors, including: (1) Continuous price reporting during the trading session; (2) Daily reporting of trading volume and open interest; and (3) Monthly reporting of deliveries against the futures contract.

NYMEX futures contracts trade by open outcry on the Exchange floor during the day and during the evening on NYMEX ACCESS®, our after-hours electronic trading platform. Transactions are executed in a transparent and competitive environment between NYMEX members who are registered futures industry professionals. The daily settlement price for each contract is calculated pursuant to Exchange rules, which generally is the average price for all outright transactions during the closing range.

NYMEX energy futures markets are highly liquid and transparent, representing the views and expectations of a wide variety of participants from every sector of the energy marketplace. Customers from around the globe can call into a broker on the NYMEX trading floor to place buy and sell orders. On behalf of the customers, buyers announce their bids and sellers announce offers. The price agreed upon for sale of any futures contract trade is immediately transmitted to the Exchange's electronic price reporting system and to the news wires and information vendors who inform the world of accurate futures prices.

Price signals are the most efficient transmitters of economic information, telling us when supplies are short or in surplus, when demand is robust or wanting, or when we should take notice of longer-term trends. NYMEX futures markets are the messengers carrying this information from the energy industry to the public. The wide dissemination of futures prices generates competition in the establishment of current cash values for commodities.

GASOLINE

Gasoline is the largest single volume refined product by volume sold in the United States and accounts for almost half of national oil consumption. It is a highly diverse market, with hundreds of wholesale distributors and thousands of retail outlets, often making it subject to intense competition and price volatility.

NYMEX trades, among other things, New York Harbor leaded and unleaded regular gasoline futures contracts. The New York harbor gasoline futures contract trades in units of 42,000 gallons (1,000 barrels). It is based on delivery of petroleum products to terminals in the New York harbor, the major East Coast trading center for imports and domestic shipments, from refineries in the New York harbor area or from the Gulf Coast refining centers.

Average daily trading volume in these contracts has hit record levels in recent months and prices have been volatile. These market conditions reflect the basic market fundamentals where there is an imbalance of supply and demand. Tight gasoline supplies due to lack of refinery capacity, compounded by the impact of hurricane Katrina, which resulted in the closing of 9 refineries, has driven prices upward dramatically in the cash and futures market.

The importance of the Gulf Coast refineries as a key supply source for the New York Harbor via Colonial Pipeline directly impacts the physical gasoline market and the futures gasoline market. During the one-week period prior to hurricane Katrina, the cash market price for Gulf Coast gasoline averaged $1.82 per gallon (using the Platts wholesale assessment at the Colonial Pipeline), which was $.08 per gallon lower than the weekly average NYMEX futures settlement price. After the supply disruption due to hurricane Katrina, the Gulf Coast gasoline cash market rose more than one dollar to $2.84 per gallon for the daily average on August 30 (one day after the storm), $.37 higher than the NYMEX futures settlement price on August 30. This differential between the cash and futures prices represents the free market price that is derived in light of the extreme supply disruption and reflects a new equilibrium in the marketplace in response to the shock to the demand and supply balance.

NYMEX has closely monitored the gasoline futures market during this recent period of price increases in the aftermath of hurricane Katrina and has initially concluded that the market behaved rationally and the market participants acted responsibly in their futures and options trading.

SURVEILLANCE

Hurricane Katrina has had a devastating economic impact. Nine refineries in the Gulf of Mexico have been damaged beyond immediate repair and critical petroleum supplies have been lost. Prior to Hurricane Katrina, the U.S refineries had already been running at maximum capacity for years, struggling to keep up with rising gasoline demand. This huge natural disaster in a key refining region only further exacerbated an already growing problem.

The NYMEX Market Surveillance staff routinely follows trends in the cash markets, focusing on whether the futures markets are converging with the spot physical market as the NYMEX contract nears expiration. In light of the market uncertainties that resulted from hurricane Katrina, the NYMEX staff also monitored the supply and demand fundamentals in the underlying cash market to ensure that NYMEX prices reflect cash market price movements, that there are no price distortions and no market manipulation.

After analyzing events and developments over the past week, NYMEX staff believes that price increases experienced were due to fundamental market factors tied to supply disruptions in the wake of hurricane Katrina. The NYMEX system worked according to design, and added a level of economic stability to the situation by providing a viable price discovery and risk management forum.

SPECULATORS

It is widely, yet inaccurately, theorized that speculators can drive prices up. Placing blame on speculators may grab the attention of the media, but does not accurately reflect the realities of how markets work. With hundreds of commercial participants and instantaneous price dissemination, any "speculative" price would be met with an equally strong "commercial" reaction. If markets move in a direction inconsistent with actual market factors, there is a vast number of participants including energy producers, wholesalers, retailers, and government agencies that have comparable access to information. These participants will respond to ensure that prices rapidly return to where the industry consensus believes they should be.

Speculators do exist and they actually play a valuable, even necessary role in the market. They add liquidity to the market and enable commercial traders to get in and out of the market when necessary. By the nature of their role, speculative traders seek to take advantage of price trends, but because they lack the real product to back up their investment, they cannot control the price. They create virtually no impact on daily settlement prices, the primary benchmark used by the marketplace.

The Exchange has been scrutinized in the past on the role of hedge fund participation in causing market volatility. The effects of hurricane Katrina further emphasize the minimal impact hedge funds and speculators have on futures prices when compared to the real impacts of true market factors. hurricane Katrina is a natural disaster that severely disrupted the U.S. supply system and in effect drove prices higher.

Hedge funds do not account for anywhere near enough volume to affect prices.

According to a NYMEX study on the participation of hedge funds in the energy markets over a one year period beginning in January 2004, hedge funds only accounted for 4.6% of overall futures volume. Of this total, the crude oil futures market had 3.07% hedge fund participation and, its products, heating oil and unleaded gasoline, had 3.62% and 3.26% hedge fund participation, respectively.

MARKET IMPACT OF KATRINA

NYMEX directly felt the disruptive effects of Katrina in our energy futures markets. The Exchange experienced several unprecedented market events in the aftermath of Katrina. Significant price moves occurred in the energy complex on Sunday evening during the NYMEX ACCESS® trading session which commenced at 7:00 PM. During this session (which is effectively the commencement of the Monday business day) gasoline moved upward due to severe concerns around the immediate and longer term effect to refineries in Louisiana, as well as pipeline distribution systems in the region.

During regular trading hours on Tuesday, August 30, the September 2005 unleaded gasoline contract traded to its maximum upward price limit, resulting in a temporary trading halt. Exchange rules impose a price fluctuation limit of $0.25 per gallon of unleaded gasoline above or below the previous day's settlement price. When that limit is hit, a five minute temporary trading halt is triggered. This limit was reached last Tuesday when the September 2005 contract traded at $2.31. In accordance with NYMEX Rules, the market was halted at 11:15 AM and re-opened after 5-minutes with an expanded limit of $0.50 cents above the previous day's settlement.

In response to the price volatility, NYMEX increased margins on several occasions for a variety of the energy futures contracts, including gasoline and crude oil. Margin is the money or collateral deposited with the clearinghouse to protect the clearinghouse against loss on open futures or options positions. In all cases, NYMEX required additional margin to maintain the integrity of the clearinghouse. Margin is vital to ensuring the financial integrity of the Exchange and provides the clearinghouse with the ability to protect customers against counterparty credit risk. On August 30, 2005, NYMEX managed and cleared the *greatest single intra-day variation margin call scenario, when it moved nearly $2 Billion.*

During the August 30 trading session, NYMEX set daily volume records for overall Exchange volume and for gasoline and crude oil futures, as well as for the Exchanges electronic clearing platform NYMEX Clearportsm. The following day, August 31, Exchange-wide options, NYMEX Division options, and NYMEX ClearPortsm clearing once again reached record volumes. These record volume numbers, clearly reflect NYMEX's importance as a transparent trading forum where customers can effectively manage their price risk. It is precisely during such times of market volatility and uncertainty that the Exchange's vital role in facilitating price discovery and risk management is most crucial to our customers.

During the entire week following hurricane Katrina, NYMEX Compliance and CFTC officials have had a heightened presence on the trading floor overseeing all markets. All activity has been thoroughly reviewed utilizing all available electronic tools to detect any abusive activities.

CONCLUSION

At all times during this period of extreme uncertainty in the market, NYMEX has been the source for transparent prices in the energy markets. Our price reporting systems to the world's vendors have worked flawlessly and without delay. Our trading systems during regular trading hours and during after hours trading on our electronic platforms have performed flawlessly.

Even though as consumers we may not like the result, the NYMEX marketplace performed its responsibility to create open, competitive and transparent energy pricing. We can only imagine the market uncertainty and further devastation to consumers if NYMEX were unable to perform its duty and prices were determined behind closed doors.

I thank you for the opportunity to share the viewpoint of the New York Mercantile Exchange with you today.

Chairman BARTON. Thank you, Mr. Newsome.

We now want to hear from Mr. Cooper. And Mr. Cooper is the Executive Director of the Association of Oil Pipelines. You are recognized for 7 minutes, sir.

STATEMENT OF BENJAMIN S. COOPER

Mr. BENJAMIN COOPER. Thank you, Mr. Chairman. I am Ben Cooper with the Association of Oil Pipelines, a nonprofit trade association of oil pipelines. We very much appreciate the opportunity to be here today. I filed a full statement with the committee and will summarize here, and I will give you the short summary first.

Oil pipelines affected by Hurricane Katrina were rapidly restored to service, are now in service, and are able to carry oil from imports, offshore platforms, and refineries that provide supply. Second, oil pipeline transportation rates are a few cents per gallon and have not changed during the hurricane, so oil pipeline rates have had no role in the recent petroleum price increases.

Hurricane Katrina affected the operations of several major oil pipelines and facilities in the gulf coast. Today, the capacity of these pipelines has been substantially restored. The good news is that all of these pipelines weathered the hurricane with little damage and no spills. However, in the aftermath of the hurricane they were taken out of service, among other things, by the loss of grid electric power. Oil pipelines operate using large electric pumps. The electricity needed to run even one pump is enough to supply a small town.

After the hurricane, transmission and generation in south Louisiana and Mississippi were shut down, yet 3 days later our pipelines began to come back on line. Within 6 days, most were at or could anticipate full operation. The extraordinary efforts of pipeline employees, of the employees of the electric utility companies that supply power to these pipelines, and of some very dedicated public servants has restored the capacity of these pipelines. One pipeline operator, for example, located several large, many large portable generators all over the country, and with the help of the Department of Transportation's Pipeline and Hazardous Materials Safety Administration, moved them to the affected areas to enable key pumps to restart. Another operator actually rewelded bypass lines to allow pumps on either side of a shut-in facility to operate to restore partial service.

Pipelines are motivated to get their systems operating as soon as possible. The interests of the public and the pipelines are aligned in this. Of course, the shutoff of major offshore platforms and refining capacity in the storm's wake means that supply may still be affected even after the pipeline transportation system is fully restored. But when the supply does become available, oil pipelines will be ready to transport it.

Let me talk about oil pipeline rates, because we have been asked to address whether oil pipeline companies have contributed to the sudden increase in gasoline prices by raising the rates charged for transportation. The facts are that pipeline rates did not change during the past week. The Federal Government regulates the rates of interstate oil pipelines. We are the only part of the petroleum supply system that is under Federal regulation.

Our member companies deliver petroleum safely to nearly every region in the United States for a few cents a gallon. A typical rate for transport of petroleum product from the gulf coast to the Southeast is about 2 cents a gallon, to the Northeast about 3 cents a gallon, and to Chicago for about 2½ cents a gallon. Oil pipelines provide transportation services to customers. The customers are the ones who decide what to ship, where to ship, and when to ship. The decision of how much to ship of each commodity and to which destination is made by our shipper customers, not by pipeline operators.

I would like to share a couple of lessons at least for our industry from this experience. Federal policy should assign a leadership role from within the Federal family to address oil pipeline problems during these events. In the wake of Hurricane Katrina, DOT's Pipeline and Hazardous Materials Safety Administration performed highly useful services in coordinating and addressing bottlenecks as oil pipeline operators sought to locate and deliver emergency equipment and specialized generators to particular pump stations.

The Pipeline and Hazardous Material Safety Administration is the Federal agency most knowledgeable about oil pipelines and is an excellent choice for the role of assisting oil pipelines during emergencies. Legislation may be required to authorize this.

Second, restoration of grid electric power is absolutely critical to the resumption of pipeline service and needs to receive the highest priority during these events. We have a new appreciation of the interdependency of pipelines with electric power. The Federal Government should be doing everything in its power to assist the electric utility industry generally and utilities individually to harden facilities to overcome threats and to rapidly recover when power is lost despite all efforts.

A final note. Today oil pipeline capacity is near full under normal conditions. Oil pipeline infrastructure will soon require expansion to meet the needs of consumers, to accommodate changing supply patterns, for example, such as the growth of Canadian tar sands production, to meet stricter requirements for product quality such as ultra load sulfur diesel fuel, to meet stricter requirements for product composition such as boutique fuels, and to provide infrastructure security.

A support of public policy, including continuation of flexible rate treatment, permitting assistance, and creative approaches to acquiring pipeline rights-of-way will be required to ensure that oil pipeline expansions are made when needed, are there to meet expectations that the committee may have about refinery capacity.

AOPL looks forward to working closely with the Department of Transportation, the Federal Energy Regulatory Commission, this committee, and the rest of Congress to ensure that the oil pipeline industry is able to meet the challenges in the future, and we thank you for our opportunity to appear today.

[The prepared statement of Benjamin S. Cooper follows:]

PREPARED STATEMENT OF BENJAMIN S. COOPER, EXECUTIVE DIRECTOR, ASSOCIATION OF OIL PIPE LINES

INTRODUCTION

My name is Benjamin S. Cooper. I am the Executive Director of the Association of Oil Pipe Lines. AOPL is a 501 (c) (6) non-profit trade association of interstate oil pipelines, which includes pipeline transporters of crude oil, refined petroleum products, liquefied gases and anhydrous ammonia. Our Association's 53 members transport about 85 percent of the crude oil and refined petroleum products delivered by pipelines. AOPL members include pipelines that transport crude oil from production and import points to refineries and pipelines that transport the refined products produced in those refineries to end users and distributors (retailers, wholesalers, airports, railroads, etc.). AOPL's membership is comprised of domestic U.S. oil pipeline companies and two Canadian oil pipeline companies.

My testimony will first discuss the impact that hurricane Katrina has had on oil pipeline operations and lessons learned during the past week. I then will cover the role played by oil pipelines in petroleum supply, describe government oversight of that role and sketch the challenges faced by the industry in providing sufficient capacity to meet our nation's current and future petroleum transportation needs.

Impact of Hurricane Katrina

As the Committee knows, the major impact of the hurricane was felt in Louisiana and Mississippi. Four effects of the storm have been important to oil pipelines with operations in these states:

- The lives of local pipeline personnel have been severely disrupted;
- Key pipeline facilities have been flooded;
- Electric power has not been available; and
- The supply of crude oil and products to ship in pipelines has been disrupted

The major affected pipelines have been Colonial and Plantation, which together account for a major share of the refined petroleum products transported along the eastern seaboard, as much as 60% of the supply in some areas of the southeast. Both pipelines were shut down in an orderly way to maintain product quality and pipeline integrity in anticipation of the storm. They then were prevented from restarting by the severity of the storm's impact, in particular, by the loss of electric power. Both companies were able to resume limited service on Wednesday, August 31, when they were able to arrange for alternative power sources. As of September 3, both were receiving some utility electric power. Colonial was running at 80% of capacity, and Plantation was running at 95% of capacity.

Capline, a crude oil pipeline that transports crude oil from the Gulf of Mexico to refineries in the mid west, and one of the pipelines that would carry oil from the Louisiana Strategic Petroleum Reserve sites, was also shut down. As of September 3, service on Capline was restored to 80% of capacity after the integrity of the pipeline was established and utility electric supply to some pumps was re-established.

Dixie, a propane pipeline serving markets in the Southeast could not reopen after the storm due to loss of power. Dixie has also partially resumed service with the restoration of some utility electric power, and as of September 3 was operating at 50% of capacity.

Finally, Louisiana Offshore Oil Pipeline, which operates facilities for receipt and transport of crude oil imported in large tankers was also shut down, but as has since resumed operation at 75% of capacity.

The common denominator in these shut downs is the location of key facilities in areas in the direct path of the storm where flooding was extensive and electric power was out for considerable periods of time. All have substantially recovered as facilities formerly isolated by flooding are reactivated and electric power comes on line. The impact of the shut down of Colonial and Plantation continues to be felt in areas where alternative supply, for example, from imports or waterborne carriers, is not feasible. Of course, the massive shut down of refining capacity in the storm's wake meant and will mean for some time that quantities of supply from these sources will be limited, even after the transportation system is fully restored.

Some questions have arisen regarding whether these pipeline companies were economically advantaged by the hurricane and contributed to the sudden increase in gasoline prices by raising the rates charged for transportation. The facts are that pipeline rates did not change during the past week. For example, Colonial Pipeline's tariff from Pasadena, Texas to Atlanta Georgia (82.82 cents/barrel—less than 2 cents/gallon) was set on July 1st and remains unchanged. In fact, several pipeline companies were negatively impacted by the loss of revenue and extraordinary costs incurred to bring their operations back in service as soon as possible.

Lessons Learned from Hurricane Katrina

- The decision by the EPA to act quickly to waive temporarily area specific fuel requirements under the Clean Air Act in the widest possible area allows the petroleum distribution system to make the most effective use of existing supplies. Several pipelines serving the Midwest immediately began receiving nominations of alternative gasolines to move north and east. This was an important action that was taken in a timely manner.
- Federal policy should assign a leadership role from within the federal family to address oil pipeline problems during these events. In the wake of hurricane Katrina, DOT's Pipeline and Hazardous Materials Safety Administration performed highly useful services in coordinating with the Federal Emergency Management Agency and addressing bottlenecks as oil pipeline operators sought to locate and deliver emergency equipment and specialized generators to particular pump stations. PHMSA is the federal agency most knowledgeable about oil pipelines, and is an excellent choice for the role of assisting oil pipelines during emergencies.
- Hoarding and panic buying exacerbate petroleum fuel shortages. Officials need to be active early and continuously to discourage, to the extent possible, these reactions. In addition, dissemination of false information by the media can make hoarding and panic buying worse and generally has a negative impact on markets.
- Restoration of grid electric power is critical to the resumption of pipeline service and should receive the highest priority during these events. The federal government should be doing everything in its power to assist the electric utility industry generally and utilities individually to enhance the ability of utilities to overcome threats and recover rapidly where power is lost despite all best efforts.
- Finally, hurricane Katrina provides a sobering data point in the nation's understanding of the interdependency of the energy supply system and a highly painful real world experience with the impact of a loss of key energy services and infrastructure that approximates many homeland security emergency scenarios.

The Role of Oil Pipelines in the U.S.

Oil pipelines provide about ⅔ of the petroleum transportation in the U.S., measured in barrel miles. Unlike natural gas, which can only be transported by pipeline, alternatives to petroleum pipeline transportation exist and include tankers, barges, rail and trucks. However, each of these alternatives has significant limitations, and, as a result, pipelines are the primary method of bulk transportation of petroleum over medium to long distances. It is difficult to imagine how our transportation network, which is 95% powered by petroleum, could operate without oil pipelines.

Pipeline transportation has dual advantages of efficiency and safety. About 17% of the annual ton-miles of our nation's freight are carried by petroleum pipelines, at a cost of about 2% of the total U.S. freight bill. Pipelines share with tanker vessels the safest record in petroleum transportation, safer than barge, rail or truck. Deaths and injuries from petroleum pipeline transportation are rare and the environmental impact of pipeline transportation is less than any of its alternatives. Oil pipelines are able to deliver petroleum safely to nearly every region of the U.S. for a few pennies per gallon. A typical rate to transport petroleum product from the Gulf Coast to the Southeast is about 2 cents per gallon, to the Northeast is about 3 cents per gallon and to Chicago is about 2.5 cents per gallon.

Economic Regulation of Oil Pipelines

The federal government regulates the economics of interstate oil pipelines—in fact oil pipelines are the only part of the petroleum supply system that is under federal economic regulation.

The Federal Energy Regulatory Commission administers the provisions of the Interstate Commerce Act to ensure that interstate oil pipelines:

- Function as common carrier providers of transportation to any qualified shipper;
- Charge no more than publicly available rates filed in advance with the FERC, which are typically limited to a few cents per gallon;
- Assign space on the pipeline based on monthly nominations from all interested shippers and prorate access to that space among all applicants in a posted, nondiscriminatory way when the line is full;
- Exercise no undue discrimination among shippers;
- Maintain confidentiality of shipper records and not share information of any shipper with any other shipper; and
- File annual reports on pipeline company income and cost data with the FERC that are available to the public.

Oil pipelines provide transportation services and charge fees that do not fluctuate with the price of the products that are transported. Because oil pipelines do not own the products that they transport, they do not benefit from any product price increases. In fact, refined products pipelines are generally adversely impacted by high commodity prices, as higher prices increase power costs and marginally result in lower consumption levels. Even when an oil pipeline is an affiliate of a major integrated oil company, the Interstate Commerce Act and FERC oversight establishes a wall between the pipeline portion of the firm and the owners' transportation operations.

Oil Pipeline Transportation Rates

Typical oil pipeline rates range from 1 to 5 cents per gallon and are independent of the value of the oil being transported. Thus the revenue received by the oil pipeline is a few cents per gallon, regardless of the sale price of that gallon, whether that sale price is $1.00, $2.00, $3.00 or more.

Oil pipeline rates are posted in FERC-filed tariffs that normally take effect after 30 days and are subject to protest during that period. Oil pipeline rate changes must be justified using one of four rate mechanisms: indexation, a settlement rate agreed to by all affected shippers, market-basis or cost-of-service. In calendar years 2003 and 2004, there were 1096 oil pipeline tariff rate filings. Of those, 937 (88%) were index-based, and 159 were justified on another basis. Of the 159 others, roughly 49% were market-based, 30% were settlement rates, 14% resulted from pervious settlements and 7% were cost of service based.

Most oil pipeline tariffs cover a specific group of products. For instance, a "Products Tariff" would apply the same tariff rate to gasoline, diesel, jet fuel and kerosene product shipments between the same points. For instance, Colonial's tariff defines "Petroleum Products" to mean "gasolines and petroleum oil distillates", which would include jet fuel, diesel fuel and heating oil. There are also crude oil tariffs, propane tariffs, etc.

Pipeline tariffs do not tend to change frequently and, unlike commodity prices, are not adjusted as a result of short-term market circumstances. Since nearly 90% of tariffs are indexed, most adjustments are done on an annual basis and occur on July 1 of each year when the new FERC index takes effect. Even market based rate changes occur infrequently, with some changes actually rate decreases to meet competitive market conditions.

Pipelines also file rules and regulations tariffs that set forth the pipeline's conditions of service. These filings explain such things as the pipeline's tendering process, minimum batch size, allocation policy and product specifications. Such rules and regulations are required to be administered in a non-discriminatory manner. A system of checks and balances on oil pipeline behavior operates through the ability of any shipper to protest any alleged deviation from FERC requirements.

Oil pipelines are providers of transportation services for generally fixed fees for our customers, who determine what to ship, where to ship or when to ship. The decision on how much to ship of each commodity and to which destination is made by our shipper customers. Pipelines then ship multiple products on a regular cycle of products. On a normal basis, we provide transportation for all products to all destinations on a regular cycle.

The oil pipeline business is volume driven, and the incentive for pipelines from both a revenue and customer relations standpoint is to transport as much product as possible. Any inference that oil pipeline operators are purposely contributing to product shortages by reducing or shutting down capacity to cause higher product prices is simply false. In fact, the oil pipeline industry's drive to transport more volumes contributes to market liquidity, which on the margin should contribute to more competition and lower prices. The extraordinary efforts of our member companies to return their systems to service as fast as possible in the aftermath of hurricane Katrina provides ample evidence of the pipeline industry's motivation and commitment to resume business and recognition of the critically important role played by pipelines in enabling adequate supplies of petroleum products to reach destination markets..

The oil pipeline industry is not a large generator of revenue by comparison with other sectors of U.S. industry, including other sectors of the energy industry. For 2003 (the most recent data available) the entire FERC-regulated oil pipeline industry received gross revenue of $7.7 billion to deliver 13.2 billion barrels of crude oil and refined petroleum products to its various customers. A single company's revenue in many other sectors of the economy would far exceed the oil pipeline industry's revenue as a whole

Pipeline ownership is diverse, with several forms of ownership as detailed below:

- Major integrated oil companies (for example: ExxonMobil Pipeline Company, Marathon Pipe Line LLC, Chevron Pipeline Company, Shell Pipeline Company);
- Joint venture pipelines owned by shippers and other pipeline companies (for example: Colonial, Explorer, Trans-Alaska Pipeline, Capline); and
- Independents engaged primarily in oil pipeline transportation (Buckeye, TEPPCO, KinderMorgan, Enbridge, Plains All American).

A substantial percentage of the pipelines are independently owned and operated, with the current trend towards increased independent ownership of oil pipeline assets. Major integrated oil company ownership of oil pipelines has been steadily decreasing in recent years, with major oil companies now representing a minority of oil pipeline asset ownership.

In sum, the amount charged to transport oil by pipeline is controlled by either regulation or market forces and is quite small in relation to the value of oil itself. The cost of transporting oil and petroleum products by pipeline has a minimal, if any, impact on consumer prices.

Oil Pipeline Capacity

While the cost of transporting oil by pipeline has a minimal impact on consumer prices, access to adequate pipeline capacity can make a substantial difference in consumer prices. As we have seen following hurricane Katrina, when adequate pipeline capacity is not available, shortages, price increases and price volatility for petroleum consumers are the result. Even before hurricane Katrina, we saw this, for example, in Arizona in 2004 and in the Midwest in 2003 when key pipelines were out of service.

The U.S. oil pipeline infrastructure is a large system created over many years. Volumes moving on those pipelines grow only in response to increases in oil demand, that is, a few percent a year. Volumes can sometimes also increase or decrease dramatically due to changes in supply patterns such as refinery closures, new crude supplies and other significant changes. Additions to capacity often present large hurdles to individual companies in terms of capital requirements and perhaps more importantly, acquisition of right of way and required permitting. The current system, constructed principally in the 1950s and 1960s with excess capacity for that time, is quite close to full capacity at today's levels of domestic petroleum consumption, and pipelines have had to adjust to a just-in-time inventory mentality and to seasonal fuel switches that put additional strain on the system.

Oil pipelines are another component of the U.S. energy infrastructure that will require expansion in coming years to meet the needs of consumers. A supportive public policy, including continuation of the recent trend to market based and indexed rate treatment, permitting assistance and creative approaches to rights of way, will be required to ensure that oil pipeline expansions are made when needed.

Key Aspects of Oil Pipeline Operations

Oil is moved through pipelines by large pumps powered by electricity. Oil pipeline companies are large consumers of base-load electricity. Pumps are located at the origin point of the pipeline and at intervals typically 30 to 50 miles apart, depending on terrain and the location of major facilities for pickup or delivery of oil. For a pipeline of significant size, pumps at these stations of 3,000-5,000 horsepower are typically used, requiring megawatt quantities of electric power. The only feasible method for delivery of electricity in these quantities is through connection to the utility grid.

Oil pipelines maintain tanks at points along the line to facilitate the scheduling of pipeline transportation. For refined product pipelines, the need for tankage is a significant issue as the number of distinct products shipped increases. Pipeline transportation tanks hold oil that is owned and controlled by shippers. The volume in these tanks typically only represents a limited supply in relation to overall petroleum demand.

I will be glad to try to answer any of your questions, and our Association would be pleased to work with the Committee on any follow up from this hearing.

Chairman BARTON. We thank you, sir. We now want to hear from Mr. Bill Douglass, who is here representing the National Association of Convenience Stores and the Society of Independent Gasoline Marketers of America. You are recognized for 7 minutes. Welcome.

STATEMENT OF BILL DOUGLASS

Mr. DOUGLASS. Good evening, Mr. Chairman, and members of the committee. As you said, my name is Bill Douglass. I am Chief Executive Officer of Douglass Distributing Company headquartered in Sherman, Texas. Thank you for inviting me to testify before you today on behalf of NACS and SIGMA

On the impact of Hurricane Katrina on the Nation's wholesale and retail fuel supply and prices. I will concentrate my testimony on my personal experiences over the past 10 days as a marketer in Texas and on the experiences of fellow marketers and other areas during the past 10 days. In the interest of time, I will have to move through the charts I brought with me this afternoon fairly quickly.

The first chart depicts the daily movements of wholesale prices in the Dallas-Fort Worth market last week. These wholesale prices jumped an average of over 11 cents per day, for a total increase between Monday August 29 and Friday September 2 of 44 cents a gallon.

The second chart shows how my company reacted to these rack price increases in terms of our retail outlet prices. As you can see, our retail prices in general rose by a similar and in some cases lower amount than our wholesale costs.

Chart 3 provides a broader look at wholesale prices in the Dallas-Fort Worth market last week. My company's experience was not unique. These prices happen to be on branded racks, and they went all the way to $3.10 when you add the tax.

Chart 4 summarizes the changes in rack pricing in each region of the country broken down by pad.

Chart 5 provides a look at wholesale prices, that is, rack prices, last week in five randomly chosen cities: Atlanta, Boston, Dallas-Fort Worth, Detroit, and Philadelphia. All of these cities witnessed substantial increases in rack gasoline prices last week, and these figures do not include the taxes or fees or freight.

There has been widespread media reports and even some comments by congressional leaders of gasoline price gouging by gasoline marketers in the wake of Katrina. I cannot assure the committee that isolated incidents of profiteering for personal gain in the midst of this crisis did not occur last week. It is important for this committee to understand, however, before you rush to judgment on whether my or other retailers actions were proper, how I and other retailers establish our retail prices in a market with escalating wholesale prices.

Simply stated, I try to set my retail prices on the basis of the replacement costs of the gallons I have at my outlets. When the wholesale prices are rising, I know the next load of gasoline I purchase from my supplier will cost me substantially more than my last load. My sales must generate sufficient cash for me to make the next purchase and pay my supplier.

If the only thing you knew about my company was that I raised retail gasoline prices by over 40 cents per gallon last week, would you suspect that I was attempting to profit from this crisis? Maybe. But based on the information I have given you today, I trust that you would reach a different conclusion after you have investigated the facts. I urge this committee and your colleagues to gather the facts on last week's gasoline supply and retail pricing situation before reaching conclusions about my actions or the actions of other motor fuel marketers.

The enactment of the Energy Policy Act of 2005 is a good first step toward addressing the Nation's problems of shrinking refining capacity and a trend toward higher gasoline prices. I commend you, Mr. Chairman, and your colleagues for taking the lead in making this important legislation a reality after 5 long years. Specifically, your provisions gave the Environmental Protection Agency the statutory authority to waive certain gasoline and diesel fuel controls last week, providing the market with much needed flexibility to move product between markets to mitigate the supply disruptions. This is an immediate example of the positive impact this energy bill has had on the market. There are other important provisions in the 2005 energy bill that will assist in expanding domestic refining capacity and in mitigating gasoline supply dislocations and price spikes.

NACS and SIGMA urge this committee and this Congress to build on the progress made through the Energy Act of 2005 in the following ways: Assure prompt implementation of EPAC's 2005 provisions, including the joint Environmental Protection Agency and Department of Energy study on increasing gasoline and diesel fuel supplies while protecting the environment. Streamline permitting and siting procedures for expanding existing domestic refining capacity, and for the construction of new grassroots refineries. Adopt additional incentives to expand our domestic refining capacity. And, investigate the pricing policies of credit card companies, whose charges make up an ever increasing portion of the price of gasoline at retail outlets, particularly when gasoline prices are high.

Thank you for inviting me to testify here today on this important topic, and I would be pleased to answer any questions my testimony may have introduced.

[The prepared statement of Bill Douglass follows:]

PREPARED STATEMENT OF BILL DOUGLASS, CHIEF EXECUTIVE OFFICER, DOUGLASS DISTRIBUTING COMPANY REPRESENTING THE NATIONAL ASSOCIATION OF CONVENIENCE STORES AND THE SOCIETY OF INDEPENDENT GASOLINE MARKETERS OF AMERICA

I. INTRODUCTION

Good afternoon, Mr. Chairman and members of the Committee. My name is Bill Douglass. I am Chief Executive Officer of Douglass Distributing Company, headquartered in Sherman, Texas. My company operates 14 convenience stores and supplies gasoline and diesel fuel to 165 retail locations throughout the Dallas-Fort Worth area.

I appear before the Committee today representing the National Association of Convenience Stores ("NACS") and the Society of Independent Gasoline Marketers of America ("SIGMA").

II. THE ASSOCIATIONS

NACS is an international trade association comprised of more than 2,200 retail member companies operating more than 100,000 stores. The convenience store industry as a whole sold 142.1 billion gallons of motor fuel in 2004 and employs 1.4 million workers across the nation.

SIGMA is an association of more than 240 independent motor fuel marketers operating in all 50 states. Last year, SIGMA members sold more than 58 billion gallons of motor fuel, representing more than 30 percent of all motor fuels sold in the

United States in 2004. SIGMA members supply more than 35,000 retail outlets across the nation and employ more than 350,000 workers nationwide.

Together, NACS and SIGMA members sell approximately 80 percent of the motor fuel retailed in the United States each year.

III. SUMMARY OF TESTIMONY

Thank you for inviting me to testify before you today on the impact of Hurricane Katrina on the nation's wholesale and retail motor fuel supply and prices. The past ten days have been some of the most challenging in my thirty years as a motor fuel marketer and I welcome this opportunity to share my personal experiences, and the experiences and impressions of other NACS and SIGMA members with whom I have talked, with you.

As an initial matter, I would like to express my personal sympathy, and the sympathy of our entire industry, for the victims of Hurricane Katrina. Individually and collectively, our industry shares the suffering of our fellow citizens and will do all in our power to alleviate this suffering at the earliest possible date.

My testimony will touch on three broad topics today. First, I will provide the committee with as much information as I have available on the impact of Hurricane Katrina on gasoline supplies and prices. Specifically, I will share with you my personal experiences over the past ten days and summarize, to the extent possible, the information I have received from my fellow retailers.

Second, I am here to respond to allegations that I, and my industry, have taken advantage of this tragedy by "gouging" our customers by raising retail motor fuel prices. Such allegations are personally offensive to me, and in general reflect a lack of understanding of the market events that have led to the gasoline and diesel fuel price spikes of the last ten days. While it is certainly possible that some "bad actors" have sought to exploit this crisis for personal gain, I can assure you that their actions are not the actions of the vast majority of our industry.

Third, my testimony contains recommendations to the committee on steps that should be taken to lessen the likelihood that such supply disruptions and wholesale and retail price spikes will occur in the future. Unfortunately, these recommendations are remarkably similar to the steps NACS and SIGMA have been urging public policymakers to take for the last ten years. While the enactment of the "Energy Policy Act of 2005" earlier this summer was a good first step towards implementing some of these recommendations, much remains to be done.

IV. IMPACT OF HURRICANE KATRINA ON WHOLESALE AND RETAIL GASOLINE PRICES

For much of the eastern two-thirds of the nation, the impact of Katrina on wholesale and retail gasoline prices could not have been more immediate and profound. I will leave it to other witnesses here today to discuss the impact Katrina had on crude oil production and imports, crude oil movements from production to refineries, domestic refining capacity, and the movement of finished gasoline and diesel fuel throughout the country via pipeline, barge, and truck. That is not my area of expertise. Instead, I will concentrate my testimony on my personal experiences over the past ten days as a marketer in Texas, and on the experiences of fellow marketers in other areas over the past ten days.

It will be helpful for me to use several charts to graphically make these points. This first chart (Chart 1) depicts the daily movements of wholesale prices in the Dallas/Fort Worth market last week. This is the "rack," or wholesale price—the price at which my suppliers are willing to sell me, and other marketers, truckloads of 87 octane conventional gasoline. As you can see, these wholesale prices increased daily, and dramatically, last week. On August 28th, before Katrina struck, my wholesale gasoline cost was $2.36 per gallon including federal, state, and local taxes. Early last week, as Katrina struck the Gulf Coast, these wholesale prices jumped an average of over eleven cents per day, for a total increase between Monday, August 29th and Friday, September 2nd of 44 cents per gallon.

I must point out that I am a branded marketer—the stations I own and supply fly the flag of a major refiner. The wholesale prices in this chart reflect branded rack prices, not unbranded, or independent, rack prices. During this same five day period, wholesale prices for these unbranded stores rose 73 cents per gallon, or over 18 cents per day.

This second chart (Chart 2) shows how my company reacted to these rack price increases in terms of our retail outlet prices. As you can see, our retail prices in general rose by a similar, and in some cases, lower amount than our wholesale costs. In short, my company reacted primarily to changes in wholesale price increases when determining where to set our retail prices. In some cases, because of competition from other retailers in our market area, we did not pass the entire increase in rack prices through to retail. On these days, virtually every gallon we sold from our stations resulted in no or negative profit margins for our company, once our operating costs are taken into account.

My personal experience is similar to the experiences of other retailers across the nation. NACS and SIGMA obtained rack pricing data from the Lundberg Survey, an independent report on wholesale motor fuel prices, for several major metropolitan areas for the past two weeks. This chart (Chart 3) provides a broader look at wholesale gasoline prices in the Dallas-Fort Worth market last week.

The next two charts (Charts 4 & 5) indicate that my experience in Texas was not unique. Chart 4 summarizes the changes in rack pricing in each region of the country, broken down by PADD. As you can see, wholesale prices were up significantly last week in all areas of the country. Chart 5 provides a look at wholesale rack prices last week in five randomly chosen cities—Atlanta, Boston, Dallas/Fort Worth, Detroit and Philadelphia. All of these cities witnessed substantial increases in rack gasoline prices last week.

I have used these charts to provide you with detailed evidence that Katrina had a widespread impact on gasoline prices in much of the country over the past two weeks—not just in the areas devastated by the storm itself. Because crude production was reduced, refineries crippled, and gasoline pipelines were taken out of service, gasoline supply shortages began to occur, first in areas close to the areas hit by Katrina and rapidly moving outwards to areas of the country served directly or indirectly by the production, refining and transportation hub of the nation's Gulf Coast.

These statistics confirm that retail gasoline price increases last week were justified by movements in the wholesale cost of gasoline. While two months from now hindsight may provide us with additional facts that will indicate that the markets could have responded to this supply crisis differently, as we are going through this crisis, the fundamental laws of economics tend to apply forcefully—if demand remains the same or increases and supply is reduced, prices will rise. This is the situation we have experienced for the last ten days.

V. ALLEGATIONS OF PRICE "GOUGING"

Last week, there were widespread media reports, and even some comments by congressional leaders, of gasoline price "gouging" by gasoline marketers in the wake of Katrina. I can not assure the committee that all of these reports are false or that isolated instances of profiteering for personal gain in the midst of this crisis did not occur last week. I wish I could.

However, I can tell you that such actions were not the norm in our industry. The vast majority of gasoline marketers are fair and scrupulous businesses. As my testimony has shown, I personally responded to wholesale price hikes in my area in setting my retail prices. I am not aware of any credible instance in which retail price increases were not justified by the supply crisis faced by a retailer.

It is important for this committee to understand how I and other gasoline retailers establish our retail prices in a market with escalating wholesale prices. Simply stated, I try to set my prices on the basis of the replacement cost of the gallons I have at my outlets. This is an important concept which may not be readily grasped. When wholesale prices are rising, and I know that the next load of gasoline I purchase from my supplier will cost me substantially more than my last load, my sales must generate sufficient cash for me to make that next purchase and to pay my supplier.

For example, assume the gasoline at one of my retail stations cost me $2.00 per gallon yesterday. I know that the next gasoline truckload from my supplier, to be purchased tomorrow, will cost me $2.25 per gallon. I will, if I can based on competition in my area, set a retail price at my outlet today that will cover the higher price I will have to pay tomorrow. If I don't, I will be forced to borrow money from my company's banks to pay for tomorrow's gasoline. Such debt only increases my cost of staying in business and adds to the upward pressure on retail gasoline prices. It is a sound business practice for a retailer to price today on the replacement cost of gasoline at the outlet, not the cost of product actually at the outlet.

If instances of profiteering on this tragedy have occurred, federal and state officials have ample legal recourse for dealing with those bad actors, including Section 5 of the Federal Trade Commission Act. Such behavior must not be tolerated now or in the future in our industry or any industry.

However, just as such behavior must not be tolerated in our industry, neither should the media or opinion leaders react to such anecdotal reports by issuing blanket indictments of all motor fuel marketers. Such generalizations may make for good "sound bites," but they do not reflect what is actually happening across the country and unfairly damage the reputations of many companies that are struggling to meet the challenges of the current crisis.

If the only thing you knew about my company was that I raised by retail gasoline prices by over 50 cents per gallon last week, would you suspect that I was attempting to profit from this crisis? Maybe. But based on the information I have given you today, I trust that you would reach a different conclusion after you had investigated the facts. I urged this committee and your colleagues to gather the facts on last week's gasoline supply and retail pricing situation before reaching conclusions about my actions or the actions of other motor fuel marketers.

As a final point with respect to retail pricing, I have one more chart to share with you (Chart 6). This chart outlines the approximate gross revenues that several different parties in the petroleum exploration, refining, and distribution system realize from each barrel of crude oil. Simply stated:

- In August 2003, the royalty owner of the crude oil received approximately $4 per barrel; in August 2005, the royalty owner received about $8 per barrel;
- In August 2003, the crude exploration and extraction company was receiving approximately $28 per barrel of oil; in August 2005, this company received about $67 per barrel;
- In August 2003, a refiner was receiving around $11 per barrel; in August 2005, this company received about $27 per barrel;
- In August 2003, a gasoline retailer was receiving approximately $6 per barrel; in 2005, that retailer still received about $6 per barrel; and,
- In August 2003, a credit card company was receiving approximately $1.50 per barrel; in 2005, that company is receiving approximately $3 per barrel.[1]

Based on this information, I question whether it is appropriate to single retailers out for pricing scrutiny.

VI. RECOMMENDATIONS FOR THE FUTURE

In 1996, Tom Robinson, a former president of SIGMA, offered the following testimony before the Senate Energy Committee as part of a hearing on "Recent Increases in Gasoline Prices." "The federal and state governments regulate the gasoline refining and marketing industry with little or no thought given to costs, distribution difficulties, or market efficiencies. Congress must acknowledge that . . . the present course will lead to further market disruptions and higher gasoline prices at the pump." Mr. Robinson made that statement over nine years ago.

Last year, I testified on behalf of NACS and SIGMA at a subcommittee hearing of this committee and stated:

> "Our nation's gasoline and diesel refining industry is shrinking at a time when consumer demand continues to rise. Unless we collectively change course, domestic refining capacity will be unable to keep pace with demand, gasoline and diesel fuel price spikes such as the one we have experienced this year will become the norm rather than the exception, and our nation will become more reliant on imports of gasoline and diesel fuel to meet increased consumer demand in the coming years. Congress has a choice, it can either pursue policies that will encourage the expansion of domestic refining capacity, or it can turn its gaze overseas for our nation's future gasoline and diesel fuel needs."

Unfortunately, both Mr. Robinson's and my predictions have come true. Domestic refining capacity continues to shrink, wholesale and retail motor fuel price spikes have become the norm rather than the exception, and more of our nation's gasoline needs are being met by foreign sources. NACS and SIGMA assert that it is time to stop talking about these problems and do something about them.

In my opinion, the enactment of the "Energy Policy Act of 2005" (EPAct 2005) is a good first step towards addressing these problems. I commend you, Mr. Chairman, and your colleagues for taking the lead in making this important legislation a reality after five long years. Specifically, your provisions gave the Environmental Protection Agency the statutory authority to waive certain gasoline and diesel fuel controls last week, providing the market with much needed flexibility to move product between markets to mitigate supply disruptions. This is an immediate example of the positive impact this energy bill had had on the market.

There are other important provisions in the 2005 energy bill that will assist in expanding domestic refining capacity and in mitigating gasoline supply dislocations and price spikes, including:

- Repeal of the reformulated gasoline program's oxygenate mandate;
- Restrictions on creation of new "boutique fuels" which strain refining capacity and the distribution system;

[1] All information based on publicly available sources.

- Authority for retailers to blend compliant RFGs for limited periods each summer; and,
- Federal tax incentives to encourage the expansion of domestic refining capacity.

NACS and SIGMA urge this committee and this Congress to build on the progress made through EPAct 2005 in the following ways:

- Assure prompt implementation of the EPAct 2005 provisions outlined above, including the joint Environmental Protection Agency and Department of Energy Study on increasing gasoline and diesel fuel supplies while protecting the environment;
- Streamline permitting and siting procedures for expanding existing domestic refining capacity and for the construction of new grassroots refineries;
- Adopt additional tax incentives to expand our domestic refining capacity, or a federal government-led effort to site and build three new 500,000 barrels per day refineries on federal lands to augment domestic production;
- Encourage increased price transparency and lower price volatility in the nation's gasoline futures markets by increasing the number of delivery points and product types under such contracts; and,
- Investigate the pricing policies of credit card companies, whose charges make up an ever-increasing portion of the price of gasoline at retail outlets, particularly when gasoline prices are high.

None of these recommendations will result in a substantial short-term increase in gasoline supplies or retail price decreases. However, if we do not undertake these initiatives now, we will be sure to repeat the experiences of the past two weeks in the future.

VII. CONCLUSION

Thank you for inviting me to testify today on this important topic. I would be pleased to answer any questions my testimony may have raised.

Chart 1- Douglass Distributing Wholesale Price Experience
August 28 - September 2, 2005 (per gallon prices)[1]

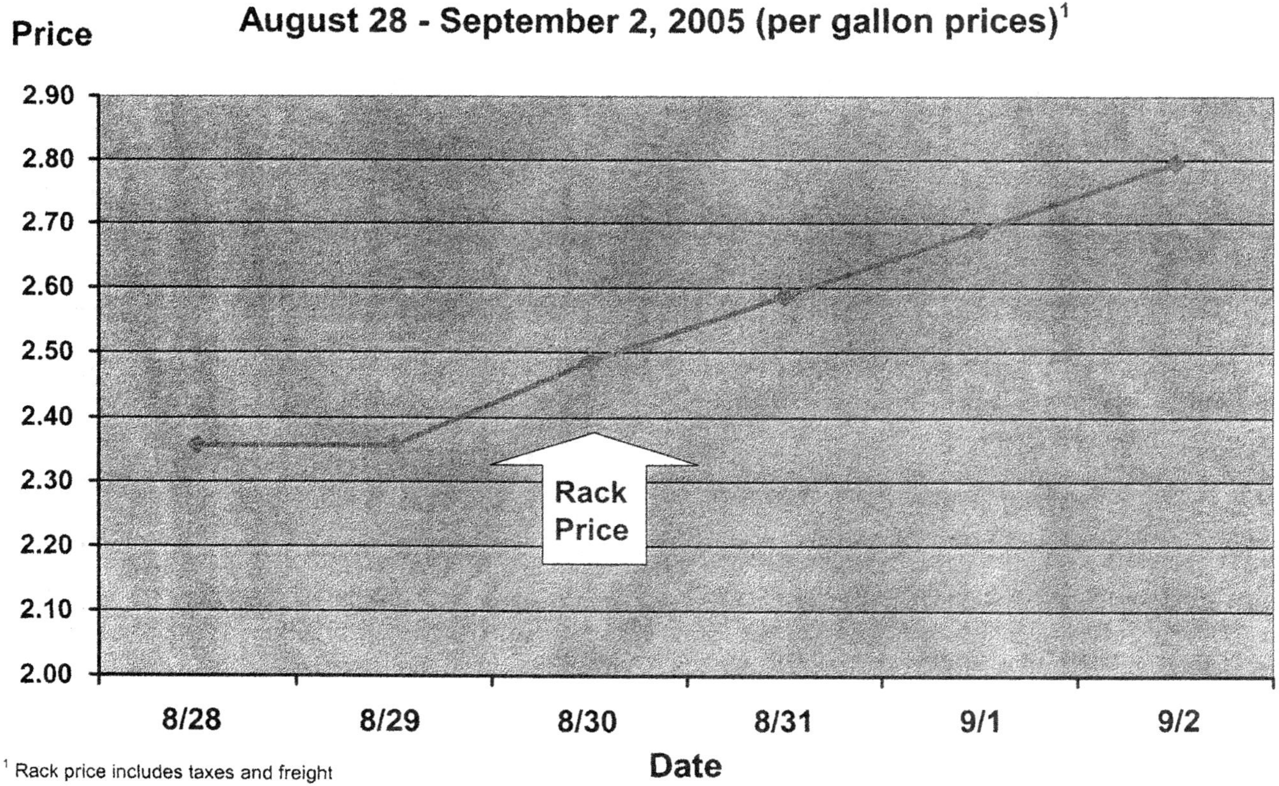

[1] Rack price includes taxes and freight

Chart 2- Douglass Distributing Wholesale/Retail Price Experience August 28 - September 2, 2005 (per gallon prices)[1]

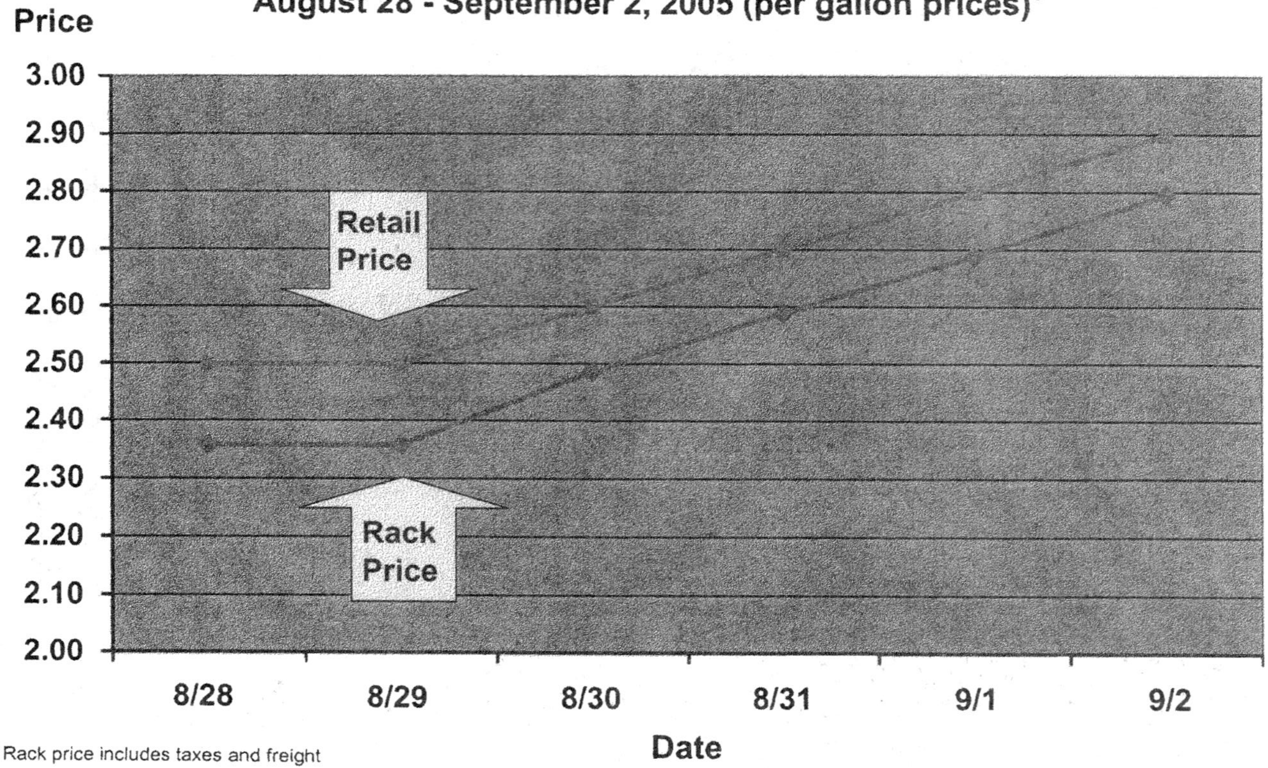

[1] Rack price includes taxes and freight

Chart 3- Wholesale Price Experience for Dallas/Fort Worth August 29 - September 2, 2005 (per gallon prices)[1]

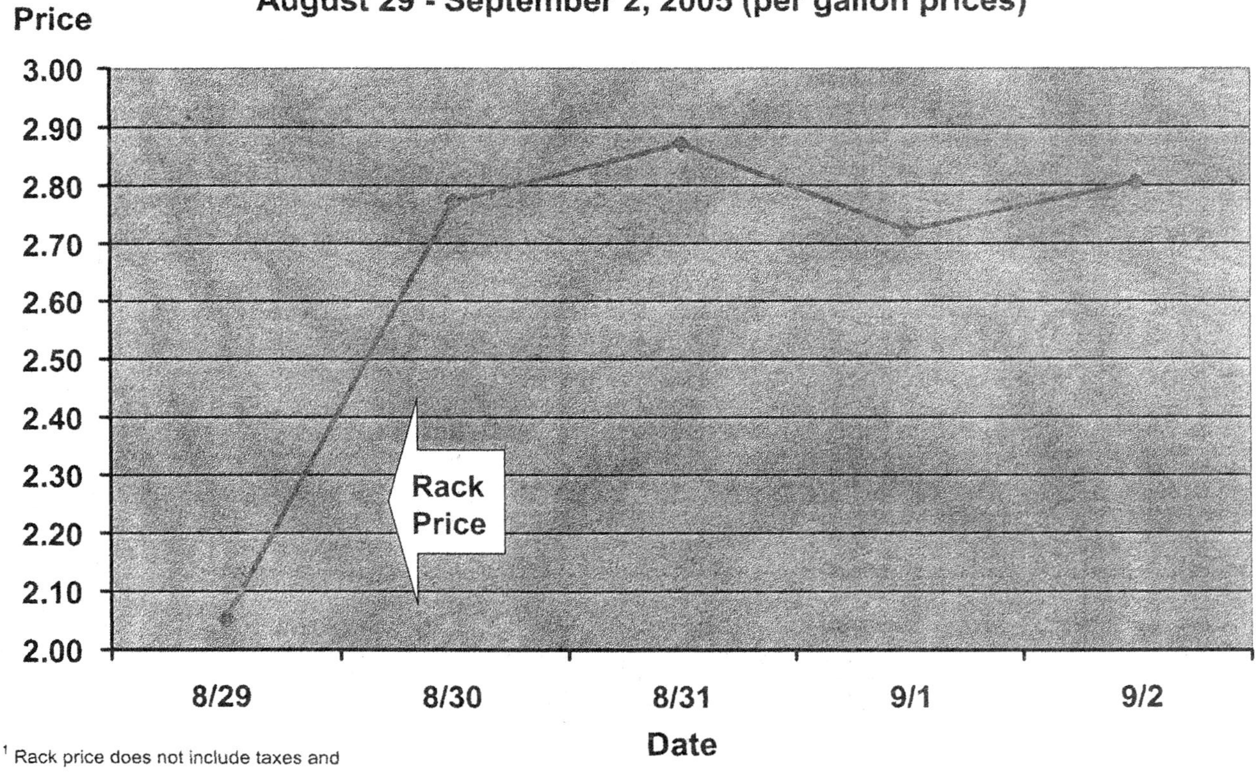

[1] Rack price does not include taxes and

CHART 4 - WHOLESALE PRICES INCREASES BY PADD AUGUST 26 - SEPTEMBER 2, 2005 (per gallon prices)[1]

PADD	Close September 2, 2005	Change from August 26, 2005
PADD I (East Coast)		
Branded	$2.5454	+55.08 Cents
Unbranded	$2.7919	+83.62 Cents
PADD II (Mid-West)		
Branded	$2.7919	+48.11 Cents
Unbranded	$2.7800	+72.55 Cents
PADD III (Gulf Coast)		
Branded	$2.4004	+43.80 Cents
Unbranded	$2.7760	+80.06 Cents
PADD IV (Rocky Mountains)		
Branded	$2.4259	+34.57 Cents
Unbranded	$2.4819	+37.60 Cents
PADD V (West Coast)		
Branded	$2.3551	+21.95 Cents
Unbranded	$2.7038	+48.25 Cents

[1] Excluding taxes and freight.

Chart 5- Wholesale Price Experience in Various Cities August 29 - September 2, 2005 (per gallon price)[1]

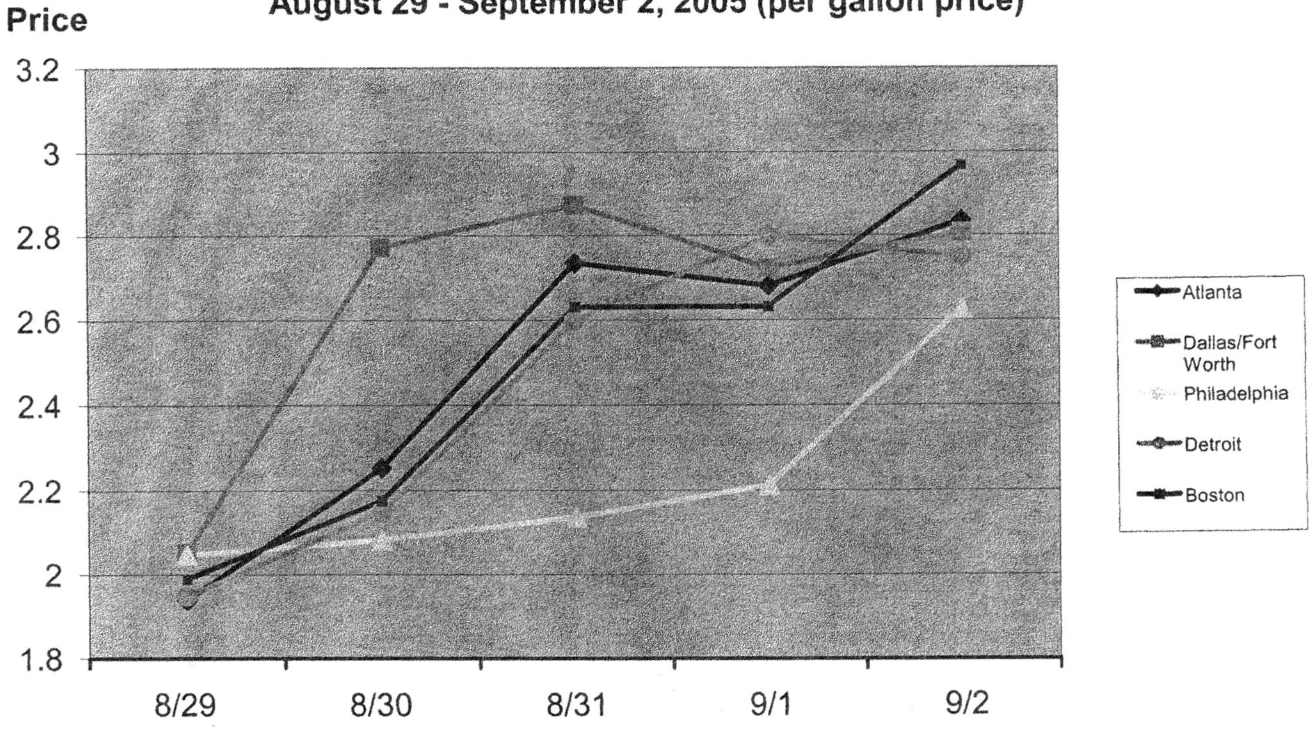

[1] Rack price does not include taxes and freight

Chart 6- Approximate Gross Revenue Received by Different Parties in 2003 & 2005 ($ per barrel)

Party	Average 2003 Revenue	Average 2005 Revenue
Royalty Owner	$4.00	$8.00
Crude Exploration and Extraction Company	$28.00	$67.00
Refiner	$11.00	$27.00
Retailer	$6.00	$6.00
Credit Card Company	$1.50	$3.00

Mrs. MYRICK [presiding]. Thank you, Mr. Douglass. Thank you very much. Mr. Smith. Welcome.

STATEMENT OF WILLIAM L. SMITH

Mr. SMITH. Thank you. My name is Bill Smith, and I am the Chief Technology Officer for BellSouth. The purpose of my testimony today is to address the impact of Hurricane Katrina on BellSouth's people and our network. I will describe the damage that Hurricane Katrina has caused, which has been unlike any other hurricane we have experienced, and I will give you a status of the restoration efforts.

Given the area that we serve, BellSouth has dealt with hurricanes for years; however, we rarely lost operational status of an entire central office. Katrina has been very different. We have lost service at some point during the storm in 24 BellSouth central offices in the impacted area. The majority of these central offices were in the New Orleans area that was flooded. These central offices failed due to flooding and logistical problems presented by the security in the area and the ability to get fuel to the emergency generators.

Our operations in Florida, Alabama, Mississippi, and Louisiana have all been impacted by Hurricane Katrina. In places like Gulfport and Biloxi and New Orleans, the impact on our customers, our employees, and our network have been catastrophic, and restoration efforts are still encumbered by flooding, debris, and security issues. In other areas of Louisiana and coastal Mississippi and Alabama, we are well under way in our restoration efforts and are progressing well. In Florida, we are actually wrapping up our restoration efforts and freeing up resources like generators and technicians to move into the areas that are still impacted.

Let me move to the impact on our people. BellSouth has approximately 13,000 employees in the States of Alabama, Mississippi, and Louisiana, and approximately 6,500 of those were in the impacted area. I am pleased to say that as of today we have located all but 65 of those employees and we are working very hard to find the others.

Immediately prior to Katrina, we took steps to ensure that supplies and services would be on hand. We knew that employees would be called upon to work around the clock, and, as Governor Barbour said this morning, many of our employees actually lost their own homes. So we established what we call BellSouth tent cities that we can actually house, shelter, feed our employees and their family, because we knew that we would be counting on them to operate to help us restore our network. And, in fact, we are currently operating six of these tent cities in the impacted area and serving over 8,000 meals a day to provide assistance for our employees and their families.

Let me move to the impact on our network. Our network operations team actually started tracking Hurricane Katrina as early as August 23. We began making preparations for landfall in south Florida as a Category 1 hurricane. That actually occurred on Thursday evening, August 25. Then we tracked Katrina as she became a Category 4, 5, and then made landfall as a Category 4 storm in New Orleans or just east of New Orleans at approximately 2 p.m. on Monday, August 29. There were reported wind speeds of over 145 miles an hour and the storm surge was reported as high as 25 feet.

I have this chart you can see that we have used to assess the impact on our network. We have categorized the damage in geographic areas as catastrophic indicated by red, severe indicated by yellow, or moderate indicated by green. We have restoration efforts well under way in the green areas and are moving into the yellow areas and the red areas as well. The unique problem that we have experienced with Katrina has been the severe flooding, particularly in New Orleans. It is not unusual in these situations for a central office to lose its commercial power and for BellSouth to continue to provide power using generators. These generators, however, require fuel and they require technician access to maintain them. With Katrina, the continued flooding and security issues severely hampered our ability to maintain our network as we would normally do.

Now, I will spare you a lot of the details, but suffice it to say that our experience in New Orleans' main central office at 840 Poydras Avenue was an example of what we saw. We actually had 82 people in that office working to man our equipment and our emergency operations center. And everything was fine until the flooding started after the hurricane. At that point and subsequently, we were advised that there was gunfire in the area, it was not safe to keep our employees there, so we actually made arrangements to evacuate those employees with heavily armed State police. We later got FBI and Federal Marshal protection back into the area to secure the central office, and had heavily armed convoys taking fuel and water back into the location. Obviously, that is not something that we normally see in normal hurricane restoration activities. I am happy to say, however, that office has maintained operational status throughout this period.

Nevertheless, with all these difficulties, we have made huge strides to restoring our network. As of yesterday morning we actually had 506,000 lines remained out of service, and that is less than one-fourth of the original number. We have restored service to all but 18 of those central offices that were impacted.

Now, let me move to what we can ask from the government for help. Overall, the cooperation and assistance from state, local, and Federal agencies has been very good. The FCC led by Chairman Martin has been extraordinarily helpful. They have reached out to offer assistance in many areas, in particular, waiving rules that would have hampered our ability to restore service in a quick manner. We will continue to need this kind of assistance. The Louisiana and Mississippi Public Service Commissions have also stepped up to provide assistance, as well as the Department of Homeland Security, the White House, and the Department of Defense, Northern Command. We have also had great cooperation from the FAA, the Bureau of Alcohol, Tobacco, and Firearms as well as the U.S. Marshal Service.

Right now we need three things. First, we need safe access to our facilities and adequate security for our personnel. Second, it will take many months for us to complete our repair work even though we will be working around the clock. We have engaged and restored 22 hurricanes since 1992, including storms such as Hugo, Andrew, and now Katrina. Congress and the private sector alike should be cautious about building unrealistic expectations about how quickly we can fully recover from the impact of this storm.

Third, the government needs to recognize that the cost to BellSouth to restore these communications infrastructure items will be significant. We have estimated those costs to be between $400 and $600 million. Now, to put this in perspective, the storms that we experienced last year, the four hurricanes, cost $200 million, and we are still in the middle of the hurricane season. So restoration of this infrastructure will require that we deploy capital not as a cost plus utility but in a very competitive industry. We will deploy this with other companies, depending on our facilities, despite the fact that we don't share the burden of this restoration between those companies.

Thank you.

[The prepared statement of William L. Smith follows:]

PREPARED STATEMENT OF WILLIAM L. SMITH, CHIEF TECHNOLOGY OFFICER, BELLSOUTH

I. INTRODUCTION

My name is Bill Smith, and I am the Chief Technology Officer for BellSouth. BellSouth is a full-service communications company providing service to customers in the nine southeastern states of Alabama, Florida, Georgia, Kentucky, Louisiana, Mississippi, North Carolina, South Carolina, and Tennessee. I have worked for BellSouth for 26 years, and in my current position I am responsible for overseeing the planning of our overall network, integrating new technology into our network, and ensuring the interoperability between our networks and those of other carriers.

The purpose of my testimony is to address the impact of hurricane Katrina on BellSouth's people and our network. I will describe for you the damage that Katrina has caused, which is unlike any hurricane BellSouth has experienced, and to give you the current status of our restoration efforts. What I will give you today is a snapshot—the situation changes literally every few minutes, as power is restored, flood waters recede, field surveys occur, and repairs are made. Furthermore, because we are still assessing the full impact of the storm on our network and our customers, our damage estimates are preliminary. It will take some time for us to know with certainty the total magnitude of the destruction caused by hurricane Katrina.

Given where our network is located, BellSouth has dealt with hurricanes for years. However, with most hurricanes, from Camille to Andrew, we rarely lost operational status for an entire central office. A central office is a building that houses the switching and transmission equipment for a geographic area; it is usually the first "building" that all of the wires coming from houses and offices go in connecting to BellSouth's regional network.

But Katrina was different. Based on what we know today, we lost service in 24 of BellSouth's central offices in the impacted area. Some of these offices were located in coastal Mississippi and were destroyed by the storm surge when Katrina came ashore. The vast majority of BellSouth's central offices that are currently out of service are located in greater New Orleans. These central offices failed due to flooding and logistical problems presented by security on the street. The flooding and security issues that BellSouth has had to confront is what makes Katrina different from other hurricanes—both in terms of impact to the network and on our ability to restore service.

Operations in Florida, Alabama, Mississippi and Louisiana, all have been impacted by hurricane Katrina. As I will describe in more detail below, we have 3 different types of restoration efforts underway. In places like Gulfport and Biloxi, Mississippi and New Orleans, the impact on our customers, our employees and our network have been catastrophic and basic restoration is still encumbered by flooding, debris and security issues. In other parts of Louisiana, coastal Mississippi and Alabama, we are well into our restoration efforts and progressing well. In Florida, we are wrapping up our restoration efforts, and freeing up resources like generators, and of course technicians, to move to the other areas where they are needed.

As is standard operating procedure for us during hurricane season, on August 23, BellSouth's network operations team began tracking Tropical Depression 12, then located over the southeastern Bahamas with 35mph winds, moving northwest at 10 mph. This is business as usual for us, but none could have imagined what was to follow. There are two integral pieces to this story: the network, and the people. I plan to first walk you through the people side of this story, because without our people, we would have no company and no network. It is our employees who make BellSouth what it is.

II. KATRINA'S IMPACT ON PEOPLE

BellSouth has about 13,000 employees in the states of Alabama, Mississippi, and Louisiana, approximately 6500 of whom were in the hardest hit areas affected by the storm. As of now, we have located or made contact with all but about 110 of those employees, and we are making every effort to locate these employees. Prior to Katrina, BellSouth already had in place an 800 number for BellSouth employees to call to report their status in the event of an emergency and a separate number employees could call to get emergency information. Immediately prior to Katrina, we also took steps to ensure adequate supplies and services were on hand, sending non-perishable food to strategic areas where employees could be stationed, setting up structural materials including tents, showers, toilets, tables, and chairs, and engaging janitorial and guard services. Our experience with prior hurricanes has taught us that our employees will be called upon to work round the clock, and they can best perform the extraordinary tasks expected of them if their basic needs for food, shelter and the safety of their family are addressed.

As Katrina hit the Gulf Coast on August 29th, we assessed potential locations for what we call BellSouth tent cities—stations where employees in affected areas could seek shelter, receive food, ice, water, showers, laundry services, air mattresses, linens and clothing, medical care and financial loans. In addition, we had on hand access to our employee assistance program to provide counseling services as needed. The first tent city was set up in Gulfport, Mississippi on August 30th, a second opened in Baton Rouge on September 1st, and a third on September 2nd in Covington, Louisiana. With the addition of tent cities in Hattiesburg and Jackson, Mississippi, and Kenner, Louisiana by the end of this week, BellSouth will be operating six tent cities that will serve over eight thousand meals daily, and provide assistance for our employees and their families, including medical care.

III. KATRINA'S IMPACT ON BELLSOUTH'S NETWORK—RESTORATION EFFORTS [1]

BellSouth has 1591 central office buildings across its region. 578 of these central office buildings are located in Alabama, Louisiana and Mississippi. As we do with every storm, our network operations team was tracking Katrina as early as August 23, and began making preparations for potential landfall in the Florida Gulf Coast area. Because our network equipment requires power to operate, our standard hurricane procedures include ensuring that generators are in working order and fuel tanks filled for our central offices and our key administration offices, closing shutters and sealing windows, sandbagging critical facilities, and making arrangements for additional generators where needed.

Despite these precautions, Katrina brought considerable damage to BellSouth's network. Katrina first made landfall in South Florida as a Category 1 hurricane on Thursday evening, August 25, and caused considerable damage to the area. Katrina, a Category 5 hurricane that dropped to a Category 4 just before landfall, made landfall in our operating area for the second time at approximately 2 p.m. on Monday, August 29, just east of New Orleans. In some areas, winds exceeded 145 miles per hour and the storm surge was reported as high as 25 feet. In assessing the impact on our network, we have categorized damage to geographic areas caused by Katrina as "catastrophic", "severe" or "moderate". The "catastrophic" areas are red on the map attached as Appendix 1; severe areas are yellow; and moderate areas are green. Much progress has already been made restoring service in the "moderate" areas of the region.

In the Gulf region of Mississippi, Alabama, and Louisiana, we had 4.9 million access lines prior to the storm. Of those 4.9 million lines, 1.6 million were in the red zone, 782,000 were yellow and 2.6 million were green. A snapshot on August 30 after the storm estimated that 2.475 million lines were actually affected. All 1.6 million lines in red zones were affected; 500,000 of the 782,000 lines in yellow zones

---

[1] Because restoration efforts in Florida are mostly complete, the network impacts in section III will focus on operations in Alabama, Mississippi, and Louisiana.

were affected; and 440,000 of the 2.6 million lines in green zones were affected. Details on a state by state basis are attached as Appendix 2.

The unique problem that BellSouth has experienced with Katrina is that the severe flooding, particularly in New Orleans, has made it difficult to get a clear assessment of the extent of the damage. Normal hurricanes have an initial surge, the water recedes, power begins restoration, and we follow power with sweeping telecom restoration resources. When the levees broke in New Orleans, the water did not recede. The flooding has greatly complicated our restoration efforts. In most hurricanes, it is not unusual for a central office to lose commercial power and for BellSouth to continue providing service using generators. Those generators require fuel, and we have to be able to get our network technicians to those central offices in order to ensure that the generators are fully fueled and operating correctly. With Katrina, we have 24 central offices that are without commercial power and are operating under generators. However, because of the continued flooding we have not been able to access or support many of these central offices in New Orleans as we would in normal hurricane restorations.

Our experience in the New Orleans Main Central Office at 840 Poydras Street gives a sense of the situation on the ground. BellSouth employees began staffing an Emergency Operations Center (EOC) on the 12th Floor of the building on Sunday, August 28. The office lost power and engaged generators when the storm hit on Monday, but occupants breathed a sigh of relief that there was no flooding. Then, the levy broke and conditions rapidly deteriorated on Tuesday. Technicians and engineers in the office were trying to re-establish service and maintain power by keeping the generators fueled and running. As the situation in New Orleans deteriorated with violence and looting, the New Orleans police and the Louisiana State Police told us to evacuate the building. There was gunfire in the area and it was therefore unsafe for our employees to remain in the area. At 3:00 p.m. CST, the Louisiana State Police arrived and provided us with an armed escort so we could leave the building. We moved to Baton Rouge and, concerned for the security of the building, we arranged for FBI agents to take occupancy of the building at approximately 9:00 that evening. By Friday morning, the Louisiana State Policy and the FBI occupied the building. At that time, we began armed and escorted caravans to the building to bring fuel for the generator, water for the coolers, and BellSouth personnel to maintain the equipment. We are not yet back to full support but have managed to keep this key switch operations despite the flooding and security concerns.

And another example heroic story rises out of the coastal town of Gulfport, MS. On September 3, a brick wall protecting the main generator keeping the central office alive started to give way. Nine workers from that central office ran from the basement, where they had been working while riding out the storm, to the rooftop room and fortified the walls with whatever they could find—plastic tarps, plywood and even the cardboard from a science project of one worker's son.

Nevertheless, we have made huge strides towards restoration of communications capabilities. As of the morning of Tuesday, September 6, 506,000, less than one-fourth of the original 2.475 million lines, remained impacted. Only 4,900 of the 440,000 lines remained impacted in the green areas; 23,200 of the 500,000 lines in the yellow zone remained impacted; and 478,500 of the 1.6 million lines in the red zones remained impacted.

As of September 6, we have restored service to all but 20 central offices. This restoration is due to the tireless efforts of our employees on the ground who are working around the clock with a single minded mission of restoring communications to these hard hit areas, and to the efforts of our wireless and wireline industry colleagues who have partnered with us with an unwavering commitment to enable communications.

Of course, other carriers rely upon BellSouth's network in order to provide service to their customers. We have an Emergency Control Center in Atlanta, which is the control center from which we are coordinating our hurricane response, overseeing network restoration efforts, and working with other carriers to restore communications. We are coordinating a contingent of impacted carriers with one mission—to make communications work. We collaborated with other carriers, without regard to ownership or jurisdiction, and brokered capacity and worked through technical issues. We conduct two daily calls—one with wireless carriers and one with wireline carriers. Using our network data and resources we assisted in developing a joint wireless restoration plan, now underway, bringing competitive service providers together to serve a single goal of restoring communications. A joint industry team has agreed on a list of prioritized sites and are working together to restore wireless service to these sites in the New Orleans area. This has included traveling by boat to several of the sites in order to survey what equipment is needed to restore service. They traveled by boat to survey sites on Sunday and Monday, and may have already enabled communications from some of these towers while I am speaking with you. This has been a remarkable collaborative effort.

In terms of restoration priority, we have been and continue to focus our support on public safety concerns, including hospitals, E-911 centers and law enforcement. Following the storm, in Florida and Alabama, there were no E-911 centers that incurred outages. For Mississippi, service was impacted to 43 E-911 centers, and service to all 43 centers has been restored on site or by re-routing the calls to other centers. In Louisiana, 35 E-911 centers were impacted, and 28 of these are back in service, either partially on site or through re-routing of calls to other centers. Seven of the centers are out of service. Of the 7 E 911 centers that remain out of service, all are located in the New Orleans area and are served by the Franklin tandem, which is flooded. Four of the centers are located in Plaquemine and St. Bernard parishes, low lying parishes along the Mississippi River below New Orleans which were in the eye of the storm as it came inland.

We are continuing to work around the clock to restore service. Our restoration efforts involve surveying the damaged area. That activity is approximately 80% complete. Next we concentrate on restoration of highest priority circuits, specifically those which support public safety including hospitals, E911 centers and law enforcement. Then we focus on supporting other carriers, including the wireless industry.

IV. NEEDS FROM GOVERNMENT

What can the government do to help? The cooperation and assistance from local, state and federal agencies overall has been good. The FCC, led by Chairman Martin, has been extraordinarily helpful. The FCC has reached out to offer assistance in many areas: waiving rules that will help customers who are without service; taking actions that have and will allow for the quick restoration of network facilities (including the emergency routing of traffic over whatever facilities are available for use); and helping with the publication of "find me" numbers to help locate BellSouth employees. We will continue to need this type of help, particularly related to the efforts to restore communications in Louisiana and the Mississippi Gulf Coast areas. The magnitude of the damage will present unique issues that will need to be resolved quickly and efficiently in order to restore service.

The Louisiana and Mississippi Public Service Commissions have also stepped up to provide assistance to the industry in efforts to assess damage, maintain the operation of the remaining network, and restore service to impacted areas.

BellSouth has been extremely engaged with the Department of Homeland Security's Infrastructure Protection Directorate, headed by Bob Stephan, and most specifically, DHS's National Coordinating Center (NCC). The NCC, which is a division of the National Communications System of DHS, provides a focal point for industry and the Federal government to share operational information and coordinate needs to respond to crises just like this. BellSouth maintains an office at the NCC headquarters, along with many other major wireline, wireless, and satellite providers.

Our representatives there work around the clock to facilitate response efforts for FEMA, DHS, the National Guard, State Emergency Management Agencies and Operations Centers, NORTHCOM, and many other organizations. On industry's behalf, the NCC works through a myriad of concerns, with security and fuel at the top of the list, along with abatement of the flood waters. Industry has also worked together to coordinate fuel convoys, search and rescue, network impacts, and logistics. It's been an outstanding example of the public-private partnership in action.

Through the NCC, and through direct sources, BellSouth has been in communication with the Department of Energy and the Nuclear Energy Regulatory Commission (NERC), which has provided status information on power and electricity. The White House Executive Office of the President has also been strongly supportive in responding to specific issues that required support. We have had good coordination and information from the FAA and DHS on aviation needs. The Department of Defense's Northern Command has also been very helpful, providing information, support, and logistics as well. BellSouth has also had outstanding security support from the Bureau of Alcohol, Tobacco and Firearms, as well as the US Marshal Service, which were coordinated through FEMA. Keith Hennessey—Deputy Director of national Economic Council at the White House Executive Office of the President, helped BellSouth get the employee hotline number that I described earlier publicized on Fox, CNN, and MSNBC Cable networks, as well as Direct TV satellite network.

Right now, we need several things. First, we need safe access to our network facilities. This will require the abatement of the flooding in New Orleans, which I understand is underway. Once the flood waters have receded, we need adequate security measures to ensure the safety of our technicians trying to assess and conduct repairs.

Second, it will take many months for BellSouth to completely repair the damage caused by Katrina. We will continue to work around the clock to restore service to our customers as they have rebuilt and are ready to be served. BellSouth has engaged and restored 22 hurricanes since 1992, storms such as Andrew, Hugo and now Katrina. Congress and the private sector alike should be cautious about building unrealistic expectations about how long it takes to fully recover from a storm packing the furor of a Katrina.

Third, the government needs to recognize that the cost to BellSouth to restore the communications infrastructure will be significant. BellSouth has estimated that the cost to restore our network as a result of hurricane Katrina will be between $400 and $600 million. By comparison, the cost to BellSouth of the damage caused by the four hurricanes that hit Florida last year was approximately $200 million. And, of course, we're still in the middle of the hurricane season, and the long term impacts of the flooding in New Orleans are hard to estimate.

Restoration of our near ubiquitous infrastructure will demand that we deploy capital, not as a cost plus utility, but as a company in a very competitive industry. We will be expected to rebuild without knowing what our ultimate demand will be. And, we will rebuild this network in an environment where many companies depend on our network for providing service to their customers, but policy doesn't equally distribute the burden of restoration among all players. The FCC has been very helpful in waiving rules that hamper restoration. We will need continued focus from the policy community on rules and regulations that hamper access to capital. Timely restoration requires that we spend this money now, well in advance of knowing what people and businesses will actually return to affected areas, and when.

Mrs. MYRICK. Well, thank you for what you have done. And also, please thank your employees for their commitment as well. It is important to all of us, and I know it means a great deal.

Mr. Lashof.

STATEMENT OF DANIEL A. LASHOF

Mr. LASHOF. Thank you very much, Madam Chairwoman, and members of the committee, I appreciate the opportunity to participate in today's hearing. And first let me add my voice to those of the other witnesses and members of the committee in expressing deep sympathy for the victims of Hurricane Katrina and great appreciation for the emergency responders who are on the ground struggling to restore service and provide emergency services.

While Katrina has produced a horrendous catastrophe along the gulf coast, its impact has also rippled across the country. And, for many Americans, that has been most evident in the price of gasoline. Both immediate and long-term responses are needed to address the fundamental vulnerability that Hurricane Katrina has revealed. But let me start with the Hippocratic Oath: In searching for the right responses, let us first do no harm. Let us be sure that we avoid counterproductive actions that don't actually address the real problems, and would needlessly expose people to higher pollution levels and harm to the environment.

First let me address fuel standards which we have discussed here. Certainly EPA's prompt action to temporarily waive certain fuel requirements has ensured that these standards are not responsible for the increases in gasoline prices that consumers have seen during the last week. It also shows that EPA has the statutory authority that it needs to respond to supply disruptions and other emergencies. No permanent changes in the Clean Air Act can be justified based on the aftermath of Katrina, and responsible policy and law require the clean air waivers should not be extended any longer than necessary to respond to the immediate supply disruptions. If Congress does wish to reduce the number of fuel specifications to provide additional flexibility in the market, it should harmonize these standards upwards by making it easier for States and regions to opt in to Federal formulated gasoline programs and protect their citizens with clean air.

Turning to refinery capacity, similarly, while it may be desirable to increase refinery capacity, particularly outside the gulf region, there is no justification for relaxing environmental requirements in order to site new refineries. There simply is no credible evidence that environmental requirements have played a significant role in the economic decisions that refiners have made to consolidate and to reduce spare capacity. We have heard other testimony that it is primarily an economic driver with respect to margins. In fact, in response to an inquiry from the ranking member of this committee, EPA has said that there are no pending applications to restart any of the refineries that have closed since 1980. And with respect to new refineries, the only application that I am aware of, which is the Yuma facility which has been mentioned here in Arizona, it has actually already received an air permit which was granted less than 1 year after a complete application was submitted.

Now, let me turn to the Arctic Refuge, where we have heard, since Katrina, renewed cause to open the Arctic National Wildlife Refuge to oil exploration and production. And these are also impossible to justify based on the short-term supply disruption caused by Katrina. Even if you take EIA's optimistic estimates of potential annual production from the Arctic Refuge, drilling would affect by their estimate gasoline prices by less than 1.5 cents per gallon and then not until 2025. As shown in my exhibit, oil from the Arctic Refuge would be a drop in the bucket—it is the red curve there just hugging the bottom—compared with the oil demand reduction we could achieve with a national commitment to oil savings which I will address in a minute.

We need a national commitment to reduce our dependence on oil because currently our dependence is very dangerous to our security and economy. With only 3 percent of the world's oil reserves and 25 percent of the world's oil demand, there is no way for the United States to drill its way into energy security. The only effective way to reduce our vulnerability to oil price shocks is to significantly reduce our dependence on oil.

This is true for family budgets as well as for the national energy security. For example, for an average family driving 2,500 miles a month, a $1 gallon run-up in gasoline prices as we have seen in recent weeks takes $120 out of their monthly budget if they are driving vehicles that average 21 miles per gallon; but it would only take $60 out of their budget if those vehicles average 42 miles per gallon, which is within our technical capability.

Turning to short-term action, to respond to the short-term disruption in oil, I believe the President should call on the Nation to conserve gasoline. And I don't think it is enough for him to simply say people shouldn't buy gasoline if they don't need it. The President should specifically ask for consumers to pitch in by taking five immediate and relatively simple steps that would save 7 percent of our gasoline demand: First, by keeping tires inflated; second, by sticking to the speed limit; third, by reducing engine idling; fourth, by using car pools, transit, and telecommuting; and, fifth, by keeping cars tuned and using efficient engine oils. These are sensible steps that all Americans can take that can help us all through this short-term problem.

To reduce our vulnerability and increase our security in the future, a broad coalition called Set America Free, which involves national security organizations, religious leaders, and energy experts, calls on Congress to establish a minimum national commitment to reduce our oil commitments by saving 2.5 million barrels of oil a day within a decade and 10 million barrels of oil a day by 2025. We can achieve that with a combination of diversifying our supplies away from petroleum, using biofuels that Mr. Shimkus has mentioned earlier in the hearing, as well as American know-how and technology to make sure that every gallon of fuel that we use is used with the utmost efficiency. By doing that, we could save more than 15 times as much as the production from the Arctic Refuge could potentially deliver cumulatively over the next 20 years.

Equally important, in contrast to oil savings, Arctic Refuge drilling would do nothing to insulate our economy from the effects of future oil supply disruptions, because those would ripple through the economy and affect the price that everyone pays regardless of how much crude oil we import or how much comes from domestically because we have national and global markets.

So, in conclusion, Mr. Chairman, I believe that there are short-term measures that we should call on all Americans to take to pitch in to help us through the immediate supply disruption. In the longer term, we need a real commitment to oil savings, and we should move forward with approaches that really respond to the problems we have and not use the short-term crisis to justify permanent changes that are inappropriate.

Thank you.

[The prepared statement of Daniel A. Lashof follows:]

PREPARED STATEMENT OF DANIEL A. LASHOF, SENIOR SCIENTIST, NATURAL RESOURCES DEFENSE COUNCIL

INTRODUCTION

Thank you Mr. Chairman. My name is Daniel Lashof and I am a senior scientist at the Natural Resources Defense Council. I appreciate the invitation to participate in today's hearing.

Mr. Chairman, it is now clear that hurricane Katrina is among the worst natural disasters in American history. My deepest sympathy goes to the victims and their families and my deepest respect goes to the emergency workers who are struggling to provide relief in almost unimaginable conditions.

While Katrina produced a horrendous catastrophe along the Gulf Coast its impact has also rippled across the country. For many Americans this is most evident in the price of gasoline. For some of us this is an annoyance that means that our Labor Day trip to the beach was a little more expensive than we had anticipated. But for millions of low-income Americans higher energy costs have thrown carefully balanced family budgets out of whack, creating real hardship.

With tempers running short as some motorists have watched the price of gasoline increase as they were waiting in line to fill up, it is natural to look for someone to blame. I urge that we resist the temptation to offer simplistic explanations or simplistic solutions. Where there is evidence of price gouging it should be investigated and prosecuted to the full extent of the law. But we also need both immediate and long-term responses that address the fundamental vulnerability that hurricane Katrina revealed.

FIRST, DO NO HARM

Some argue that America should open its wild lands for oil exploration and drilling or relax environmental safeguards to reduce gasoline prices and U.S. dependence on imported oil. But these are inappropriate, wasteful, and ineffective responses to the aftermath of Katrina.

EPA's prompt action to temporarily waive certain clean fuels requirements has ensured that these standards are playing no role in the gasoline price increases that consumers have seen during the last week. EPA's action also demonstrates that current law already provides the necessary authority to respond to short-term supply disruptions. No permanent changes to clean air laws can be justified based on the aftermath of Katrina, and responsible policy and the law require that clean air wavers should be extended no longer than necessary to respond to the actual supply disruption. If Congress wants to reduce the number of different fuel specifications it should make it easier for states and regions to adopt the federal reformulated gasoline program, and not lock in the use of dirtier conventional fuels.

Some have cited a decline in the number of refineries operating in the United States as evidence that environmental regulations have discouraged investment in new capacity, driving up gasoline prices. The facts do not support this claim, however. While the total number of refineries has declined, total capacity has increased as refiners have found it to be more cost effective to expand capacity at existing facilities than to operate small refineries or build new green field plants. Refiners have also consciously sought to reduce excess capacity to improve refinery margins. Environmental permitting has not played a significant role in these decisions. In response to an inquiry from the Ranking Member of the Committee, EPA has said that there are no pending environmental permit applications from any of the U.S. refineries that closed since 1980.[1] With regard to new refiners, the record shows that in the case of the proposed facility in Yuma, Arizona, an air quality installation and operating permit was granted by the Arizona Department of Environmental Quality less than a year after a complete application was received.[2]

Similarly, renewed calls to open the Arctic National Wildlife Refuge to oil exploration and production are also impossible to justify based on the short-term supply disruption caused by Katrina. Although drilling advocates claim there is potentially 16 billion barrels of oil in the Arctic National Wildlife Refuge, this figure is an upper bound estimate (one-in-twenty chance) for the amount of oil that is potentially recoverable, regardless of extraction costs. Using a price-adjusted mean estimate (which better represents the basis for production decisions regarding potential future discoveries), the actual amount of oil that is economically extractable would be far less. Investment decisions would be made based on expectations of long-term average prices, which are far lower than current peaks. For example, at $40 per barrel the economically recoverable total would be about 6.7 billion barrels. Moreover, it would take 10 years for any oil from the Arctic Refuge to reach the market. Even during the predicted production peak in 2027, the coastal plain would produce about 3 percent of America's daily oil demand.[3] Even with EIA's optimistic estimate of potential annual production from the Arctic Refuge, which is much higher than can be justified by actual experience with North Slope fields, drilling would affect gasoline prices by less than 1.5 cents per gallon in 2025.[4]

A national commitment to oil savings could yield more than 15 times as much as production from the Arctic Refuge cumulatively over the next 20 years (see exhibit). Equally important, in contrast to oil savings, Arctic Refuge drilling would do nothing to insulate our economy from the effects of future oil supply disruptions, which would ripple through the oil market and affect the price of domestic and imported crude equally.

DANGEROUS DEPENDENCE

Our fundamental vulnerability is rooted in America's dangerous dependence on oil. Thirty years after the first Arab Oil Embargo our transportation sector remains

[1] Letter from Charles Ingebretson, EPA Associate Administrator, to Congressman Dingell, dated September 29, 2004.

[2] The permit was granted on April 14, 2005. Letter from Nancy Wrona, Director Air Quality Devision, Arizona Department of Environmental Quality, to Jeff Donofrio, Committee on Energy and Commerce Democratic Staff, dated July 29, 2004 shows that the complete application was received on July 14, 2004.

[3] Arctic National Wildlife Refuge production analysis conducted by Richard A. Fineberg (Principal Investigator, Research Associates), January 2005.

[4] U.S.DOE/EIA. Impacts of Modeled Provisions of H.R.6 EH. h. EIA estimates that allowing drilling in the Arctic Refuge will reduce world oil prices by $0.57 per barrel in 2025. Assuming a one-to-one impact on gasoline prices, this translates into $0.57/42 = $0.014 per gallon.

97 percent dependent on oil; imports account for over half of our supply; and our vehicle fleet remains woefully inefficient. In fact, after increasing from 13.1 to 22.1 miles per gallon between 1975 and 1987 the average fuel efficiency of new personal vehicles has actually declined to 21 miles per gallon in 2005, according to the latest government report.[5]

As a result of rising global demand, particularly in the United States and China, and unrest in the Middle East and other major oil producing areas, oil markets were already tight before Katrina struck. Refinery acquisition costs for crude oil had more than doubled from $24 per barrel in 2002 to almost $53 per barrel in July 2005.[6] China's 32 percent, or 1.6 million barrel per day, increase in oil consumption between 2001 and 2004 was the largest single factor increasing global demand, but the United States was not far behind. Although U.S. consumption grew by only 5.5 percent over this period, that represented more than a 1 million barrel per day increase due to our much larger consumption base.[7]

With only 3 percent of the world's oil reserves and 25 percent of the world's oil demand, there is no way for the United States to drill its way to energy security. The only effective way to reduce our vulnerability to oil price shocks is to significantly reduce our dependence on oil. For example, if the fuel efficiency of our personal vehicle fleet was 42 miles per gallon today, rather than 21 miles per gallon, U.S. oil demand would be lower by 4 million barrels per day, oil markets would have spare capacity, and the impact of any gasoline price spike would be far smaller. For an average family driving 2500 miles in a month, a $1/gallon run up in gasoline prices takes $120 out of their monthly budget at 21 miles per gallon, but only $60 at 42 miles per gallon.

Unfortunately, neither the energy bill enacted last month nor the fuel economy standards proposed on August 23rd will achieve substantial oil savings.

The United States needs to make a national commitment to reduce our oil dependence, through both immediate conservation measures and through investments that increase our efficiency and diversify our sources of fuel.

IMMEDIATE CONSERVATION MEASURES

During the Second World War, Americans met our nation's energy challenges with an unprecedented spirit of conservation, using every gallon of gasoline wisely. Californians showed again during the electricity crisis in 2001 that the conservation spirit is alive and well today, responding by cutting their power demand by 10 percent without any draconian measures.

The President should announce a "National Emergency Gasoline Conservation Program" to respond to the short-term supply disruption caused by Katrina. There are five simple steps American consumers and businesses could begin taking immediately to reduce gasoline consumption. These steps could cut gasoline consumption by several percent, helping to relieve gasoline shortages, save money, and cut pollution at the same time.

In contrast to drilling in the Arctic National Wildlife Refuge, which would not begin to produce oil for many years, these measures would yield immediate benefits.

1. Check tire pressure.

- More than a quarter of all cars and nearly one-third of all SUVs, vans, and pickups are driven with tires at least 8 pounds below their proper levels, according to a new survey by the Department of Transportation.
- If all Americans kept their tires properly inflated, our nation would cut its gasoline use by 2 percent.
- Maintaining the correct tire pressure also would save lives. Under-inflated tires are more prone to tread separation and blowouts, which can cause fatal accidents.
- Congress should help by authorizing the president to require all service stations to offer free air and to post prominent signs and stickers that say, "Check your tire pressure every time you fill up—For your safety and America's energy security."

[5] Light-Duty Automotive Technology and Fuel Economy Trends: 1975 Through 2005. EPA420-R-05-001. July 2005.

[6] U.S. Department of Energy, Energy Information Administration. http://www.eia.doe/pub/oil_gas/petroleum/data_publications/petroleum_marketing_monthly/current/txt/tables01.txt Accessed September 2, 2005.

[7] U.S. Department of Energy, Energy Information Administration. http://www.eia.doe.gov/emeu/ipsr/t24.xls Accessed September 2, 2005.

2. Obey the speed limit.

- Slowing down from 75 to 65 miles per hour would reduce highway gasoline consumption by about 10 percent.
- If Americans followed the speed limit on our nation's highways, we would cut total national gasoline use by about 2 percent.
- Slowing down also would save lives.
- Congress should provide extra funding for states that strictly enforce speed limits and post signs that encourage slower driving: *"Drive 65—for your safety and America's energy security"*

3. Turn off the car engine while waiting in line.

- Americans who run their engines while they are parked or waiting in line waste as much as 4 million gallons of gasoline every day, according to the U.S. Department of Energy.
- Drivers cannot avoid idling in traffic jams, but they should turn off their engines while parked or waiting at drive-in windows. If the wait is longer than 30 seconds, starting up a car again uses less gasoline than leaving it running.
- If drivers turned off their engines while parked or waiting in line, we would cut national gasoline use by about 1 percent.
- Congress should help by authorizing the president to require parking lots, banks, fast-food restaurants, and other drive-through stores to post signs stating: *"Turn off your engine while you wait—for cleaner air and America's energy security"*

4. Use car pools and public transit, and telecommute.

- If each commuter car carried just one more passenger once a week, we would cut gasoline consumption by about 2 percent. That would translate into big savings for the average American worker. Someone with a daily commute of 10 miles each way and a 20- mpg vehicle would save 236 gallons of fuel per year by opting to carpool, telecommute or use transit, according to the American Public Transportation Association.
- A study in Minneapolis-St. Paul found that more than one in 10 employees shifted from driving to some other way of commuting when offered tax-free commuter benefits equal to those provided in the form of free parking.
- Congress should promote commuter choice with a tax-free benefit for employees who car-pool, use transit, bike to work, or telecommute (currently limited to $100) equal to that provided in the form of free parking (currently limited to $175). The federal government also should support and promote Web sites that help commuters find drivers traveling similar routes at similar times. Posters at workplaces could say: *"Car pool or ride the bus—for America's energy security'"*

5. Keep cars tuned and use fuel-efficient engine oil.

- A poorly tuned or poorly maintained engine can increase gasoline consumption by as much as 10 to 20 percent.
- Following the recommended maintenance schedule in your owner's manual will save drivers fuel and cars will run better and last longer.
- Motor oils with additives that reduce friction may increase a vehicle's fuel economy by 3 percent or more. Fuel-efficient oils are marked with an "Energy Conserving" label by the American Petroleum Institute (API).
- Congress should authorize the president to require service stations to post prominent signs trumpeting the benefits of keeping cars tuned and using fuel-efficient oil. Signs could say: "Keep your car tuned to save gas for America's energy security" and *"Use fuel-efficient motor oil to save gas for America's energy security"*

A NATIONAL COMMITMENT TO REDUCE OIL DEPENDENCE

To reduce America's vulnerability to future oil supply disruptions, whether from natural disasters, war, or terrorist attacks, we need to make a national commitment to invest in reducing our dependence on oil.

While there are many views of the energy bill enacted last month, everyone agrees that it does not represent such a commitment. In fact, the administration strongly opposed the Senate-passed measure that would have required the president to develop and implement a plan save at least 1 million barrels per day of oil and this critical proposal was not included in the final bill. Yet the conference report retained a provision that effectively lowers fuel economy standards by extending a loophole that allows automakers to claim credit for producing "dual fuel" vehicles, boosting their fuel economy numbers on paper by as much as 1.2 miles per gallon, even though these vehicles use gasoline more than 99% of the time.[8] While biofuels have great potential to reduce our oil dependence, rather than promote use of alternative fuels this provision will increases gasoline consumption by 15 billion gallons over the life of its 10-year extension. Wasting 5 billion gallons of gasoline more than the estimated fuel savings from the administration's proposed light truck fuel economy standards.

The fuel economy standards proposed by the administration on August 23rd miss a critical opportunity to seriously address America's oil dependence. Despite record oil prices and mounting instability in oil producing countries such as Iraq and Iran, the new administration plan actually calls for a slower increase in light truck standards than the modest 1.5 mpg increase adopted by the administration in 2003 when oil was selling for less than $30 a barrel. The proposal also exempts the heaviest SUVs and pickup trucks that weigh over 8500 pounds, such as the Hummer H2 and Ford Excursion, and does not address the car standard, which hasn't been updated in nearly 20 years. As an example of how out of touch this proposal is, its benefits were calculated assuming that the average price of gasoline over the next 25 years would be less than $1.60 per gallon.

Technologies and fuels exist today that can reduce wasteful use of oil in vehicles, industry, aviation, and buildings, delivering savings of at least 3.2 million barrels of oil per day (mbd) by 2015. By 2025 we could save at least 11.2 mbd, cutting our demand in half. We can reach these goals while enhancing the competitiveness of U.S. automakers and farmers by combining efficiency standards with incentives to retool factories, accelerate the production of gasoline-efficient vehicles, and deliver alternative fuels to consumers. Because our economy and national security are tied to America's dependence on oil, smart energy policies that deliver near term results would reduce America's vulnerability, stimulate our domestic economy, and help keep our nation safe

The Set America Free coalition has brought together national security and religious leaders, as well as energy experts, in calling on Congress to take immediate action and establish a national commitment to save 2.5 million barrels per day by 2015—as much as we currently import from the Persian Gulf—and at least 10 million barrels per day by 2025.

Saving oil requires mobilizing American ingenuity, factories, and farms around a clear goal. The first, most critical, step is for Congress to establish a national commitment to cut oil expenses and reinvest the resources—otherwise sent to oil producing countries—in American factories and farms. During World War II, American factories converted in just months from building cars to building tankers and bombers that became the arsenal of democracy. And after the first oil crisis in the early 1970s, America cut its oil demand to keep our economy strong. Although some may doubt the ability to turn this ship around, history shows us that American efficiency and ingenuity can meet the challenge. Given technologies and fuel available today we know that saving 2.5 mbd by 2015 and at least 10 mbd by 2025 is an achievable, practical goal that would deliver near term benefits in the next 5 to 15 years, while also starting the United States on a new path toward significantly greater energy independence and security thereafter. An analysis of how these savings can be achieved is attached to my testimony.[9]

Failure to take these steps would perpetuate unacceptable risks for our economic and national security, American jobs, and consumers. Rising oil prices have placed a devastating and disproportionate burden on U.S. automakers, according to a report released last month by NRDC and the University of Michigan. Without serious action to improve fuel economy performance, Detroit automakers will continue to lose thousands of jobs and millions in earnings, leaving them at a sharp disadvantage to their Japanese competitors. This report is also attached to my testimony.[10] Rather than exporting billions of dollars more to oil regimes with every rise in the prices of oil, the United Sates should be investing those dollars at home to support domestic industries and jobs, and leading the world in reducing global demand for oil.

CONCLUSION

Katrina has highlighted the vulnerability of our energy system due to our dangerous dependence on petroleum to fuel our transportation system. The best way to

[8] Department of Transportation. *Effects of the Alternative Motor Fuels Act CAFE Incentives Policy*. Report to Congress. March 2002.

[9] Bordetsky, A. et al., *Securing America: Solving Our Oil Dependence Through Innovation*. NRDC and IAGS, 2005. http://www.nrdc.org/air/transportation/oilsecurity/plan.pdf

[10] McManus, W. et al., *In the Tank: How Oil Prices Threaten Automakers' Profits and Job*. NRDC and OSAT, July 2005. http://www.nrdc.org/air/transportation/inthetank/contents.asp

reduce our vulnerability—both immediately and in the longer term—is to reduce demand by becoming more efficient with every barrel of oil we use and to diversify our supply by relying more on homegrown biofuels. A national commitment to saving oil is long overdue. If we make the commitment now America's oil dependence could be reduced by 2.5 million barrels per day by 2015 and by at least 10 million barrels per day by 2025. Meeting such a commitment will reduce our vulnerability to catastrophes like Katrina, protect the environment, and make us more secure.

Chairman BARTON. We thank you. Last but not least, a good friend to the committee who has testified numerous times, Mr. Mark Cooper, the Research Director for the Consumer Federation of America. You are recognized for 7 minutes, sir.

STATEMENT OF MARK N. COOPER

Mr. MARK COOPER. Thank you, Mr. Chairman. There is a real benefit to going last today, and you will see why.

Chairman BARTON. You know what everyone said, for one thing.

Mr. MARK COOPER. That is right, and I have gotten some interesting numbers today. Again, let me stress as all of us have, we have had a catastrophe of immense proportions, human proportions, physical proportions, economic proportions. And it is extremely important I think to act very, very quickly and seize this moment to reorient the way we address these fundamental problems. And as my testimony points out, we have been saying the same thing for 4 years. Maybe folks will start listening.

Public policy cannot prevent accidents or catastrophic acts of nature, but it can build systems that are resilient, robust, and flexible to minimize the impact of those inevitable accidents on our society. We believe it is quite clear that the business practices and public policy in the oil industry have combined to create a gasoline sector that has difficulty coping with even minor events, not to mention the disaster that struck the gulf coast. If the measure of performance of an economic sector is adequate supplies at stable prices, then this industry and the public policy under which it operates has failed, not just in the wake of Katrina, but also repeatedly over the past 5 years.

The bulk of my testimony today presents word for word the policy conclusions that we reached in a report over 4 years ago. There is no Monday morning quarterbacking here. After analyzing the industry structure conduct and performance, we urged policymakers to move aggressively in five areas. And we gave specifics. We said: Restore reserve margins by developing efficiency. The first and most important thing is to get the fuel efficiency of our fleet increased. The pitiful proposal of increase that we got last month was based on a price of gasoline of $1.80 a gallon, including taxes, in 2012. That number is economically irresponsible and socially irrational. They should have done the analysis with a much higher price, with the social value of gasoline, and we would have asked for much more efficiency.

Second, we ask for expansion of refinery capacity by redeveloping the 50 sites that had been closed. And as you have heard today, not one of them has applied for reopening. They were closed for business reasons, not environmental reasons. They should have been reopened. We wanted a list, we wanted a study. Those are sites that you can't hide behind the environmental laws for.

Second, we call for increasing flexibility in storage and stock policy. We talked about strategic product reserves. Mr. Chairman, our friends in the IEA who are giving us product are drawing down their own strategic product reserves to help us. We don't have a strategic product reserve. We should have one. We should also have a policy requiring storage. This is not an industry that can function without storage. In the electric utility industry we have reserve margin requirements of 15 to 20 percent above peak precisely because they are high capital, high intensity, inflexible industries. We need to think about that here as well.

I said discourage private actions that make markets tight or exploit markets. And you have the numbers before you today. I have done this quickly, but here are the numbers you heard today about gouging.

Mr. Caruso told us that the price of gasoline went up the equivalent of $19 a barrel. Mr. Douglass told us that his rack price went up $18 a barrel. That is the wholesale price he pays. Mr. Cooper told us that there were no increases in transportation costs because they are regulated. Mr. Caruso also told us that the WTI price went up $5. The difference between that WTI price and the rack price is $13. That is the refiner margin. And if you look over at the charts that were taken down, unbranded went up $26 equivalent per barrel. That is the refiner margin against unbranded gasoline. If you want to look for gouging, take a hard look at those numbers. Get the record, find out what happened. There is a good case there that it wasn't crude and it wasn't the retailer and it wasn't the transport; it was the refining sector that sets the wholesale price of gasoline.

Finally, we said promote a competitive industry. Almost every wholesale market in this country is concentrated, every regional retail market is concentrated, every regional refining market is concentrated. They got that way over the past 15 years as a result of mergers. Now, the FTC says they are not highly concentrated. In this industry, you get a lot of market power with a little bit of market share, and these firms have a sufficient market share to affect the price. The GAO discovered that, found that, showed that, and the FTC attacked them for showing us the facts.

I have attached to my testimony three graphs, one which shows the growth of gasoline since I last testified or since we did our report of 6 percent, one which shows an elimination of oil spare capacity and refineries, and one which shows no increase in our stocks.

Essentially, we were living on the razor's edge, and Katrina pushed us off. But Katrina is only the last in a series of accidents and interruptions that have hit this industry. It was the worst possible single event that we could have suffered, yet its impact is only an extrapolation of what has been going on. Four times in the last 5 years we have had these price spikes, and everyone said: Stocks were low, refinery capacity was stretched to the limit, and we had an incident, a pipeline broke, a refinery had a fire, a storm. Once is a surprise, twice is bad luck, but three times is a systematic pattern, and it is incumbent upon this Congress and our government officials to take action to insulate us against these kinds of events because they do happen and they will happen.

So it is time for a dramatic change in our approach to policies. I offer the same advice I gave 4 years ago but only with greater urgency. We should pursue each of these options twice as fast, aim twice as high, because we have wasted a critical 4 years in providing the American consumer with the gasoline sector that they need and deserve.

Thank you.

[The prepared statement of Mark N. Cooper follows:]

PREPARED STATEMENT OF MARK N. COOPER, DIRECTOR OF RESEARCH, CONSUMER FEDERATION OF AMERICA

Mr. Chairman and Members of the Committee, my name is Mark Cooper. I am Director of Research at the Consumer Federation of America (CFA). CFA is a non-partisan, non-profit association of 300 pro-consumer groups, which was founded in 1968 to advance the consumer interest through advocacy and education. We have been analyzing the petroleum industry for decades and have issued numerous reports in the past five years, as the seeds of the underlying conditions for the current crisis became apparent.[1] I greatly appreciate the opportunity to share our views with the Committee today.

Public policy cannot prevent accidents or catastrophic acts of nature, but it can build systems that are resilient, robust and flexible to minimize the impact of the inevitable accidents on our society. It is evident that the business practices of the oil industry and public policy in this country have combined to allow a gasoline industry that cannot respond to even minor incidents, not to mention the disaster that struck the Gulf Coast last week. If the measure of performance of an economic sector is adequate supplies at stable prices, then this industry has failed the consumer, not just in the wake o Katrina, but also repeatedly over the past five years.

The bulk of my testimony today presents word-for-word the policy conclusions that we reached in a report released over four years ago. After analyzing the structure, conduct and performance of the oil industry, we urged policymakers to move aggressively in five areas.

- Restore reserve margins by developing both efficiency (demand-side) and production (supply-side).
- Increase market flexibility through stock and storage policy.
- Discourage private actions that make markets tight/or exploit market disruptions by countering the tendency to profiteer by withholding of supply.
- Promote a more competitive industry.
- Address the disproportionate burden that rising energy price place on lower income households.

Unfortunately, these policy recommendations were not included in any significant way in the recent energy legislation and actions by Federal agencies. For example, the single most important mid- and long-term policy we advocated, improving the fuel efficiency of the vehicle fleet, could have taken significant pressure off of gasoline markets, if it had been embraced four years ago. In the past four years, gasoline consumption in America increased by about 6 percent.

Regrettably, just last month the Department of Transportation proposed timid improvement in fuel efficiency standards, gerrymandered the calculation to let more gas guzzlers escape scrutiny, and exempted some of the worst gas guzzler from fuel consumption standards altogether. Simultaneously, it threatened to preempt states like California from imposing stricter standards. That proposal, which would take ten years to lower consumption by the equivalent of less than what we use in one month, does not address the problem in any meaningful sense.

Similarly, the key short-term policy we recommended of increasing reserve margins and stocks has been neglected. I have attached to my testimony three updated graphs from our July 2001 analysis. The first shows gasoline consumption. The second shows refinery capacity compared to demand and the second shows gasoline stocks above minimum operating levels. Together these show that we were living on the razor's edge, with rising demand, little excess capacity and small stockpiles. Katrina pushed us off the edge.

Hurricane Katrina is just the latest in a series of accidents and interruptions that have hit the oil industry and consumers hard. It is the worst possible single event

[1] I have submitted for the record four reports on the oil and gas industry that we have prepared over the past four yours. Our policy discussion from the first of these in July 2001 are included as part of this statement.

one could imagine from the point of view of the domestic gasoline sector. Yet, its impact reflects the same underlying problems that have afflicted the industry for the past half-decade. At the start of each price spike in the past few years we hear the same refrain. "Stocks were low, refinery capacity was stretched to the limit." A refinery fire here, a pipeline breach there, or a storm triggers a spike.

Once may be a surprise; twice may be bad luck; but by the third time, it is a pattern that demands a systematic response, not hand wringing. Moreover, if many different accidents keep happening, if many different things can go wrong, and if running an overtaxed system makes outages more likely, it is incumbent upon policy makers to do something about it. The industry and public policy have failed to create a system that can meet the needs of the American consumer and America as a nation is paying the price for that inaction.

It is time for a dramatic change in the approach to policy. I offer the same advice we gave four years ago. The only thing I would change is the urgency in the recommendations. We should move twice as fast and set our goals twice as high because past inaction has made the problem we face even more critical.

ENDING THE GASOLINE PRICE SPIRAL

MARKET FUNDAMENTALS FOR CONSUMER-FRIENDLY POLICIES TO STOP THE WILD RIDE

JULY 2001

V. POLICY RESPONSES

A. ECONOMIC FUNDAMENTALS AND POLICY PRINCIPLES

Public policy responses must reflect physical and economic reality. Since the laws of physics cannot be repealed, public policy must be cognizant of the increased likelihood and severe impact of accidents in energy industries, like refineries and pipelines. Physical and institutional structures must be prepared to deal with accidents in this industry.

The low short run elasticity plays a critical role in price volatility and the exercise of market power. The extremely low elasticity of demand is one of the key characteristics of the gasoline market. Suppliers are well aware of the rigidities in the market and can take advantage of them under the right circumstances. Because the gasoline market is so large, even small and short term pricing abuse imposes substantial costs on the public.

Under these circumstances, firms with relatively small market shares can increase profits by withholding supplies, unless the elasticity of supply is high. Unfortunately, petroleum product markets do not exhibit very elastic supply. Reserve margins and stocks are crucial.

1. Supply

Avenues for increasing supply are available, but they may not be pursued, if left to industry business decisions. Since short-term elasticities are quite low, a variety of resources that can be called upon to meet demand quickly are necessary to prevent price volatility. Having reserve margins of production and transport capacity would dampen price volatility. Stockpiles and storage are the best option when demand shifts or supply is interrupted. Import of product is an important option when refinery capacity is not available or, depending on geographic location, when pipeline capacity is not available.

The recent closure of refineries also suggests an avenue for expanding capacity. The most readily available path to expanding capacity may be to identify existing facilities that have been shuttered, or sites that have been recently abandoned to expand capacity while minimizing environmental impact should be explored. Each of these options should be considered, particularly in markets where capacity is tight and ownership is concentrated.

The behavior of small refiners in response to the elimination of programs that supported their existence makes it clear that public policy can affect the number and geographic distribution of refinery capacity. If we want geographically dispersed refinery capacity to promote local responses to supply problems, we just have to pay for it.

2. Demand

In the long run, reducing the size of the market, without imposing deprivation on consumers, is the major policy challenge.

The consumption patterns deeply embedded in spatial relationships lead us to conclude that increased fuel economy is the more readily achievable approach to reducing gasoline consumption than changing living patterns. Reducing fuel use per vehicle allows existing mobility patterns to be preserved, while consumption is reduced.

Shifting preferences for vehicles (toward higher efficiency vehicle types) requires greater change in social behaviors. It is also vulnerable to changes in taste. Moreover, it requires a change in the stock over a substantial period of time, perhaps a decade. While policies to affect these behaviors should be pursued, their complexities and difficulties should be recognized.

Attempting to overlay mass transit on existing living patterns may be pursued as a long-term strategy. However, given consumer preferences and the spatial distribution of activity, this is a substantial task. The increasing suburbanization of living patterns frequently results in relatively low densities and high costs for mass transit. Changing the geographic distribution of work, home and play, requires the greatest amount of social change.

3. Distributional Effects

Equity impacts of rising energy prices, particularly as they affect low and lower middle income households, must be dealt with directly. Neither general tax cuts nor existing energy assistance programs, such as the Low Income Home Energy Assistance Program (LIHEAP), address the problem of rising or volatile transportation energy costs. Even if it could be argued that LIHEAP addresses the general energy needs of groups, ad hoc efforts to increase programs like LIHEAP tend to fall short and come long after the impacts of rising energy prices have been felt.

B. POLICY TARGETS

It is time for public policy to seek permanent institutional changes that both reduce the chances that markets will be tight and reduce the exposure of consumers to the opportunistic exploitation of markets when they become tight. To achieve this reduction of risk public policy should be focused on achieving five primary goals.

- Restore reserve margins by developing both efficiency (demand-side) and production (supply-side).
- Increase market flexibility through stock and storage policy.
- Discourage private actions that make markets tight/or exploit market disruptions by countering the tendency to profiteer by withholding of supply.
- Promote a more competitive industry.
- Address the disproportionate burden that rising energy price place on lower income households.

1. Expand Reserve Margins By Striking A Balance Between Demand Reduction and Supply Increases

We have earlier identified the hierarchy of policies to reduce demand. Increasing the fuel efficiency of the fleet through increased standards for mileage and use of hybrid vehicles should be given top priority. Shifting preferences for vehicle types and modes of transportation through taxes and incentives are a second category to be considered.

A goal of achieving an improvement of vehicle efficiency (reduction in fleet average miles per gallon) equal to economy wide productivity over the past decade (when the fleet failed to progress) would have a major impact on demand. It would require the fleet average to improve at the same rate it did in the 1980s. It would raise average fuel efficiency by five miles per gallon, or 20 percent. This is a mid-term target. This rate of improvement should be sustainable for several decades. This would reduce demand by 1.5 million barrels per day. This would return consumption to the level of the mid-1980s.

Expanding refinery capacity by 10 percent equals approximately 1.5 million barrels per day. This would require 15 refineries, if the average size equals the refineries currently in use. This is less than one-third the number shut down in the past ten years and less than one quarter of the number shut down in the past fifteen years. Alternatively, a ten percent increase in the size of existing refineries, which is the rate at which they increased over the 1990s, would do the trick, as long as no additional refineries were shut down.

Placed in the context of redevelopment of recently abandoned facilities or expansion of existing facilities, the task of adding refinery capacity does not appear to be daunting. Such an expansion of capacity has not been in the interest of the businesses making the capacity decisions. Therefore, public policies to identify sites, study why so many facilities have been shut down, and establish programs to expand capacity should be pursued.

Once the magnitude of the task on the supply-side is placed in perspective, and given the objective analysis of the environmental costs involved, the call to overturn environmental laws loses its force. It seems that expansion of supply-side capacity can be accomplished within the current confines of environmental laws. To the extent that the costs of compliance can be demonstrated to be a significant problem, then underwriting compliance (directly through financial subsidies or indirectly through research) rather than relaxing standards should be pursued.

This combination of demand-side and supply-side policies to improve the long run market balance would restore the supply/demand balance to levels that typified the mid-1980s.

2. Expanding Storage And Stocks

It has become more and more evident that private decisions on the holding of stocks will maximize short term private profits to the detriment of the public. Increasing concentration and inadequate competition allows stocks to be drawn down to levels that send markets into price spirals. While the strategic petroleum reserve has been developed as a strategic stockpile and companies generally take care of operating stocks, the marketplace is clearly not attending to economic stockpiles. Companies will not willingly hold excess capacity for the express purpose of preventing price increases. They will only do so if they fear that a lack of supply or an increase in brand price would cause them to lose business to competitors who have available stocks. Regional gasoline markets appear to lack sufficient competition to discipline anti-consumer private stock policies.

Public policy must expand stocks. Participants in the distribution of gasoline can be required to hold stocks as a percentage of retail sales. Public policy could also either directly support or give incentives for private parties to keep storage. It could lower cost of storage through tax incentives by draw down stocks during seasonal peaks. Finally, public policy could directly underwrite stockpiles. We now have a small Northeast heating oil reserve. It should be continued and sized to discipline price shocks, not just prevent shortages. Similarly, a Midwest gasoline stockpile should be considered.

3. Taking The Fun And Profit Out Of Market Manipulation

In the short term, government must turn the spotlight on business decisions that make markets tight or exploit them.

Withholding of supply should draw immediate and intense public scrutiny. It needs to be backed up with investigations. Since the federal government is likely to be subject to political pressures not to take action, state government should be authorized and supported in market monitoring efforts. A joint task force of federal and state attorney's general could be established on a continuing basis. The task force should develop databases and information to analyze the structure, conduct and performance of gasoline markets.

As long as huge windfall profits can be made, private sector market participants will have a strong incentive to keep markets tight. The pattern of repeated price spikes and volatility has now become an enduring problem. Because the elasticity of demand is so low—because gasoline is so important to economic and social life—this type of profiteering should be discouraged. A windfall profits tax that kicks in under specific circumstances will take the fun and profit out of market manipulation.

Ultimately, market manipulation could be made illegal.

4. Promoting A Workably Competitive Market

Further concentration of these industries is quite problematic. The Department of Justice Merger Guidelines should be rigorously enforced. Moreover, the efficiency defense of consolidation should be looked on skeptically, since inadequate capacity is a market problem.

Restrictive marketing practices, such as zonal pricing and franchise restrictions on supply acquisition should be examined and discouraged. These practices restrict flows of product into markets at key moments.

Markets should be expanded by creating more uniform product requirements. These should not result in a relaxation of clean air requirements.

5. Low-income assistance

Rather than fight repeated battles over supplemental appropriations, it would be more effective to index assistance payments to energy prices. It may be time to consider new programs that deal directly with transportation fuel costs. Transportation energy is a necessity in the 21st century.

Chairman BARTON. Well, thank you, Mr. Cooper.

The Chair is going to recognize himself for the first 5 minutes of questions. I want to ask Mr. Cavaney—and you may have said this in your testimony; I was having to go to vote and come back—is there evidence that any of the oil production or distribution gathering or transportation pipelines or production platforms in the Gulf of Mexico have not performed as desired in terms of preventing spills during the hurricane? In other words, I have not seen in the popular press any story about a rupture or a spill, so I am assuming that part of the system has performed as designed. Is that your understanding also?

Mr. CAVANEY. It is. Daily, Mr. Chairman, the MMS reports all the activity that they have done through their overflights and the searches in the area. There have been no spills related to production.

Chairman BARTON. So we have had a Category 5 hurricane take a huge swath out of three of our coastal States, but in terms of the offshore oil and gas production and delivery system, while it has been shut down, it has not spewed any environmental damage into the gulf coast?

Mr. CAVANEY. No, Mr. Chairman. Another point, too, is these platforms out there, particularly those in the deep water, are designed to withstand the roughest of conditions, conditions like what they experienced. There are over 4,000 total operating platforms out in the Federal section of the OCS area, and of those we have had reports of 41 of them which were all very close inshore and basically first generation platforms with very little production receive damage or were effectively destroyed. But of all the rest in deep water, only four platforms have experienced damage. We learn lessons when we go through these things, and those will help us going forward. So I think fairly much, if you look at 95 percent of the production was shut in because we abandoned all those and shut them down properly, now we are at a stage back where only 57 percent is shut in within a week after the storm went through there. So I think the design parameters and the safeguards performed as expected and as anticipated, but we do learn some new things each time and we will from this one.

Chairman BARTON. Now, we have before us on my far right two of the watchdogs for the consumer, and to the left, with the exception of the telephone gentleman and the Governor of Louisiana's representative, I have got the full panoply of the oil and gas industry from Mr. Cavaney of API who goes out and finds and hopefully produces the oil, to the gentleman on the New York Mercantile who helps create a market for the crude, to Mr. Slaughter who helps purchase and refine the crude, and Mr. Cooper who helps transmit both the crude and the refined products, to Mr. Douglass who actually sells them at retail—not you individually, but the groups that you represent.

My first question: Do any of you gentlemen have any knowledge of any attempt to collude, to set a price because of this shortage? Do you have any personal knowledge of attempting to coordinate an orchestrated increase in the price that has resulted in the gasoline prices being as high as they are right now across the country?

Mr. CAVANEY. Mr. Chairman, I can speak for API and its members. We have no knowledge. We have cooperated, as I mentioned in my testimony, for decades and decades on dozens of investigations by the FTC, and we are daily monitored, as Bob Slaughter had mentioned, by the FTC in 360 different cities. This industry operates, because it is incredibly, incredibly efficient and it is highly diverse particularly as you get closer to the retailer. And I think that is our protection.

Chairman BARTON. Now, Mr. Cooper has pointed out in his testimony that if you have consolidation and if you have a concentration, even at what would normally be considered a low level, 10 to 15 percent of a given market, that in and of itself can result in prices going up disproportionately. I am going to ask the indulgence of the full committee. I want to start with Mr. Cavaney and then work my way through the system and have each of you try to explain how your segment sets the price or how it is created. In other words, I want Mr. Cavaney to explain to the best of his ability where this world price, the raw material price comes from, and Mr. Newsome to explain if there is any group of people like the Hunt brothers or somebody who could create a position to move the market artificially, and just work our way down. Because I am like everybody else, I don't like going from 2.47 a gallon Tuesday morning to 2.79 a gallon Tuesday night, which I had to do down in Texas, and I just thought I was getting ripped off big time until I heard from Mr. Upton in Michigan that the price went to 3.50, and then when I watched on TV and saw it Georgia it was at $6. So I felt a little bit better about only having to pay 40 cents a gallon more when I found out about the rest of the country. But I think it is a fair question that the American people ask, is why, when we have such a little—and 2 million barrels a day in terms of production and 4 million barrels a day in terms of refinery capacity and two major pipelines being shut down is not little, but in the overall scheme of things why that would cause prices to go up everywhere in this country.

So, Mr. Cavaney, can you briefly explain to the best of your knowledge how the market price is set for crude that your people produce?

Mr. CAVANEY. Yes. Mr. Chairman, I might just comment, we share the same concern; we don't like the prices, either. But we just went through an experience down in the gulf that is a once in a generation or once in every several generation experience.

Directly to your point, the products that we produce, natural gas and crude oil, those are set and traded in worldwide markets. Crude oil is the world's largest commodity, it is highly transparent, and the prices are established off of what are called benchmarks, which are certain quality parameters, certain viscosity, certain amounts of sulfur. And then those that are considered to be more attractive have an established differential that floats around. The more attractive lighter grades, which are more efficient, you can get higher yield, trade a little higher. The lower ones trade a little lower.

Chairman BARTON. But can any company——

Mr. CAVANEY. Everyone pays the same price.

Chairman BARTON. Can any producer, ExxonMobil, the biggest in the world, can they or the Saudis last week, could they have moved the market, the crude price higher by something that they did or colluded with other producers?

Mr. CAVANEY. Not in that short period of time. Nobody could do any of that, because the market is so transparent and so large.

Chairman BARTON. Do you agree or disagree with that, Mr. Cooper? I am going to use you as my referee here.

Mr. MARK COOPER. Well, collusion is not the issue. And you——

Chairman BARTON. Well, I am trying to get to——

Mr. MARK COOPER. But you posed the question properly. It is that, it is not the crude market. I mean, the crude price is set by a cartel primarily, and it is a political price, it is not an economic price. The refinery market——

Chairman BARTON. If you have a surplus, I agree with that. But we don't have a surplus of production.

Mr. MARK COOPER. But the quantity of supply, the amount of production capacity is a strategic variable. When you have a small number of people who decide what to do, then supply is not out. It doesn't happen in nature, it comes out of political decisions.

Chairman BARTON. I am just asking, do you have any—I am going to give you a chance after each person speaks, because you are a smart guy. Okay? But on the crude price, do you stipulate that any producer in the world tried to raise the price above a market level in the last 2 weeks?

Mr. MARK COOPER. Certainly not in the last 2 weeks. The conditions were set over a period of time by political decisions among the people.

Chairman BARTON. Now, Mr. Newsome, is there anybody on the New York Mercantile Exchange or any other commodity exchange that, in a market this big, has the ability to move the price higher on a purely speculative play?

Mr. NEWSOME. No, sir. As you know, I come from the regulatory side of the equation, not as a trader or anything else, and so monitoring markets, market surveillance is something that is very near and dear to me. That is one of the reasons that NYMEX is considered the benchmark in many energy products is because you have hundreds of companies, representatives that all come together to compete in a very open, transparent environment to come up with a price.

Chairman BARTON. So there is nobody out there that could try to corner the oil market or the natural gas market and move the price higher for purely speculative purposes in the last 2 weeks?

Mr. NEWSOME. No.

Chairman BARTON. Do you agree or disagree with that, Mr. Cooper?

Mr. MARK COOPER. I would like people to look at that one very carefully. You would have gotten the same answers, and you probably did, about the electricity market in California 4 years ago. There are a lot of people chasing a lot of oil out there, a lot of financial transactions, but very little physical product changing hands. That is a concern to me. He has got the data, you can take a look at it.

Chairman BARTON. In your opinion, how big of a position would you want or a group colluding have to take to move the market?

Mr. MARK COOPER. In a market that is this tight, you know, the tighter the market, the smaller the position you need to have that influence. But I am not asserting that. But that is a good question and there is data that you can look at these transactions, ask how many times this oil, this single barrel of oil changed hands, just like we asked how many times the electrons changed hands.

Chairman BARTON. You believe, while on the crude—the producers, while you think there are political considerations over time that have tightened the market or prevented production coming on line, on the commodity exchanges do you think it is possible somebody could have moved the market differently than pure market pressures moved it?

Mr. MARK COOPER. The amount of speculation going on is a source of concern. We have seen it in the Wall Street Journal and a number of other places. It is an objective study. Absolutely.

Chairman BARTON. We will put a question mark there. Now we go to Mr. Slaughter, who takes the products that have been produced and purchased, and it is your job now to refine those products. You take the price coming in that is set on the exchanges. The refined products that are coming out, your refiners have the ability to set a price. How is that going out price set?

Mr. SLAUGHTER. It is basically set from market conditions. There are two factors: One is the cost of production, and the second is the market that it is being sold in.

Chairman BARTON. Well, when Mr. Douglass, I think he had the chart showing the rack price going up everywhere in the country over a period of I think 3 days. It went up faster in some regions than it did in others, but that is a refinery setting that price. Why did it? Your crude input didn't really go up that much. It went from about $65 a barrel to $70, and then when the President announced the release of the SPR the crude price went back down. Unless it went up big time today, it is $65, $66, $67. But the price that Mr. Douglass is paying went up and it stayed up. Now, what happened there?

Mr. SLAUGHTER. First of all, there are 48 refining companies and 149 refineries. Each sets prices, of course, individually. But what happened was that with what happened in the gulf last week, as Mr. Cavaney pointed out, was a once in a lifetime event. It knocked out 20 percent plus of the Nation's refining capacity, and it also knocked out the major supply pipelines for about one-third of the country to a half of the country, as well as the Midwest as well as the East and the South. When that happens, basically prices go up. Prices went up. If you look at what happened to gasoline on the exchanges, gasoline prices went way up because people were anticipating that there would be a supply shortage as a result of what had happened with Hurricane Katrina.

Chairman BARTON. I am not being argumentative, Mr. Slaughter, but I am trying to understand. Your input price did not go up. There is obviously some supply disruption to these pipelines that got shut down, and we know that overall there is going to be a shortage because we have less refined product because the refineries themselves are down. But I am still trying to understand why the refiners' posted prices went up everywhere as quickly as they did. I understand raising a price to alleviate a shortage. If I don't have any gasoline coming into Atlanta, I know why that price is going up: Because you are not getting any more gasoline. I don't quite understand why it is going up in Texas and California where the refineries are still operating, the distribution systems are still operating; you just have to allocate across the 21 million barrel a day market, you have got to allocate a shortage overall of somewhere between 2 and 3 million barrels once you go through your inventory.

Mr. SLAUGHTER. What happens in the event of a supply disruption is that basically, you know, prices throughout the country will basically go up to try to allocate the product through market forces. And——

Chairman BARTON. So is there a marketing manager at each refinery who takes it upon himself or herself to set the price? Does it come down from the corporate office? I mean, how do you get to that price?

Mr. SLAUGHTER. Basically, the companies have their own pricing strategies and they respond to market conditions and what they also perceive as replacement cost and what the marketplace is basically going through in terms of general supply problems.

Chairman BARTON. Mr. Cooper, do you want to comment on that?

Mr. MARK COOPER. They have market power. There is a line in the movie, A Beautiful Mind, about John Nash. And the key line he says is: Adam Smith was wrong. When he realizes that nine people can cooperate, a small group of people, maximize their profits by observing each other's behavior. Right? It is called the Nash equilibrium. He won the Nobel Prize for it and he spent 25 years learning how noncollusive games, noncooperative games—every refinery market in this country is concentrated. Several of them are highly concentrated. So he told you the right answer; he says: We will charge whatever the market will bear. And in this particular market, there is only a small number of other players out there to observe. So they all put their prices up at the same time. And people call me all the time and say they all did 15 cents overnight. How can that happen? Because most experience is that someone in that market would say, hey, I will eat a little bit of that to expand my market share.

The refinery industry doesn't behave that way. The RAND study showed that coming out of the mergers of the 1990's. You put your finger on it. They have the market power to charge whatever the market will bear.

Chairman BARTON. I am not stipulating they have the market power. I am trying to find out what they are really doing.

Mr. MARK COOPER. Well, his answer is that is where the wholesale price is set. And the numbers you see today are damning numbers, that $18 a barrel and $13 a barrel.

Chairman BARTON. Well, it is not automatically a damning number if it prevents a shortage. If you raise the price to prevent a shortage so that willing buyers have an opportunity to purchase, that is not damning. If it is artificially created for whatever reason, that is damning.

Mr. MARK COOPER. It is damning if there are not enough willing sellers to prevent excess profits.

Chairman BARTON. Well, do you know of any refinery that held refined product off of any given market?

Mr. MARK COOPER. Well, they created a market in which there was no surplus. That is the long-term problem of letting those 50 refineries get closed.

Chairman BARTON. We are going to encourage cooperation to get some of them reopened and build some new ones.

Mr. MARK COOPER. But it would be real important to get some other players, some independents in those markets. If those same folks own those refineries, you are at the same point.

Chairman BARTON. Let us go to Mr. Cooper. I am going to stipulate, since you are regulated and you testified that the pipelines didn't change their price.

Mr. BENJAMIN COOPER. That is described on page 4 and 5 of my testimony.

Chairman BARTON. So we are going to pass you over. Now we go to Mr. Douglass. You are getting these higher prices from Mr. Slaughter's group, but you didn't just pass those through. The retail price went up too.

Mr. DOUGLASS. Correct.

Chairman BARTON. Okay. Do you want to explain how that happened?

Mr. DOUGLASS. Yes. May I be allowed to put up chart 2?

Chairman BARTON. You may.

Mr. DOUGLASS. If we can put up the second chart that we have that was an attachment to our statement that was filed, you will see that refiners—in this particular case I am a branded marketer and this is a branded supplier, my biggest supplier. They raised the price 13 cents the first day. By the way, they lagged the market. I had other suppliers who raised it 37 for their branded stores and 70 cents for the unbranded business that I use commercially. So this is a very restrained marketing company.

And for the benefit of those who have called them names today, it is ExxonMobil. So they were dragging. They were actually being very conservative in their price moves compared to the others. However, they did raise my price 13 cents a gallon. And the way it works in our business is you have to pay—the next truckload, you have to pay that extra 13 cents. So you immediately identify that as part of your inventory. So we price on the basis that what is in the inventory at the time of an escalating market, we better move consistent with that, or if we sell that other inventory and don't get back the 13 cent increase, we will end up losing money.

Chairman BARTON. You fast-forward to replace your inventory.

Mr. DOUGLASS. Correct. We are definitely trying to put a replacement cost——

Chairman BARTON. And you are stipulating that the retail price just went up whatever the wholesale price you had to pay went up.

Mr. DOUGLASS. Actually, we lost 5 cents in the transaction over time.

Mr. MARK COOPER. He lost because the credit card companies charge him a percentage of the sale, and that went up as well even though the cost of that transaction didn't go up at all, and that is 3 percent. So if you went from $2 to $3, he lost 3 cents out of that. They took a bigger bite out of it.

So those are the numbers that suggest where the price increase came from. That doesn't mean there weren't individual stations who may have gone hog wild.

Chairman BARTON. You think the retail guys are okay?

Mr. MARK COOPER. As far as I can tell, and these numbers nailed it, and I have been talking with people, and the rack price was just going up. And this gentleman had to come here and put it on the screen.

Chairman BARTON. If we had any black hats in the room—and I am going to stipulate that everybody here has at least a gray hat on, and I want to say everybody has a white hat on, but if we had to go from white to shades of gray, you are going to say the speculators on the commodity markets and the refiners are the ones that are not lily white?

Mr. MARK COOPER. The latter I have seen some numbers; the former I would like to see some analysis.

Chairman BARTON. I am going to give those two guys a chance to rebut, and then I am going to deal the next person.

Mr. Slaughter or Mr. Newsome.

Mr. SLAUGHTER. Yes, Mr. Chairman. The point that Mr. Douglass raised, the antitrust laws prevent collusion in our industry, as you know.

Chairman BARTON. I don't think anybody is alleging collusion based on what I have heard today.

Mr. SLAUGHTER. The spot price, because of the anticipation of supply shortages for products, went up considerably during this period of time when there was the greatest disruption. A number of our companies froze prices or were selling prices well below spot price during that period of time. So, I mean, there is different behavior by each individual company.

The point I had before, the tremendous amount of our cost of our product is involved with a cost accrued, plus the cost of taxes. Different people behave differently, but everybody could see that that situation that we had this week was a cataclysmic event threatening so much of the market for our products. The people responded accordingly. And, you know, basically it is a market situation. It was a cataclysmic event. It knocked out 20 percent of the refining supply, and everyone responded accordingly. A number of them took the steps that Mr. Douglass has suggested and froze prices or actually sold below spot.

Chairman BARTON. Mr. Newsome.

Mr. NEWSOME. Thank you, Mr. Chairman.

As I commented in my testimony, speculators have become easy fodder to pick on by the media, but they play an important role in the market.

Chairman BARTON. I am not picking on them. I am trying to give an explanation to the average citizen, if they are lucky enough or bored enough to watch this entire hearing, about what the best experts in the country explained is our pricing. We will politicize it, but I at least want to try to get on the record what it is.

Mr. NEWSOME. The reality of how the marketplace operates keeps speculators from pushing the price. It is a very open, competitive marketplace. And in one pit you have got commercials, you have got the banks and brokers, you have got the locals or the speculators that are all bidding and competing one for another.

The markets move based upon fundamentals. If a speculator tries to take the market in a direction that is not fundamentally sound, there is someone always there ready to make him pay, saying, yeah, do it, and we will turn around, and it will cost you a lot of money. The reality of the market is that everyone can't move in one direction whether that is high or whether it is low. The futures markets are a zero sum game. For every winner, there is a loser. So, that, just the way that the markets operate don't allow it to get away from the fundamentals.

Chairman BARTON. I want to thank the committee for indulging me. I don't normally take that much time, but this is such an important issue that I wanted to try to get on the record at least how the pricing system works or doesn't work depending on your point of view. And I want to again thank the committee for letting me do that.

Mr. Rush is recognized for 5 minutes.

Mr. RUSH. Thank you, Mr. Chairman. Mr. Chairman, I don't expect to take my 4 or 5 minutes. I just have a question that is not in line with your line of questioning.

I want to ask Mr. Angelle, I had occasion this morning to be at a meeting with the president of the American Red Cross as it relates to the problem that we are experiencing in the gulf coastal region, and I asked her a specific question. And I asked her, why didn't the Red Cross go into New Orleans earlier? And her response to me startled me really. And she said that the Governor did not—ordered her not to go into New Orleans to conduct rescue and recovery or relief work, for her to conduct relief efforts in New Orleans. And that astounded me. And I expected to—I wanted to ask the Governor. I thought that maybe she would be here, but you are a representative. Are you aware of that?

Mr. ANGELLE. No, sir. In my role as secretary of the Department of Natural Resources, I obviously have primary jurisdiction over managing our energy resources in Louisiana, and I am not qualified to answer that question.

Mr. RUSH. Thank you.

Mr. Chairman, I yield back.

Chairman BARTON. Did you yield back? Wow, thank you.

Mr. Hall, 5 minutes.

Mr. HALL. Thank you, Mr. Chairman.

All of you probably sat through the first session and heard our questions and answers, and I guess, Mr. Cavaney, I will start with you. And I thank you for being here and sharing your knowledge with us.

What is needed—what we need to do is to bring some stability to the energy market, and that is easy to say and hard to do. It is going to require long-term investments and strategies.

What, in your opinion, will be the most effective strategy for dealing with the spiraling cost in the short term first? We got an idea on the long term, but what can we do tonight or before milking time tomorrow, or within the reasonable foreseeable future?

Mr. CAVANEY. I think the steps that were undertaken by the administration and by the IEA in terms of making available some finished product, which is being brought over from their reserves, did about as much as one could expect. We have seen over the last several days both drops, pretty significant, in the futures markets of both crude oil and finished product. I am not forecasting here, but we may well look back and see that the peak here occurred the last several days ago, because with the passage of time, what we are finding is both the refineries are, in fact, coming back; the pipelines are up, as we mentioned; and the refined product that is coming over is helping fill a void. So we are on the right track in the near term.

What we ought to focus on is trying to help work together instead of actually working against one another. And by that I mean we in the industry as we are, we ought to be out there getting supplies we need, getting the repairs, getting our people down there and cooperating with the other entities which are so important.

Utilities have done a terrific job of getting power down there. Government ought to be in place to be able to provide help when needed, and that is quick approval on permits, helping to expedite things that need to go through the process of getting approval, and in some cases possibly even helping coordinate closely the various different entities within the government itself.

And finally, consumers, and I think this is the part where the biggest savings can be made. And I don't think we as a country have done anywhere near the job we can, and that is in energy conservation. We can give you on our Web site api.org a whole litany of things that a driver can do in his or her automobile that can actually improve the miles per gallon by as much as 10 or 12. That is fairly significant if you are driving a car that gets about 25 miles per gallon. And there is also, very obviously, that if you have two cars, one gets good mileage and other gets less so, try and plan your longer trips for the other. That is the equivalent of actually adding product instantaneously.

The problem is most people think, I am just an individual, and my little contribution won't mean much. But if you do a few simple things and multiply that by several million, you are adding more production than we can ever find overnight.

So the answer to your question is individuals can look at this and begin to create, I think, an approach that won't only serve us well during this immediate postrestoration period, but can serve us well going forward, and that is energy conservation not only in transportation sector, but in all sectors.

Mr. HALL. Would you repeat that?

Mr. CAVANEY. I will do it a little more passionately next time.

Mr. HALL. Seriously, I know you are a little disgusted that you don't have a better audience. Everything is being taken down by the reporter, every Member of Congress, and we will read and reread it. That is the way we make decisions. I was kidding you. I don't expect you to repeat it.

But Mr. Cooper made a great case for more refineries, and the chairman, and the chairman's committee, those that work with him and support him and advise him, worked with him, and in his—I guess in his good judgment, he put subtitle H, sections 391, 392, their aim toward the Governors of this State to make it more possible and more likely that they could pursue the construction of a refinery. And I am not sure, I don't think we had any report language or anything telling him exactly how to do it, because probably it will be done slowly and by people that are more knowledgeable about the business certainly than I am. But if we have the support of the EPA, and we set that out in the bill itself, there is a lot of "thou shalts" in there that ought to make the acquisition of a refinery a little more attainable for the people, Governors of the State, with a lot of their finances; that they have to reduce the time that it takes to build one and might make it more attractive for investors to shorten the time for them to have some expectation of their payback.

But what are the—do you know—I asked someone who represented the Federal Government if they had any anybody from any of the 50 Governors tie into them on this, and I think maybe, Mr. Chairman, our Governor has discussed it with you, and you discussed it with several other Governors. Chairman BARTON. I asked the President to send a letter to all 50 Governors asking them if they want to consider using this particular provision, and I talked directly with Governor Perry of Texas, who is considering strongly doing just that.

Mr. HALL. Maybe I am talking to the wrong bunch, Mr. Chairman. When did you send that letter?

Chairman BARTON. I haven't sent the letter yet. I just made a phone call. I asked the President of the United States——

Mr. HALL. I will follow up and I will send Red a copy so he can help me answer that and help us work that out. But it is up to—how much more time do I have?

Chairman BARTON. You have been out about a minute and a half.

Mr. HALL. Could I ask one real quick question of Mr. Douglass? He is my constituent and my good friend. He has been an effective witness before this committee before.

Bill, would you clarify how individual stations make adjustments in pricing in times of market fluctuations; do suppliers give you anticipated prices?

Mr. DOUGLASS. When you say anticipated, we get our price changes every day. But in a volatile market like we are presently in, they may be twice a day, anticipated in the sense that if they told us at noon, they would raise the price at 6; if they told us at 6, they would raise it at 12.

Mr. HALL. Who is they?

Mr. DOUGLASS. The oil companies.

Mr. HALL. Do you deal with the same one all the time?

Mr. DOUGLASS. We have eight different suppliers, but we have one principal supplier, a branded company. But what you have to do is look at your replacement costs, because they draft you as soon as they load the truck. They literally have you on the computer, and when that truck is loaded, they draft your account. So you have to pay when you buy, and if you don't pay, obviously they don't let you buy.

Mr. HALL. What are the prices based on?

Mr. DOUGLASS. The price that we pay?

Mr. HALL. Yes.

Mr. DOUGLASS. Or the price that we charge?

Mr. HALL. The price that you set. You set a price to sell.

Mr. DOUGLASS. Essentially we try to recover our costs, because when I used my chart, one, I had a 14-cent margin at that point, which is a gross margin before any expenses are taken out. And when the truck picks up that fuel, and it is 10 cents, in this case, 13 cents higher on the next day, we have to raise our price because our costs remain. And as they pointed out, the credit card costs alone range from 5 to 7 cents a gallon of that 14 that I had. But if you will notice on that chart, my margin went down to 10, so I had a net margin, if you will, of 3 cents to pay all the expenses associated with it.

So you must move forward. You must take your inventory.

Mr. HALL. My time is really up. I will write you a letter and ask you to give some more answers, really to follow your invoice right from the time you got it and the time you pay it, and how much you raised your prices compared to how much the people that sold to you raised theirs.

Did you ever raise yours in excess of what——

Mr. DOUGLASS. No, sir. No, sir. In fact, we lost margin, if you look on this transaction. We lost margin as it went up, and our expenses went up, as I indicated, through credit card expense and so on.

Mr. HALL. The only way you could have kept from losing that was to arbitrarily raise yours above what you were paying directly for it?

Mr. DOUGLASS. That is correct.

Mr. HALL. You didn't do that?

Mr. DOUGLASS. I did not do that.

Chairman BARTON. Before I recognize Mr. Stupak, I want to understand something. You just said if I come into one of your stations, and I use my Visa credit card, and I buy 1 gallon of gasoline, you are charged 7 cents?

Mr. DOUGLASS. Yes, sir.

Chairman BARTON. But if I buy two gallons, you are charged 14 cents? And——

Mr. DOUGLASS. Yes, sir.

Chairman BARTON. And if I charge 10 gallons, you are charged 70 cents?

Mr. DOUGLASS. Yes, sir.

Chairman BARTON. So the more I buy, the more you are charged?

Mr. DOUGLASS. Absolutely.

Chairman BARTON. Even though it doesn't cost any more for a transaction if I buy 10 gallons than if I buy 1.

Mr. DOUGLASS. That's correct.

Chairman BARTON. That doesn't make a whole lot of sense.

Mr. DOUGLASS. In my testimony I asked that we investigate the pricing of the credit card companies.

Chairman BARTON. Interesting.

Mr. Stupak.

Mr. STUPAK. Thank you, Mr. Chairman.

Mr. Newman—I am sorry, Newsome—the risk premium, what is it right now on a barrel of oil?

Mr. NEWSOME. I think it would vary depending upon what analyst you talk to.

Mr. STUPAK. All right. Most papers are saying 15 to 20 dollars; is that correct?

Mr. NEWSOME. I am not an oil analyst, so it would be difficult for me to say, but my opinion would be that is very high.

Mr. STUPAK. What do you think it is?

Mr. NEWSOME. I would say it is much lower than that, below $5.

Mr. STUPAK. So Bloomberg, Kiplinger, all these people are just wrong?

Mr. NEWSOME. Well, they all have their opinions.

Mr. STUPAK. Okay. Thank you.

Mr. Douglass, certainly I enjoyed your testimony. I really found your charts quite interesting. The chairman was just asking you about if you take a look at your chart number 6 about the credit card companies, as you can see right there, they went up twice, or 200 percent; average 2003 revenue was a buck 50. They are up to $3 now, 200 percent increase, royalty owners. They have doubled, if you will, their cost to a retailer?

Mr. DOUGLASS. Sure.

Mr. STUPAK. Royalty owners, they doubled, crude exploration and extraction, two and a half times, if my math is correct; refiner, two and a half times; and retailer stayed the same.

Mr. DOUGLASS. Correct.

Mr. STUPAK. Is that a set margin you work off?

Mr. DOUGLASS. Well, actually, I was reflecting incidentally on this chart. It says average 2003 revenue. My staff didn't understand my cursive. It was August 2003 revenue and August 2005. And, in fact, what I did, in the event I got challenged, I brought invoices with me.

Mr. STUPAK. So everybody in the industry has gone up at least 200 percent in the last 2 years except the retailer?

Mr. DOUGLASS. That is correct.

Mr. STUPAK. If you look at your chart, go to chart number 4, if you will. You start on August 26 and go to close of September 2. That is a week's time. Katrina hit land on August 29. Shortly thereafter the President released an SPR, or oil out of SPR, as I and others have been advocating for some time to try to bring some stability here. And Mr. Caruso testified earlier that once the oil was reduced from the SPR, the cost of a barrel of oil actually went down.

Mr. DOUGLASS. Yes, sir.

Mr. STUPAK. But the price keeps going up.

Mr. DOUGLASS. Well, we have a chart, chart 3 here, which does not reflect taxes or, if you will, freight, but it is the price for unbranded fuel, and you will see that it did drop off from its high of 831 for unbranded fuel. What happened with the branded companies, they were lagging that particular pricing strategy, so they were still catching up. They were still passing through the——

Mr. STUPAK. But then the next day on September 2, it is going back up again.

Mr. DOUGLASS. Yes, sir.

Mr. STUPAK. Where would it be on the 3rd, 4th? Has it gone up, stabilized? I am just asking you to guess. It is just on your chart.

Mr. DOUGLASS. It has actually dropped.

Mr. STUPAK. Dropped again?

Mr. DOUGLASS. Yes, sir.

Mr. STUPAK. We appreciate your charts and appreciate the suggestions you have.

Mr. Slaughter, I got a number of questions for you. And the chairman is turning on the clock, so I am going to have to move quickly, so my time is limited. If you answer yes or no, I certainly would appreciate it.

Allow me to begin by quoting a 1995 industry document in which the oil industry noted that—I am quoting now—that if the U.S. Refining industry does not reduce its refining capacity, it will never see any substantial increase in refining margins.

And that is an internal Chevron document dated November 30, 1995. That suggests to me that the industry itself has played an important role in the lack of refining capacity in the U.S. So my question is in your testimony, you indicated that in 19—you indicated that 177 refineries have shut down since 1977. Is it true that there are no requests for environmental permits to reopen any of these facilities, yes or no?

Mr. SLAUGHTER. To my knowledge, there are not. Many of them have been sold or moved, or many of them could not get another permit.

Mr. STUPAK. But there is no request to reopen these 177 refineries; the answer is no?

Mr. SLAUGHTER. There may not even be a refinery on that site any more, Congressman.

Mr. STUPAK. Isn't italso true that since 1995—I should say in 1995 alone, 30 refineries have shut down, right?

Mr. SLAUGHTER. Many of them shut down because of the cost of the Federal programs that I displayed on my third chart, Congressman.

Mr. STUPAK. Sure. So 30 of them shut down. So now we are up to 207. Of the 30 that have shut down, are any of them seeking permits to reopen?

Mr. SLAUGHTER. It is funny you had mentioned that. There was a refinery in California, the Powerine refinery, two summers ago when there were supply problems in California applied to reopen, and the application was denied.

Mr. STUPAK. Denied by California?

Mr. SLAUGHTER. Denied by the local authorities.

Mr. STUPAK. Well, there have been no new refineries built since 1976, and over the past 13 years only one refinery, the Arizona clean fuels facility, has sought a permit to rebuild a new refinery; isn't that correct?

Mr. SLAUGHTER. That is correct.

Mr. STUPAK. Isn't it also true that in 1992, the State of Arizona granted a permit to the Arizona facility for construction and operation, and the company sat on the permit for nearly 8 years without actually building a refinery?

Mr. SLAUGHTER. Not to my knowledge, no. They had to remove the refining site——

Mr. STUPAK. They didn't move the refinery site until 2003 from Maricopa County to Yuma, Arizona. So from 1992, after it was approved, it sat 8 years without actually building a refinery.

Mr. SLAUGHTER. There was a lot of activity on that in the meantime, Congressman.

Mr. STUPAK. They got the permit, and they still to this day have not built a refinery.

Mr. SLAUGHTER. It is more than just one permit, Congressman. They did just get an air permit, which is a good sign, but you know there is a problem because it costs $3 billion to build a refinery of that size today. There is a lot of upfront money you have to pay, and you may have to wait 15 years until you know whether you will ever get a drop out of that refinery. There is a lot of uncertainty in making an application to build a new refinery.

Mr. STUPAK. They have all the permits they needed. What permit did they not have after 1992?

Mr. SLAUGHTER. They need air permits, and they need other permits.

Mr. STUPAK. Tell me the ones specifically they did not receive for this one. I don't want general answers. I am talking about a very specific refinery, because I found—and today's Washington Post backs up the stories here. You claim you can't get permits because of environmental reasons, yet everything we have examined finds just the opposite. And when you started off with the first quote I gave you, which was 1995, in which the refining industry says, if you don't reduce your refining capacity, you will never see any substantial increase in refining margins.

Mr. Douglass' chart here, which shows just in the last 2 years you were able to increase your revenue by decreasing refining capacity by 2½ percent, from $11 in August 2003 to $27 per barrel in 2005.

Mr. Douglass' charts sort of indicate what we are driving at here. The issue isn't environmental laws. The issue is there is no need to take refinery capacity because if you do, your margins are going to go down.

Are you familiar with the Government Accounting Office report of May of this year which says exactly the same thing?

Mr. SLAUGHTER. I am aware of that GAO report. I am also aware of a FTC report that rebuts it. You keep citing a 1995 document by one person in one company which I have not seen, and there is no evidence at all that anyone ever acted on whatever suggestion is there.

Mr. STUPAK. Really, since that time you have closed 203 refineries. I think that is quite a bit of action; 177 and 30, unless my math—you are right, it is 207.

Mr. SLAUGHTER. That is right.

Mr. STUPAK. One hundred seventy-seven——

Mr. SLAUGHTER. Actually refining capacity has increased over that period.

Mr. STUPAK. In refineries that are left.

Mr. SLAUGHTER. It is still capacity.

Mr. STUPAK. Actually it is increased by 13 percent only since change is made. I keep hearing this, and even the questions of the chairman and Mr. Hall was if we would build more refineries, and even in the energy bill, which I supported, there was movement to waive environmental laws to make the Department of Energy, the Secretary, the head person who would decide the environmental laws of this Nation. I think it is a lame excuse for putting forth when you see prices and profits like this by the refineries.

Mr. SLAUGHTER. Crude price remains the biggest determinant of product prices, Congressman. The EIA testimony today demonstrated that, what happens in the marketplace demonstrates it. The industry has been adding capacity at existing sites consistently over the time period that you have mentioned, and we have increased the actual refining capacity in the United States over the last 10 years. A number of the refineries that are cited in the larger number were inefficient, small refineries that were basically supported by the price control system of the 1970's——

Mr. STUPAK. Or, as Mr. Cooper says, the bigger ones bought up these little refineries, put them out of business; therefore, it is easier to control the price when very few are controlling the process. That is what Valero and others have done.

Mr. SLAUGHTER. Valero has bought a number of refineries in the last several years and they have added 400,000 barrels of capacity to them, Congressman.

Mr. HALL [presiding]. Mr. Stupak, have you got the answers you want?

Mr. STUPAK. No. Mr. Chairman, can we be allowed to follow those up, though, with more written questions?

Mr. HALL. I will allow you an additional 2 minutes. I will allow you an additional 2 minutes if you are trying to close up because we are not going to have a second round.

Our chairman took 20 minutes, and we will yield you 2 or 4 of those. I am not mad at him about it. Let the record show that.

Mr. STUPAK. Mr. Cooper, I brought up about the refineries being closed, more people control it, and therefore they can control also the price or the profit margin. Would you care to comment on my question to Mr. Slaughter on that line?

Mr. MARK COOPER. It is a simple proposition that we have understood for quite some time that when the number of actors in the market gets small, they can exercise market power.

In this industry, because supply and demand are so weak, market forces are weak, people can't come back, you can't increase supply, you get more market power at a lower level of concentration. And we submitted for the record a very detailed analysis of this debate between the FTC and the GAO. The FTC has a theory that mergers that don't increase the market power—the concentration ratios above a certain level don't hurt. The GAO analysis shows they do.

And let me be clear. The FTC and the GAO do not treat capacity as a strategic variable. If you look at the GAO's analysis, as capacity utilization goes up, price goes up. If you cut back on capacity, you can expect the price increase. As stocks go down, price goes up.

The GAO considers those to be exogenous; that is, they are not part of the market power problem.

Your memo, the behavior over the past 4 years shows—or 10 years—shows quite clearly an industry that has got concentrated, and as a result of concentration at the refinery level, they have power over price. That is what we call market power.

I think the numbers you have seen today suggest that they set the price at what the market will bear—you can put it that way—

but there is not enough supply side competition, which is what we like in our markets, to protect consumers from the abuse of market power. I think that is a good case, and you have good empirical evidence of that today during this crisis.

Mr. STUPAK. Then what should the Congress do to take—if I use the word we have heard a lot today—the U.S. Off the razor's edge of tight refining capacity? What should we do?

Mr. MARK COOPER. Well, we tried to emphasize the demand side. So Mr. Hall's question, what can I do before I have to milk the cows tomorrow, I talk about three Ts. Two of them I can do before tomorrow; that is, trips and tires.

You heard the suggestion. You inflate your tires, and you think about your trips. And it is not deprivation. I think about how many times I am going to the store and which car I am going to drive to the store. I agree on something with Mr. Cavaney—I hope this place doesn't get struck by lightning—and then the third one is tune-ups. Those are near-term things. In the long term, we have to get the efficiency of our fleet increased.

Because reducing our consumption affects the world market—Mr. Lashof showed you why it is a big impact on the world market. It affects the domestic market. It alleviates the refinery problem. Producing more domestic oil will not alleviate the pressure on the refineries.

Those are the things to do. I think the Government agencies that worry about gouging should stop saying there is no collusion and start saying consumers aren't being treated fairly. We should worry more about the unilateral exercise of market power.

Mr. HALL. All right. Mr. Cooper took the 2 minutes. Now, you want to——

Mr. STUPAK. I thank you for the extra time.

Mr. HALL. You yield back?

Mr. STUPAK. Yield back.

Mr. HALL. Thank you, Bart.

Chair recognizes Mr. Shimkus for 5 minutes, which will probably be 7, 8 or 9 or 10 minutes.

Mr. SHIMKUS. Thank you, Mr. Chairman.

First of all, I want to thank Chairman Barton for actually using a lot of time, because going through the chain process is really helpful because the average consumer just gets lost, and it is a great educational process.

We used to have on this committee jurisdiction over financial services. We lost that with Gramm-Leach-Bliley, and I think we have lost some of our financial service expertise on markets and futures in the issue.

So, Mr. Newsome, could you briefly explain and answer this question: If futures causes us to hedge risk, by looking at the futures market, should we have been able to predict the high prices we saw this summer? I am not talking about the hurricane, but the escalation. By looking at the futures market, should that have told us where we were headed?

Mr. NEWSOME. Well, certainly futures markets are no crystal ball, but high energy prices are not new. I mean, the market has been reacting to market fundamentals for over a year, and if you start looking at those fundamentals of record high usage from

China, India and the U.S., China is still considered potentially the 800-pound gorilla because most of their usage is industrial. And as they move to normal usage as in the U.S. Is the middle class, I think their demand for energy is going to blow through the roof.

You combine that with a political unrest in the Middle East, political unrest in Venezuela, and you get in the situation that Mr. Cooper has talked about where you have got such a tight margin that anything that happens around the globe—because these market are global—anything that happens has an impact. A refinery goes down in Venezuela, it has an impact on our market. I mean, that is how thin the margin is.

So certainly the markets have been moving in that direction over time.

Unfortunately, the hurricane became a market fundamental very quickly, and the markets reacted to that fundamental.

Mr. SHIMKUS. Thank you.

Mr. Douglass, I really appreciate this chart because I have gone to a lot of retailers, and they have talked to me about price inversion in this issue, and I think it is really helpful.

The other thing, based upon financial services aspects, they also raised the credit card issue. And I have heard that in my district, and that is part of the financial services background that we have kind—of our expertise on this committee we have kind of lost. But there are some savings. The consumers choose to use credit cards; you choose to accept credit cards instead of cash transactions. No one is being forced to either use a credit card or pay for their gas by credit card; is that correct?

Mr. DOUGLASS. That's correct, Congressman.

Mr. SHIMKUS. There are benefits to that because you get immediate payment versus if someone comes in to the quick mart and wants to write a check that may clear or may not, and you have to decide whether—at what level are you going to accept the check, $20 or $50 or—use of the credit card hedges some of your risk.

Mr. DOUGLASS. The credit card does take some of the risk, but the obvious costs of the credit card have accelerated.

Mr. SHIMKUS. And I think that—one of the issues here is that we need more competition across the board in the markets. We need more competition in the refinery industry. We need more competition for you to have selected credit cards that would have a different type of standard by which to, in essence, loan money to individuals and then pay the retailers. But how do you—government, how does government intervene to get involved in the market? And that is a huge challenge.

I was asked by my friend Vito Fossella to ask this question, and I looked at it, and I did think it is a pretty good question. It is you, Mr. Douglass, from August 29 to September 2, your testimony says that prices jumped 44 cents for branded stations—those flying the flag of the major refiners—but for unbranded, independent folks like the one I—the picture I showed earlier, Rock stations in my district, it went up 73 cents per gallon. Can you explain the difference?

Mr. DOUGLASS. I think essentially the suppliers started to be concerned that they couldn't meet their contractual obligations. And the branded stores, we are contractually obligated to our supplier, and they to us, to supply us at the historical average, and they have us on allocation currently. We are restricted in ourselves. We can't buy more each day than we bought the same day last year. So they look at that and they say, this is going out of control; I am going to be able to supply the branded people. So then they take the unbranded, if you will, or the surplus sale and take it off the market by virtually pricing it so high, it squashes the transaction.

Mr. SHIMKUS. Thank you. And this has been a long day, and I appreciate you all being here. I think we have learned a lot. We have a lot of work to do. I am for more competition across the board, and that is what makes America best. And we have to move somehow incentively to get more refineries, more independence. Of course, as I said earlier, and I was going to try to go without saying anything, ethanol and biorefineries are growing, and I encourage that also.

Thank you. I yield back, Mr. Chairman.

Mr. HALL. I thank you.

Chair recognizes the gentlelady from California Mrs. Capps for what we hope will be 5 minutes.

Mrs. CAPPS. Thank you, Mr. Hall. And I was going to offer my thanks to the person whose chair you succeeded here, Mr. Barton, because I thought the exchange that he had early in the hearing was very instructive, especially, from my point of view, that Mr. Cooper, Dr. Cooper, was allowed to respond each time, and that gave me the opportunity to learn.

Mr. Cavaney, I want to congratulate the people you represent for the lack of any accidents in this massive assault on the oil platforms. I come from California. We have a lot of oil platforms on our coast. It is not hurricanes we are terrified of, it is earthquakes. And I have several in my district—off my district—that sit on earthquake faults. And I want to say a word about taking advantage of technology that is the very best there is, and this is an example. I hope you will carry back to the industry that this happens.

Now, at the same time, I want to say, I believe—and I am going to say it to you folks in the industry—that these high gas prices and these disruptions in the oil market that we have been talking about all afternoon—all day bring home the point that we are too dependent on oil as a Nation. I believe that America needs to diversify our energy portfolio. I have actually heard more about conservation and diversification today than I heard during our preparations for our energy policy. I wish we had had more of this language that I have heard today in this hearing in that markup.

I also want to say that I think it is really unfortunate that some Members of—colleagues of mine, and some in the industry, may use this tragedy to push unpopular plans for new drilling in protected areas, and that they say we need to drill in these areas in order to deal with the high gasoline prices. And I understand there is a number of businesses and trade groups that are planning to send a letter to the Republican House leadership asking them to allow new drilling in protected areas. Their letter specifically cites Lease 181 in the eastern gulf as a priority area.

This is nothing new. Many of us have experienced this before in this body. These businesses and groups have been asking for this for years, and for years the House has rejected these drilling plans, most recently during House consideration of the Interior bill.

What is new, however, is the desire to use Hurricane Katrina as a motivating factor. And I think this goes right to the heart of profiteering motivation in the face of human tragedy, and I hope we can avoid it.

We know MMS does an inventory every 5 years. We have within our area, or confines, something like 3 percent of the known oil and gas reserves, while we are responsible for 25 percent of the world's demand.

So my question to you, Dr. Lashof, I want to ask from your perspective and your studies, would drilling off the west coast, off the east coast or in the eastern Gulf of Mexico, where there are prohibitions currently, do anything for gas prices now, or would it do also anything to lower oil prices in the future?

Mr. LASHOF. No, I don't believe it would.

Mrs. CAPPS. Maybe you could include ANWR in the same.

Mr. LASHOF. The chart I had specifically related to the Arctic National Wildlife Refuge, but it is much the same story with regard to protected areas off the coast.

In fact, if you look at the areas that are already available for oil and gas development, they contain 80 percent of the estimated resources. So the areas that are protected are a small fraction of the total resource base offshore, and, you know, overall, as you mentioned, the United States only has 3 percent of the world's reserves. Two-thirds of the world's reserves are in the Middle East. So any supply side strategy that we might adopt that involves petroleum simply can't change the balance in a significant way. And the EIA analysis estimating the effects of the energy bill, for example, even including the Arctic Refuge, which wasn't in the final bill, suggested that its maximum impact would be less than 1.5 cents in 2025. This is a trivial number compared with the kind of price movements we have seen.

Another point, as we have heard, the price movements that we have seen in the short term are mostly related to the price coming from the refineries rather than crude oil. So any drilling for crude oil has no impact on that. The price impacts we are seeing are at the refinery gate rather than due to increases in crude prices.

As Mr. Cooper said, and as I said in my testimony, improvement in efficiency, both the short-term conservation effort and the long-term commitment to real efficiency improvement and diversifying supply away from petroleum itself, can have a big impact, and it affects both the crude supply and the refinery supply, whereas a drilling response will not do that.

Mrs. CAPPS. And diversifying our energy portfolio is something we could be emphasizing more in our energy policy?

Maybe I will use my last minute to ask Dr. Cooper, last month the administration proposed a restructuring of light truck fuel economy rules claiming the new policy would save the country 10 billions of gallons of gasoline over the lives of vehicles bought from 2008 to 2011. I know there is not much time, but could you comment on this? Was this move enough? Would you recommend we de lower? And this—I am talking about regulations from the administration or from Congress.

Mr. MARK COOPER. Well, as I mentioned, you might not have been in the room, the price of gasoline used in that cost-benefit analysis was $1.80 a gallon in 2012, including taxes. So they have undervalued gasoline dramatically.

Mrs. CAPPS. What should we do instead?

Mr. MARK COOPER. Well, the answer is you should adopt a much higher standard. And we called for that last week at a press conference, and, you know, the off-the-shelf technologies can dramatically improve efficiency. People calculate to double it in new cars, and then over the course of the life of the cars, we will have a dramatic increase in efficiency.

And if we take a leadership position, the question comes up, the automobile industry says, it will kill us. That is what they told us about airbags, which is in the purview of this committee. They said, airbags is the end of the industry, and all these safety measures. And, of course, they adopted it. They adjusted to it.

So from our point of view, the most important thing we can do for our automobile industry is reorient their thinking. Instead of giving people discounts to keep them hooked on fuel-inefficient automobiles—I mean, they have lowered the private cost of the automobile sufficiently to make it economic for individuals to buy it on a private basis. There is no doubt about that. I think that is clear.

But as a society, this is suicidal, because those vehicles will be in this fleet guzzling gasoline for a decade. So there is a big gap. And Mr. Garman really all day he does not understand the gap between the private costs of gasoline and the social costs of gasoline.

The price at the pump does not reflect the economic slowdown, the trade deficit, the currency instability, the political vulnerability, all of the environmental harm. There is a social cost to gasoline. These guys couldn't even get the economic cost right.

We have to start viewing gasoline as a critical public problem. The example I use, the analogy I use, is cigarettes. And maybe there are some people from tobacco-growing States. Forty years ago this room would have been filled with people who were smoking. Half the people in this room would have been smoking. They all kind of knew it was bad for them, and they couldn't kick the habit. We as a society, through a combination of education and regulation, changed that behavior.

Gasoline is a bigger threat to our national health and welfare, and we need that same sort of commitment to education and regulation to change behaviors. And that is where I think we have to go as a society. It can make a big difference for our industry and our Nation.

Mr. HALL. A little relief now.

Mr. CAVANEY. Mr. Chairman, I would like to just take exception to that comment about comparing some of the dangerous health effects that come from smoking with our products. Most of the various pharmaceutical company inputs come from hydrocarbons that we produce and make the various drugs that save people's lives in an immense amount. So to equate one with the other is really not a fair comparison in that same sense.

Our products save lives. The plastics that come from our products that are found throughout hospitals, the drugs that we use are incredibly valuable, and so it is just not a fair comparison to equate us with some of the problems that were associated with the other industry.

Mr. HALL. Gentlelady has the right to close. I will give you a minute to close, Mrs. Capps. Is that enough?

Mrs. CAPPS. Less than that. And I want to wear another hat now as someone who has worked in health care all my life and to say I think the industry that you represent has enormous products that benefit our health very much, and the use that Mr. Cooper—I know I am putting words in his mouth when I say this, but you gave the time to me, Mr. Hall—the uses that he is talking about are very different uses of gasoline and oil.

I think actually you could make the case if we saved the product—because it is fossil fuel, it is limited—over a lifetime if we saved it for its more beneficial therapeutic uses and life-enhancing uses and did that by—you know, we are not going to stop using gasoline in our cars anytime soon, but over the long haul, I have heard advertisements from the oil industry on television talking about diversification as part of their portfolio, and that is what we are talking about here.

Mr. HALL. All right. I thank you. Does gentlelady yield back the balance of her time?

Mrs. CAPPS. I do yield back, finally.

Mr. HALL. I recognize the gentleman from Oregon.

Mr. WALDEN. Thank you very much, Mr. Chairman. I have got a number of questions I would like to work through here, so I appreciate your indulgence today and your testimony. It has been most helpful.

Mr. Slaughter, you made a comment about the spot price and how prices get adjusted because the spot price goes up. What percent, though, of the market is traded in the spot market? You know, we saw with energy pricing a spot market that went to $1,900 a kilowatt hour at one point.

Mr. SLAUGHTER. I can't give you a number on that. Mr. Walden. It is essentially a wholesale price at the rack. It generally is a higher price because it is something that someone basically is going out in the marketplace and buying, who doesn't have a contractual relationship.

Mr. WALDEN. Is there anyone on the panel that can tell me how much is contracted out and how much is bought on the spot market? Don't you maintain portfolios of different——

Mr. NEWSOME. If you look at, Congressman, the way the market is set up, you have the futures, which is a slice, and then you have got cash, which is the bulk of the marketplace.

In futures, it varies by contract. And natural gas, most of our trading is in the back months; less of it is in the spot. Gasoline and crude oil, it tends to be a little more even between spot and back months, but it depends very much on which markets you are talking about and whether it is futures or cash.

Mr. WALDEN. So in gasoline, am I correct, then, hearing what you have said, about half of it would be in the spot market? Is that a daily or hourly——

Mr. NEWSOME. No. It is not going to be half of it. But again, it depends if you are just talking spot market or cash market.

Mr. WALDEN. One of the things we heard was that the price went up at the rack—I don't know all your terminology—based on the spot market going up, and you were somewhat under the spot market. And I guess my question is if you have already got a reserve of gas or crude oil or whatever part of the business you are in already purchased and handled, then what percent, how big an impact does the spot market really have on your actual cost?

Mr. CAVANEY. In our testimony that we submitted in a written form, there is a little explanation. As the oil companies utilize different arrangements with their distribution chains—there are company-owned stores, and there are stores that are leased, and there are three different ways you can work under those, and then there is the independent. Under each of those scenarios, there can be arrangements to either go under a contract, which was mentioned here, or if someone has a financial difficulty or somebody has a new rip-roaring contract, you might say, let's work on the spot market. Let's go for a while and see if we can develop a credible history. It is changing all the time. There isn't one place you can go to get a data point that would tell you that.

Mr. WALDEN. I am trying to get a handle on this. I sat through here as my colleague from Washington—we heard virtually the same sort of panels telling us why the energy market and electricity was all fine, and there was no problem out there. And then we hear the tapes of the traders. And I will tell you what, I am not from Missouri, but a lot of companies lost a lot of credibility with this Republican business member because I had to sit here and listen to the testimony, and I believed it. And then I found out the truth. And I will tell you as vice chairman of the Oversight Investigation Subcommittee, if we find that traders have been manipulating the market, we are going to go after them, because we found that to be the case long after the fact, and it cost ratepayers in my State an enormous sum of money.

And I have already asked the Government Accountability Office to look into the trading issues. I am sure you know, Mr. Newsome, to make sure that isn't happening because consumers deserve that right to know if these markets are being operated in an honest, above-board, direct way. And I am not casting aspersions, but, you know, it is like Mr. Cooper said, you know, first time or second time, you learn from it. Third time, I am not taking any prisoners.

Mr. NEWSOME. But only thing I would say when you talk about traders, again, there are traders on future markets, traders on over-the-counter market. I was chairman of the CFTC during the whole Enron situation, brought multiple charges against energy companies. Attempted manipulation did not take place in the regulated futures market. It was an attempt through the false reporting of prices to move those prices and then take advantage through the positions they had in the over-the-counter——

Mr. WALDEN. They did the round-tripping.

Mr. NEWSOME. Round-tripping, the false reporting of prices to the indexes. But just to say traders are manipulating, I want to clarify that it wasn't traders on the regulated exchanges.

Mr. WALDEN. All right. But then I read this quote, as you probably heard me read earlier today, on the Dow Jones news wire from Mr. Addison Armstrong, manager of the exchange trade markets, TFS Energy Futures, LLC, in Stamford, Connecticut, and I quote the wire here: There are oil commodities traders who made so much money this week following Hurricane Katrina, they will not have to punch a ticket for the rest of the year.

What does that mean?

Mr. NEWSOME. Traders on the floor can make money from two different methods; one through brokerage fees, they are trading the accounts of customers. Obviously we had record volume last week, so brokerage fees were the highest they have ever been.

The second component would be the position that the traders had coming into the trading week. Typically on a Friday, traders try to get flat their positions so they don't have to take that risk over the weekend. But all traders cannot get flat. Some are long. Some are short coming into Monday. Obviously conditions changed drastically from the end of the trading day Friday.

Mr. WALDEN. But they knew the hurricane was coming.

Mr. NEWSOME. Well, they knew it was coming, but they didn't know the path it was going to take, they didn't know how hard it would hit the oil refinery section, and certainly had no idea it was going to be as strong as it ended up being. So depending on what your positions were on Monday, some were big winners, but some were big losers.

It is a net zero sum game. So for every guy that made it, there was an equal and offsetting person who lost it.

Mr. WALDEN. So as the regulator—you do, NYMX, right?

Mr. NEWSOME. Yes.

Mr. WALDEN. And do you regulate that? What is your role?

Mr. NEWSOME. I am the president of NYMX.

Mr. WALDEN. Who is your regulator?

Mr. NEWSOME. The CFTC.

Mr. WALDEN. I will follow up with them.

Mr. NEWSOME. They have been on the floor every morning with physical surveillance. They look at our large trader reports every day. They also have analytical programs that look at how traders trade, how they handle their customers' accounts, whether they are trading their own accounts, et cetera, et cetera.

Mr. WALDEN. I want to go to just a couple other quick points here. One is we have talked about a lot of—about the lack of supply because of the shutdown of the refineries and the pipelines and everything else. I haven't heard anything today about a drop in demand in that region.

I am curious about that, because if we had half a million or a million people displaced that aren't driving, that aren't moving—did anybody see a demand reduction?

Mr. CAVANEY. Yes. There is demand reduction. We can't really quantify because we collect data, and weekly data, and put it out. Tomorrow morning we will be putting out a report, and we will be glad to send you something on that.

But we have already seen in the previous month before the hurricane was coming up that higher prices were having an impact, and, typically, demand had flattened out year-over-year basis. So trading nationally was already down, and when you see something like this, you can fully expect that it is going to be down pretty significantly.

Mr. WALDEN. One final question for Mr. B. Cooper—it says B. Cooper and M. Cooper. This is a question I get a lot. I was out 851 miles around part of my district. My district is bigger than any State this side of the Mississippi. A question I would ask is the gas station set this price in the afternoon, and that price—I can find 2.79 gas today—when you talk about forward pricing and how when they call and tell you at noon it is going to go up 13 cents or whatever, and you immediately then raise prices, right? No.

Mr. BENJAMIN COOPER. You got the wrong guy.

Mr. WALDEN. I am sorry. Mr. Douglass, I am sorry, because you just do the pipelines, and you are not party to any side of this because you didn't do anything.

Mr. Douglass, I apologize. The question that comes in is, okay, you move your prices up immediately. Do they go down immediately when they call and say tomorrow it is going to be 13 cents less? Does it take the same time to go down as it is to go up?

Mr. DOUGLASS. Actually it is driven by several factors. First obviously is the cost. When the cost went up 13 cents, we attempted to erase 10 cents; you know, the next day we attempted to go up 10 cents. The market then determines whether you get to keep that 10 cents or whether you have to roll it back to zero. You may have no margin.

Mr. WALDEN. Market being competitors and the community?

Mr. DOUGLASS. Correct.

Mr. WALDEN. So when the supplier comes to you and says, I am going up 13 cents, you automatically go up. But it is the community that decides how soon you go down?

Mr. DOUGLASS. Absolutely. They will discipline you in a hurry. If you are overpriced, your business will drop in half.

Mr. WALDEN. All right. Mr. Chairman, thanks for your indulgence. Gentlemen, thank you for your testimony.

Mr. HALL. We recognize Mr. Inslee, the gentleman from Washington, for 5 minutes.

Mr. INSLEE. Thank you.

I am assuming, in talking to the gentlemen from the petroleum industry, that anything that reduces demand or reduces the rate of increase of demand has the capacity to reduce the price. We have been talking to you about supply, but it is a supply and demand they taught me in economics at UW.

I want to ask you about what the Federal Government has done or not done to reduce demand, which could have the capacity of reducing the price you charge; and I want to refer you to a chart over here to your left. This chart shows what has happened with average fuel economy since 1975 to 2005. You will see the middle line there—the top line are cars, the bottom line are trucks. The middle line is the average of both.

You will see that the average in 1975 was about 14 miles per gallon. The Congress moved at that time to increase the CAFE, the Corporate Average Fuel Economy standards, to significantly increase it. As a result of that action by the U.S. Congress, that went up in about 1986 to about 23 miles a gallon, a very, very significant increase; and that had the result of reducing demand or reducing the rate of increase in demand, which had a capacity to reduce the price.

Since then, Congress fell off the wagon of fuel efficiency in about 1985; and, in fact, the fuel efficiency of our cars and trucks have gone done on average since 1985, have actually been reduced in 2005 now to about 21 or 22.

So while we have created the Internet, we have mapped the human genome, because the U.S. Congress has been asleep at the switch, we have gone down on our efficiency of cars. As a result, the demand has skyrocketed and one of the reasons our price has skyrocketed.

If the U.S. Congress wants to be serious about reducing the price of fuel, would you not suggest to us to get back on the wagon of increasing the efficiency of the cars and trucks we drive by increasing the CAFE standards? I ask Mr. Cavaney and Mr. Slaughter that question.

Mr. CAVANEY. I think your initial point is exactly correct, and that is where I differ from my colleague from the NRDC, is extra supply, whether it comes from drilling or from conservation, is beneficial to the consumer. It is going to help. And if you believe it in one area, then you have to believe in another.

Now the answer to your question is we feel very strongly that not only individuals but industries and companies ought to practice conservation and energy efficiency, and our record as an industry is very strong in that regard. We do not know—it is not our business—how to make cars and how to make trucks. But the idea that everybody ought to become more efficient, as efficient as they can, consistent with safety and the like, is something that we can support.

So I can speak to it on only the most general of terms. But you are absolutely right on in terms of reduced conservation helps people.

Mr. SLAUGHTER. I would concur in that, Mr. Inslee. We are definitely in favor of anything that creates additional supply. The Association has always advocated trying to maximize supply of fuels. We have always advocated that we do need additional refineries in the United States; and conservation will definitely help that, too.

Mr. INSLEE. Now let me move to the supply side as far as price for a minute. We went through the Enron and other energy debacle. We saw an administration that was callously indifferent, did nothing while Enron and other companies took over a billion dollars out of the Pacific Northwest in their rapacious behavior. So we are a little sensitive to supply pricing issues up in the Northwest and on the West Coast.

We have heard discussions today that the Federal Government really is very impotent in dealing with gouging issues in the absence of collaborative behavior by various suppliers, that if there is gouging done unilaterally or by one, we really do not have a tool in our tool box federally to enforce this, and there is 23 states that have anti-gouging laws specifically in emergency context, but only 23 States.

Some of us think that we need a Federal tool to deal with gouging, and we have proposed—today, I have introduced a bill with a bunch of Members that would essentially give the Federal Government an anti-gouging tool which—and it is just very summary fashion—would amend the Clayton Act to bar charging prices that are unconscionable in comparison to recent prices.

It would take into account the upstream components that you all have to pay for the people ahead of you in the stream. It would apply to necessary goods, which include gasoline, and would apply during times of natural or man-made disasters, pretty tightly woven bill, fairly consistent with some of the State approaches.

Now assuming that you all want to demonstrate the effort to show good corporate citizenship, which I believe—I hope that you do, is this something that you think Congress ought to at least consider, to have some type of anti-gouging regulation to give us the authority to deal with this type of issue?

Mr. CAVANEY. Well, first of all, we feel very strongly and condemn any kind of gouging. I think just the discussion of it and the fact that people are aware of it is going to have a quiescent effect on it. Besides the 23 States that actually have regulations that the Governor can enact, also any time a Governor typically declares a state of emergency he can also include gouging in there and do it. So the States do have some authority if they choose to exercise it.

One of the challenges, I think, in approaching this from a Federal level is going to be, like what we heard earlier today, is really what is the definition of gouging? If it is a one-person behavior? Does it mean, if you sell above a suggested retail price, is that gouging one industry, or do you sell within a range? And who determines it?

One of the things you need to sort of, I think, keep in mind as you look at these things is the people who are closest to where the occurrence may have happened are probably in the best position to really get all of the facts and decide what is there. So that would say at least there ought to be, in any discussion, some consideration of making sure that you work closely with the States and look at them as you go forward.

I cannot speak to whether Federal law is good or bad, but I think the discussion certainly can inform people.

Mr. HALL. The gentlewoman from North Carolina, Mrs. Myrick.

Mrs. MYRICK. Mr. Smith, you made comment that it was going to cost between $400 and $600 million for getting service back up again. But that is a cost that BellSouth absorbs as part of what you do.

Mr. Slaughter, with the refineries—it is the same thing with the refineries. You absorb the cost of whatever it costs you to rebuild, et cetera? Am I correct? Because we have been talking a lot today of what is the Federal responsibility and all, but I am in my own mind clarifying that is your cost of doing business, correct?

Mr. SLAUGHTER. To rebuild the facilities that have been damaged?

Mrs. MYRICK. Yes.

Mr. SLAUGHTER. That is what usually happens. Yes.

Mrs. MYRICK. We appreciate that. Thank you.

And thank you all again for your patience in staying this long and giving us a day of your time. We did not mean it to end up this way. But I do have just two things.

Mr. Newsome, this is not at all—I wanted to ask you, because this is not at all adversarial, so please help me. Explain two things. First one was about the hedging and short-selling that goes on and has gone on over the last, say, 4-year period. Does that have anything to do with exacerbating the price of oil by the barrel?

Mr. NEWSOME. No. Hedging is used by the market participants who actually own the product that are trying to set a price or a floor on what they think that product may be worth in the future.

Mrs. MYRICK. So that is produce moving? They would then be moving product?

Mr. NEWSOME. Well, our contracts are deliverable. Most do not go to delivery. They trade out of the contract at a moment that is advantageous. But we maintain delivery of the contract, so it gives the commercial participants an advantage over a speculator or somebody who does not own the physical product that they could be forced to make or take delivery. So unless you have the ability to make or take delivery, you are not going to go to expiration with a contract.

Mrs. MYRICK. Then you said that there are a couple of ways that traders get paid. One is a brokerage fee. Do they make more money based on sales volume?

Mr. NEWSOME. Yes. Absolutely.

Mrs. MYRICK. So if the market is volatile, then they personally benefit from that?

Mr. NEWSOME. Well, it depends. You know, for every winner there is going to be a loser. So, you know, some will benefit; and there will be an equal and offsetting loser. Because, at the end of the day, it is a zero sum game. But the speculators, which I know you mentioned earlier in your opening——

Mrs. MYRICK. That is my concern.

Mr. NEWSOME. It is an inaccurate theory that speculators can move the market. Because if a speculator—if you look at them, they try to operate off price trends and to profit off of trends in the market, whether that price is up or whether it is down. But if they try to move out of a fundamental range, you have got hundreds of commercial participants who are in that competitive environment with them that pull them right back into the range or the speculator faces a severe financial penalty.

Then, at the end of the day, the speculators do not own the physical product, so they cannot move the price. They won't go to the delivery of a product, and they create virtually no impact on the settlement price, which is the price used by commercial market participants.

Mrs. MYRICK. You feel that there is sufficient oversight on this trading?

Mr. NEWSOME. Yes. The Exchange itself, through its self-regulatory function, serves the front line of enforcing the rules. Our compliance staff is the second largest behind our technology staff at the Exchange. We have many, many tools, and we utilize all of those tools to oversee the market.

The CFTC serves as our oversight regulator. They have physical bodies on the floor watching the trading. They do a rule enforcement review of us every year to see what kind of actions we have brought, what the penalties were, to make sure they are in line with the crime, whether that is a fine or suspension of trading. And they make sure that we are enforcing all of our rules.

So we use a multitude of compliance and enforcement activity to monitor our markets. The integrity of the marketplace is critical to the reputation of the Exchange.

Mrs. MYRICK. Sometime I would like to talk to you further about that.

Mr. NEWSOME. I would love to.

Mrs. MYRICK. I yield back my time, Mr. Chairman.

Mr. HALL. All right. Thank you.

The Chair recognizes the gentleman from Texas, Dr. Burgess.

Mr. BURGESS. I thank the chairman.

Mr. HALL. You are going to close for us. Do a good job.

Mr. BURGESS. All right.

Well, Mr. Chairman, we have quoted to us from the other side several times today what has been characterized as an internal Chevron document. Have you seen this?

Mr. HALL. Not tonight.

Mr. BURGESS. Well, it is stamped "Chevron pricing exhibit 156." the date underneath there, as best as I can make out with my bifocals, is sometime in 1996.

Up at the top it says: Note. This product is gathered from industry publications. And the rest of the line is unreadable. But the paragraph that has been oft quoted today—and I wanted to give Mr. Cavaney and Mr. Slaughter an opportunity to respond to the second paragraph, if they would like to, talking about Unocal: Unocal is exploring the sale of three refineries in California due to high capital expenditures required to comply with stringent environmental regularities.

I would just submit that for counterbalance to what you were struck with earlier today.

Mr. SLAUGHTER. Yes. Dr. Burgess, that is fairly typical. Because the investments required to comply with the Clean Air Act particularly—Unocal was operating in California, which has the strictest environmental standards that require huge amounts of investment. The gasoline sulfur, diesel sulfur regulation the industry is complying with now each costs $8 billion across the industry; and over the last several years, several refining organizations or individual refineries were sold or shut down, in rare instances. Many of them were sold because their owners felt that they could not economically invest that much money in that particular facility.

And you have to remember that before 2004 and 2005 return on investment in our industry was only 5 percent. That is very low, well below the median for all industries; and so it is understandable that companies would respond that way.

Mr. BURGESS. Mr. Cavaney, do you have any further that you want to add?

Mr. CAVANEY. No, that is fine.

Mr. BURGESS. One of the questions that I posed to our friend from the Department of Energy earlier today, as I was driving around in my district last Wednesday, when it really became apparent the extent of the trouble that was going on in Louisiana, did the Department of Energy have a contingency plan that they could pull off the shelf to deal with this type of emergency? And the answer that I got was, basically, the Strategic Petroleum Reserve, but there wasn't much of any kind of lever that the Government could pull.

Does industry have a contingency plan, given the concentration of the refining and drilling capacity that we saw on the map? Does industry have a contingency plan? Because it is not new information that hurricanes strike the gulf coast.

Mr. CAVANEY. We do have elaborate contingency and security plans. And when you go into these kind of circumstances, as you point out, this is something that we have to deal with regularly, not to the level of seriousness that this was, and that is what differentiates it, I think, from previous experiences.

Every single one of our companies at every single location has an action plan that goes into effect. They are drilled on it. They know what it is. We had abandoned all of those rigs out in the gulf. We had abandoned and shut down properly all the refineries along the coast well before the hurricane was on top of us.

Part of these procedures are safety for the personnel. They are to put the safes or the shut-downs on the various connect points, which is why we did not have any spills, because they worked. When you think of those refineries, the fact that there are only three of them now that have power and are not in the process of soon to be brought on board speaks well for the systems and the safety that is done there.

One of the concerns we had—and we have worked with the Department of Homeland Security who has looked at our various plans and the like—is the physical security of these spaces in that environment, which was something we had not too often had to deal with. So those plans were not as far along as they would otherwise be but will be.

So our industry, the ones that operate worldwide, these are the kinds of environments, the kinds of conditions that we face all of the time and have throughout the many, many decades that we have been operating there. So it is not a totally new experience, which is why I think you have been able to see the return as fast as you have seen it, because there had been plans in place.

I do want to say one thing about the Department of Energy and about the administration and about the Governors, even up here in some of the committees of jurisdiction. When this started to happen, everybody called us. We called them, and we have checklists about various things that could be done besides the SPR, the idea of waivers for Clean Air Act to allow us to have different fuels, the idea of letting the drivers stay in the trucks a bit longer than the Department of Transportation regs allow. So we all go down those, and to everyone's credit, nobody was in opposition to these things, and I think there was a lot of positive reinforcement which helped them go very fast.

So, in the aggregate, that was a pretty good plan, the way it worked out, because everybody knew things need to be addressed. So I think the response of the Government, while in some areas may not have been that good, certainly in terms of the oil and gas industry, I think it was quite responsive.

Mr. HALL. I am not going to cut you off. You go ahead.

Mr. BURGESS. Mr. Inslee took his chart about the CAFE standards. Let me stay with you, Mr. Cavaney, if I could, on the CAFE standards.

I have only been here a term and a half, and I am just a simple country doctor. But, as I recall, back in 1980 and 1981, the days that we had gas lines, of course, then the price of gas rose to over $1 a gallon, and the industry picked up its production, and more gas became available. But what drove the mileage, increased mileage in cars, my observation, back then was not the U.S. Congress, but the simple country doctor back in Texas who was trying to save a little cash.

Now I will just tell you, because I live in a district that has severe problems with air quality, that I tried to do the responsible thing a year and a half ago and ordered a car that was a hybrid car, gets 50 miles to the gallon. So now I look positively brilliant. It is a good feeling to drive along the road with a feeling of moral superiority to everyone else, and I do enjoy doing that.

But I guess what I am saying is, I will trust the American consumer, I would trust American ingenuity before I would trust the U.S. Congress with increasing CAFE standards. Do you have a feel about that?

Mr. CAVANEY. I am inclined to agree with you to that extent. I mean—but what we do need is we can use leadership on stressing the importance of energy efficiency and conservation. Because I do think, all of us, we find it in our own facilities constantly if you go in and relook at things with new technology. Because it changes and thinks a different way, and if you can create some lighter weight and gain some, so much better.

But the idea of command and control in terms of how you design cars, that is not our business. I can tell you that we have worldwide companies that operate, and we were talking earlier about greenhouse gas emissions reduction and the various records.

We here in the United States, our company is using voluntary systems that they are doing, because they own the same kind of refineries and other places, actually can give you better marks of reduction here by allowing the resourcefulness to tackle the system than they can and this is how you do it precisely.

So my comments before were because of the concept, and I think we ought to do more. But the idea that you draw the actual blueprints for people to do that I think is probably overreaching, and the industry ought to be the ones that make those decisions.

I worked in a gas station during that period of time. It was consumers coming in, and they were the ones that stated the preference for cars. They were not going to pay those terrible prices.

Mr. BURGESS. And, of course, you can go on the Internet today onto Google and type in Prius plus under the search engine; and eggheads across the country have figured out ways of squeezing 100 miles to the gallon. I do not have enough confidence in my own engineering ability to do that.

Mr. Chairman, I am going to wrap up. I just wanted to say to Mr. Angelle, I did not ask you a question. I did read your testimony. I apologize I wasn't here when you spoke. Many of your constituents are in my district and Tarrant and Denton and Cooke Counties, and I have heard their stories this past week. And it a cliche to say it, but I truly do feel your pain. I feel their pain. We have new mothers separated form new babies, and a real big question mark as to how we are going ever going to put all of these pieces back together again. You have got a tremendous job ahead of you in Louisiana.

But, as your neighbor in Texas, we have been proud to help, and certainly wish you all of the best as you go through those next several weeks and months. I know it is going to be a challenge to you.

Mr. Chairman, with that I will yield back.

Mr. HALL. I thank you. And you missed a good bit of testimony from—Mr. Angelle's testimony. But we have a record, and you will have copies of it.

Thank every one of you. You have been great. The chairman chose well in getting both of these groups. Thank you for leaving your homes. One of you missed your anniversary tonight. You are in trouble tomorrow.

But God bless all of you, and thank you very much for what you have done for this country.

[Whereupon, at 7:50 p.m., the committee was adjourned.]

[Additional material submitted for the record follows:]

NEW YORK MERCANTILE EXCHANGE
October 6, 2005

The Honorable JOE BARTON
Chairman
Committee on Energy and Commerce
U.S. House of Representatives
2125 Rayburn House Office Building
Washington, D.C. 20515-6115

DEAR MR. CHAIRMAN: Thank you again for the opportunity to testify before the Committee on Energy and Commerce hearing entitled "Recovering from Katrina" on September 7, 2005. I am pleased to submit the following responses to questions from Members of the Committee for inclusion in the hearing record.

Please contact me if you should have any additional questions regarding these responses, or any other matter.

Sincerely,

JAMES E. NEWSOME
President

RESPONSES TO FOLLOW-UP QUESTIONS FOR THE RECORD SUBMITTED BY THE HONORABLE PAUL E. GILLMOR

Question: What oversight body currently patrols your industry for pricing abuses that directly affect the amount that gas stations pay for their products?

Response: Gasoline futures market prices do not automatically or directly reflect the amount that gas stations pay for their product. As we understand it, gas stations receive fairly frequent deliveries of their product and so the prices they pay to their suppliers are closely tied to the near term market for delivery of the product in their geographic area.

By contrast, NYMEX lists a series of separate calendar months in its New York Harbor Unleaded Gasoline futures contract. Each of these calendar months is separately traded. For market participants that stay in the market through the termination of trading in a particular listed calendar month, this gasoline contract provides physical delivery of tankers of gasoline during a specified delivery period of several days during the specified month between two commercial market participants in the harbor area adjacent to New York City. NYMEX does not list any gasoline contracts for next-day or day-ahead delivery of the physical "cash" commodity. On the other hand, it is our understanding that there are frequent transactions in the off-exchange "over-the-counter" (OTC) market for day-ahead or other near term delivery of the cash commodity in various geographic regions.

Any regulatory body that polices abuses affecting the price of gas at the pump, such as price gouging, has no connection to futures and would exist outside of the framework of futures regulation. NYMEX is under the full and direct regulatory authority of the Commodity Futures Trading Commission, which has exclusive jurisdiction over futures trading and has broad anti-fraud and anti-manipulation authority over its regulated markets. In addition, manipulation of cash market prices that impact prices on a futures market would be subject to the CFTC's anti-manipulation authority.

Question: The public debate is focused on the revenues of petroleum producers and retail gasoline outlets, however, little attention has been paid to the increasing income of oil commodities traders. What activities do you think can be meaningfully done to restrain the consistent and upward movement of gasoline futures?

Response: This question seems to be based on two premises: (1) that income for oil commodity traders is uniformly increasing; and (2) that this rise in income is somehow responsible for "consistent and upward movement" of gasoline futures prices. Both premises need to be re-examined carefully.

In discussing futures trading, it is important to understand how futures trading works and to understand the nature of futures markets. Futures markets allow market participants either to buy a contract that requires one to receive future delivery of the product (also referred to as a "long" contract) or to sell a contract that requires the company to make physical delivery of the product (also referred to as a "short" contract). The vast majority of open positions are not held through to delivery of the product. Instead, market participants close out or liquidate their open positions by buying a contract on the other side of the market that offsets their existing position (e.g., a long market participant buying a short contract to zero out its existing market exposure). For such participants who close out their positions, whether they make or lose money on the two transactions, i.e., the initial transaction and the close-out transaction, depends on the difference between the prices of their long and short contracts, which depend on where the market was at the time of those transactions.

The critical point to note is that a futures market is a zero sum type of market. In other words, for every winner there is also a loser. We believe that this basic reality needs to be kept in mind before placing too much faith in a wholly speculative statement made by someone sitting in an office 75 or 100 miles away from our trading floor. Clearly, there was a significant shift in energy prices, including gasoline prices, in the wake of Hurricane Rita. The impact of this shift in prices on the broad category of oil traders was related to a considerable extent to which side of the market they were on when the market moved. Some traders made money. On the other hand, as gasoline futures trading is a zero sum market, other traders necessarily lost substantial sums of money during the same period.

Turning specifically to NYMEX floor members, many floor members focus upon trading for their own personal or proprietary trading accounts. In general, these floor traders prefer to limit sharply their market risk exposure to overnight changes in market prices. Simply put, these traders prefer to go home at night being "flat" in the market and having neither a long or a short position. To the extent that such traders are successful in their trading activity, they do so by providing short-term liquidity to the market during the trading day and fluctuations in market price during a given trading session may increase the value of that activity. On the other hand, these floor traders have no incentive to gain from a consistent, upward movement in price that occurs over an extended period of time because, as noted, they close out their market positions at the end of each trading session.

Even with respect to other non-floor market participants who do maintain open positions over a period of time, NYMEX maintains strict position limits on their market activity and engages in extensive market surveillance to restrict sharply the ability of any market participant to engage in prohibited manipulative activities. The CFTC additionally has extensive resources devoted to market surveillance and enforcement efforts.

As to the broader question of possibly trying to restrict price movement in gasoline futures, a central role of futures markets is to provide a price discovery service for future prices. In this way, futures markets provide a neutral tool or a gauge as to these future prices. Futures markets are simply auction markets where buyers and sellers come together and as a result a market price is determined through open and competitive execution. This market price is derived through basic demand and supply fundamentals.

In particular, the gasoline futures price of an expiring contract month is designed to converge with the cash market price at expiration of that contract month. The futures market is a derivative of the cash market. Futures prices, which reflect the underlying cash market, are closely related to cash prices and are driven by the same economic factors. At expiration of the contract during the spot (delivery) month, the futures price converges to a single price with the cash price, which provides transparency and an efficient hedging tool for the market. Any thing that artificially restricts futures prices will have no effect on the cash market. Instead, such artificial restraints would remove liquidity and price transparency from the futures market (and shift it to the less-regulated and less transparent OTC market), and the futures market consequently would no longer be a reliable hedging tool.

In our view, the real key to truly meaningful attempts to address gasoline prices is to focus upon fundamental supply and demand issues. A comprehensive energy policy that focuses on rebuilding the US refining infrastructure is critically needed to address the supply issues, which are causing gasoline prices to be high. No refineries have been built in the US in 20 years and small ones keep going off-line. Congress should review this trend and consider whether any federal policies, including increasingly stricter EPA standards, play any role in the shrinking supply of refineries.

Question: Do you think that the anti-competitiveness and pricing protections in the petroleum marketing practices act should be extended to transactions involving oil commodities traders?

Response: We understand this question as directed to NYMEX to be referring to oil commodity traders in our futures markets. In response, we do not believe that it should be extended because the goals of that act do not apply to trading in our markets. The stated goals of this act are as follows:

> "[t]o provide for the protection of franchised distributors and retailers of motor fuel and to encourage conservation of automotive gasoline and competition to the marketing of such gasoline by requiring that information regarding the octane rating of automotive gasoline be disclosed to consumers."

Our gasoline futures contract is a standardized contract with established product specification terms that are consistent for all contracts. Our market participants are focused on our markets and generally have no awareness or interest or control over what may eventually happen with respect to disclosure of octane ratings at individual gas stations. This reflects a fairly basic instance of a difference between apples and oranges, and thus the goals and scope of that act are inapplicable to what happens on our markets.

RESPONSES TO FOLLOW-UP QUESTIONS FOR THE RECORD SUBMITTED BY THE HONORABLE BART STUPAK

Question 1. Do you believe "risk premium" plays a role in high oil and gas prices in our country?

Response. It may be useful to clarify NYMEX's role in connection with market activity. NYMEX provides a neutral market forum for the open and competitive execution of trades in our listed contracts. NYMEX does not itself engage in any trading activity or establish any market positions. Furthermore, NYMEX does not provide any services as a market advisor or analyst, and so our comments below are necessarily somewhat general in nature.

Academics have debated this issue for quite some time. The existence of "risk premium" can be neither proved or disproved because it is an abstract concept and cannot be observed directly and may be interpreted differently by different analysts or scholars. However, the implication is that "risk premium" would cause the expected futures spot price to be either higher or lower than the cost adjusted futures price. If a market participant is concerned about future tight supplies or concerned about the possibility of supply interruptions due to some type of extrinsic shock such as a political event or terrorist activity at some point along the supply chain and that participant is risk averse, then it is possible that the participant may be willing to pay a higher market price, than a participant that is risk neutral. These market decisions are based strictly on personal judgments and there are individuals in the market who are risk averse and willing to pay a premium. Alternatively, there may be those who are risk neutral and are indifferent to risk and may not be willing to pay a premium.

Although we cannot say a risk premium absolutely exists, NYMEX is a liquid and transparent market and provides the most efficient and reliable mechanism for price discovery as to future prices. By providing a centralized auction market for discovery of future prices, our markets allow market participants both to incorporate current known information as to demand and supply as well as to reflect their various views as to future demand and supply, including the views of the risk-averse as well as the risk-neutral.

Question 2. I've seen several articles that have said that prior to the Iraq war and prior to Hurricane Katrina, the risk premium on a barrel of oil was in the neighborhood of $5 per barrel. Since the beginning of the war and since Katrina, I have seen articles stating that the risk premium is now anywhere from $15-$30/barrel. Is this your understanding?

Response. Such an assertion (about a current risk premium of that size) is questionable for the reasons discussed below. As background, academics have argued that producers looking to hedge future production tend to push down futures prices by the nature of their selling activity. Buyers (speculators) of futures contracts incur the price risk and theoretically would be willing only to buy the contracts at below fair expected price due to the risk premium that would be incurred by this price risk. The result would be what is referred to in futures markets as a "backwardation" in the market where the futures price would decline progressively out into the future for the distant contact months in relation to the current or front calendar month of that futures contract. However, a review of recent crude oil prices at NYMEX reveals that this is **not** the case at present in the NYMEX market.

Since the passage of time that has now occurred following Hurricane Katrina and Hurricane Rita, the market has had sufficient time to absorb currently available information regarding the impact of the hurricanes on possible future demand and supply of the commodity. Prices traded in expectation of the hurricanes have adjusted since then with increasingly better information about the market consequences of those major events in the Gulf Coast region. As noted in our response to the prior question, the existence of and extent of any risk premium is open to discussion and is not susceptible to precise measurement. Instead, these topics would seem to be a matter of fairly broad estimates and conjecture. That stated, however, from our modest vantage point as a neutral market forum and in light of the improving quality of market information regarding the actual impact of the hurricanes, an estimate of a $15-30 risk premium at this point in time strikes us as questionable at best. Such an estimate suggests that the market is substantially overpriced and inefficient.

Currently, there is little difference between crude oil prices for 2005 and prices for 2006 and 2007, i.e., futures prices are approximately equal to futures spot (cash) prices. On October 3, 2005, the November 2005 crude oil futures contract settled at $64.47/barrel. The December 2006 contract was virtually the same at $65.49/barrel. The inference to be drawn is that to any extent that risk is being shifted in the market, it is being equally shifted between buyers and sellers. If this were not the case, there would be either a backwardation or contango (the opposite of backwardation where the prices for distant contract months are progressively **higher** than the current contract month) in the market. Current price trends indicate otherwise.

Question 3. Do you believe that the instability in the Middle East and Venezuela and the War in Iraq have caused the price of crude oil contracts on the NYMEX to increase?

Response. NYMEX provides a neutral market forum for open and competitive execution of trading in our products. We do not provide market analysis or market advice and our response should be considered in this context accordingly.

That stated, the United States now obtains a substantial amount of the crude oil needed for our economy from foreign imports, including the areas noted in the question. Specifically, the United States—averaged *total net oil* (crude and products)—imports of—an estimated 11.8 million bbl/d—during January-October 2004, representing around 58% of total U.S. oil demand. Crude oil imports from Persian Gulf sources averaged 2.4 million bbl/d during that period. Overall, the top suppliers of crude oil to the United States during January-October 2004 were Canada (1.6 million bbl/d), Mexico (1.6 million bbl/d), Saudi Arabia (1.5 million bbl/d), Venezuela (1.3 million bbl/d), and Nigeria (1.1 million bbl/d).

What seems clear is that the price of crude oil and other energy products has risen due to both supply and demand factors. Instability in the Middle East, Nigeria, and Venezuela, along with rapid demand growth in emerging market countries such as China and India seemingly have both contributed to the higher price levels. More temporary setbacks from recent hurricanes also appear to have further exacerbated the trend.

Question 4. If investors are concerned about possible terrorist attacks and the risk premium per barrel goes up an additionally $15-$30 per barrel, who reaps these benefits?

Response. As noted previously, we cannot state definitively that there is a risk premium and we do not have any conclusive views as to the size of any such premium. Indeed, stating that a risk premium is between $15-$30/barrel essentially translates to saying that oil is overpriced by that amount. Anyone who believes this is true would attempt to take advantage of the arbitrage opportunity available. A market participant would be able to sell oil now and cover his position later in the spot market for delivery. By the time of delivery, if the market participant's belief that crude was overvalued was correct, the price would be $15-$30 cheaper. For anyone who believes this scenario it would be an unprecedented opportunity to profit, however, the selling would in effect make the price go down more and more removing the premium.

In response to the premise of the question, though, in general and to the extent that oil prices rise, producers gain at the expense of consumers. Since much of the production occurs outside of the US, the benefits of higher prices generally would accrue to countries and companies that produce oil for export.

Question 5. To what extent do you believe that the futures prices for both crude and refined products supported or even advanced the sharp increases in crude and product prices?

Response. This question suggests a possible connection between the prices for crude oil and gasoline futures contracts for future periods of time and the prices for current delivery of these products in the cash market. Futures market prices do not automatically or directly reflect the amounts that are paid in the cash market for near term deliveries. Taking gasoline as an example, as we understand it, gas stations receive fairly frequent deliveries of their product and the prices they pay to their suppliers are closely tied to the near term market for delivery of the product in their geographic area.

By contrast, NYMEX lists a series of separate calendar months in its New York Harbor Unleaded Gasoline futures contract. Each of these calendar months is separately traded. For market participants that stay in the market through the termination of trading in a particular listed calendar month, this gasoline contract provides physical delivery of tankers of gasoline during a specified delivery period of several days during the specified month between two commercial market participants in the harbor area adjacent to New York City. NYMEX does not list any gasoline contracts for next-day or day-ahead delivery of the physical "cash" commodity. On the other hand, it is our understanding that there are frequent transactions in the off-exchange "over-the-counter" (OTC) market for day-ahead or other near term delivery of the cash commodity in various geographic regions.

However, while futures market prices do not automatically translate into the price for next-day delivery in the cash market, futures markets do provide one stream of information that may be considered by market participants in the cash market. Indeed, the availability of NYMEX's reliable and neutral market prices for future periods of time may actually **reduce** the possibility of price shocks in the cash markets.

In other words, the futures market can play an enormously constructive role in keeping prices in the cash market from unnecessary spikes. From time to time, cash markets may over react in the short term to breaking news about demand and supply based on the predicted severity of an event. As information becomes more and more available, as was the case after Katrina and Rita, the extent of damage was better understood and market prices came down very quickly. The futures market's role as a price discovery vehicle for future prices arguably prevented the prices in the cash market from going up as much as they may have in the absence of NYMEX's stable and neutral market forum.

By contrast, a historical example may further illustrate how the cash market operates less efficiently if an effective futures market price discovery service is not available. During the historical gasoline crisis that began in 1979, there was limited price transparency in US energy cash markets. Between 1979 and 1981 (before gasoline futures prices were fully embraced and accepted by the energy industry as a benchmark for future prices), the Exchange was receiving information from price reporters for cash markets that high prices would be sustained for extended periods of time, months, possibly years. During this period, US government price allocation controls were in effect and prevented an easing of the imbalance between supply and demand by restricting the market from reacting to weakening demand. Market prices went up quickly and dramatically and stayed there beyond what the current events would have been able to explain.

QUESTIONS FOR BOB SLAUGHTER, PRESIDENT, NPRA, FROM CONGRESSMAN BART STUPAK

1. During your appearance before the Committee, I asked you the following question:

"Isn't it true that in 1992, the State of Arizona granted a permit to the Arizona facility for construction and operation, and the company sat on the permit for nearly 8 years without actually building a refinery?"

You responded by stating: "Not to my knowledge, no" and you went on to state that "they had to move the refining site."

QUESTION: Isn't it true that in a response you provided this committee earlier this year, answering follow-up questions from congressman Dingell for the February

16, 2005 hearing entitled: "The Energy Policy Act of 2005, Ensuring Jobs For Our Future With secure and Reliable Energy," you indicated that:

"Maricopa Refining Company (MRC) was issued an installation permit for a 50,000 BPD refinery by the ADEQ on January 16, 1992," and you later acknowledged in that same response to Congressman Dingell that "the above permit was allowed to lapse and a new permit for a larger facility was submitted to ADEQ on December 23, 1999"?

NPRA Response and Background

The following is a chronology from Arizona Clean Fuels in response to the above questions:

1. Maricopa Refining Company (MRC) was issued an "Installation Permit" for a 50,000 BPD refinery by the ADEQ on January 16, 1992.
2. MRC (under the name of Arizona Clean Fuels-ACF) continued development of its refinery project in the early and mid-nineties. A significant financial investor left the project. The project was re-scoped as to refinery capacity and feedstock. The above permit was allowed to lapse and a new permit for a larger facility was submitted to ADEQ on December 23, 1999.
3. The ADEQ hired an outside contractor to prepare the permit. This contractor worked with ACF, ACF's contractor and the ADEQ to perform the BACT reviews, etc. required by the Clean Air Act. In September 2002, the above parties agreed that the information required to perform all of the permit reviews was complete and the ADEQ confirmed this on September 4, 2002.
4. During the summer of 2003, the EPA and ADEQ declared an expansion of the ozone non-attainment area in Maricopa County that included the site of the proposed refinery. ACF advised the ADEQ that it was considering alternate sites for the refinery outside Maricopa County.
5. On October 30, 2003, the ADEQ issued a proposed Draft Air Permit to the company only, for the refinery based on the December 1999 application and the Maricopa County site. This permit was not formally issued pending decision by ACF on location.
6. In October 2003, ACF advised the ADEQ that the company was proposing a new site for the refinery in Yuma County and the information required to revise the permit for the new location was submitted during the November 2003 to March 2004 period. This information was consolidated into a "new permit application" document that was submitted to ADEQ on June 28, 2004. The refinery facility was identical to that proposed in 1999 so the BACT analysis remained valid. Revisions required for the new site consisted primarily of new air emission impact modeling
7. The ADEQ issued the Draft Air Permit on September 14, 2004. Public meetings and hearings were held during October and November 2004 with the public notice period closing on January 10, 2005.
8. The permit is currently in review by the EPA with a formal response required by March 18, 2005

ONE HUNDRED NINTH CONGRESS

U.S. House of Representatives

Committee on Energy and Commerce

Washington, DC 20515–6115

JOE BARTON, TEXAS
CHAIRMAN

March 3, 2005

The Honorable Ralph M. Hall
Chairman
Subcommittee on Energy and Air Quality
Committee on Energy and Commerce
2125 Rayburn House Office Building
Washington, D.C. 20515

Dear Mr. Chairman:

Pursuant to rule 4(c)(2) of the Rules of the Committee on Energy and Commerce, I request that you forward the attached follow-up questions for the record from Rep. Hilda L. Solis and me to the identified witnesses who appeared before the Subcommittee on Energy and Air Quality hearing entitled "The Energy Policy Act of 2005: Ensuring Jobs for Our Future with Secure and Reliable Energy" held on February 16, 2005. The attached questions, along with the witness responses, should be included in the printed hearing record.

I would further ask that when the responses are received, a copy of each response be forwarded to Candy Butler (cb2000@mail.house.gov) of the Committee's Democratic staff. Ms. Butler will be responsible for forwarding the responses to the Members who originated the questions.

If any further information is required, please have your staff contact Candy Butler at ext. 6-3400 or by e-mail (cb2000@mail.house.gov).

With every good wish.

Sincerely,

JOHN D. DINGELL
RANKING MEMBER

Questions for Mr. Bob Slaughter
President, National Petrochemical & Refiners Association
from the Honorable John D. Dingell
Subcommittee on Energy and Air Quality
Regarding the February 16, 2005, Hearing entitled:
"The Energy Policy Act of 2005:
Ensuring Jobs for Our Future with Secure and Reliable Energy"

1. At the Subcommittee on Energy and Air Quality hearing on "The Status of the U.S. Refining Industry" held on July 15, 2004 (Serial No. 108-113), you testified as follows in response to questions from Rep. Gene Green:

   "Mr. Green, if I could, I just wanted to mention that there has been mention of one project for building a new refinery in Arizona that is going on. They have been trying to get a permit there for 10 years, and they haven't got one yet. So, I think that puts into context why people are focusing on existing sites."

   a. Is that a correct reporting of your testimony?

   b. According to information provided by Arizona state officials, a permit was issued to Maricopa Refining Co. (a.k.a. Arizona Clean Fuels) allowing construction and installation of equipment on January 16, 1992. Do you agree that such a permit was issued by the State of Arizona?

   c. Apparently the permit applicant had difficulty obtaining financing and allowed the permit to laspe. Almost eight years later on December 23, 1999, the contractor for Arizona Clean Fuels applied for a new air quality installation and operating permit according to Arizona state officials. Do you agree that a new permit application was submitted in December 1999?

2. According to Arizona officials, the Arizona Department of Environmental Quality on September 4, 2002, notified Arizona Clean Fuels that the permit application was deemed complete. Are you aware that just over one year later, on October 30, 2003, the Arizona Department of Environment Quality was notified that the permit applicant, Arizona Clean Fuels, intended to relocate the proposed project to a new location in Yuma, Arizona?

3. Are you also aware the Arizona Department of Environmental Quality received a new application for an air quality installation and operating permit on July 14, 2004, or one day before your testimony before the Subcommittee?

NPRA

Bob Slaughter
President

National Petrochemical & Refiners Association

1899 L Street, NW
Suite 1000
Washington, DC
20036.3896

202.457.0480 voice
202.457.0486 fax
bslaughter@npra.org

The Honorable Ralph Hall
Chairman
Subcommittee on Energy and Air Quality
Committee on Energy and Commerce
United States House of Representatives
2125 Rayburn House Office Building
Washington, D.C. 20515-6115

Dear Chairman Hall:

On behalf of the National Petrochemical & Refiners Association (NPRA), thank you for giving me the opportunity to respond to the questions you have submitted on behalf of yourself, Congressman Dingell, and Congresswoman Solis. I also appreciated the opportunity to present NPRA's views at your February 16, 2005 hearing, *Energy Policy Act of 2005: Ensuring jobs for Our Future with secure and Reliable Energy.*

NPRA looks forward to working with the Committee in the hope that our recommendations will become an integral part of comprehensive energy legislation. If you have any questions or concerns regarding this submission, please do not hesitate to contact me.

Sincerely,

Bob Slaughter

The Honorable John D. Dingell:

You have raised a number of questions regarding the status of permitting for new refinery construction in Arizona. As a preliminary matter, we would offer the following observations regarding the timeline:

1. Maricopa Refining Company (MRC) was issued an "Installation Permit" for a 50,000 BPD refinery by the ADEQ on January 16, 1992.
2. MRC (under the name of Arizona Clean Fuels - ACF) continued development of its refinery project in the early and mid-nineties. A significant financial investor left the project. The project was re-scoped as to refinery capacity and feedstock. The above permit was allowed to lapse and a new permit for a larger facility was submitted to ADEQ on December 23, 1999.
3. The ADEQ hired an outside contractor to prepare the permit. This contractor worked with ACF, ACF's contractor and the ADEQ to perform the BACT reviews, etc. required by the Clean Air Act. In September 2002, the above parties agreed that the information required to perform all of the permit reviews was complete and the ADEQ confirmed this on September 4, 2002.
4. During the summer of 2003, the EPA and ADEQ declared an expansion of the ozone non-attainment area in Maricopa County that included the site of the proposed refinery. ACF advised the ADEQ that it was considering alternate sites for the refinery outside Maricopa County.
5. On October 30, 2003, the ADEQ issued a proposed Draft Air Permit to the company only, for the refinery based on the December 1999 application and the Maricopa County site. This permit was not formally issued pending decision by ACF on location.
6. In October 2003, ACF advised the ADEQ that the company was proposing a new site for the refinery in Yuma County and the information required to revise the permit for the new location was submitted during the November 2003 to March 2004 period. This information was consolidated into a "new permit application" document that was submitted to ADEQ on June 28, 2004. The refinery facility was identical to that proposed in 1999 so the BACT analysis remained valid. Revisions required for the new site consisted primarily of new air emission impact modeling
7. The ADEQ issued the Draft Air Permit on September 14, 2004. Public meetings and hearings were held during October and November 2004 with the public notice period closing on January 10, 2005.
8. The permit is currently in review by the EPA with a formal response required by March 18, 2005

As a general matter, the refining industry has successfully gone through a major effort over the past decade to respond to changes in product fuel quality mandated by Clean Fuels requirements. During this time, the industry has met the growing domestic demand for petroleum products by limited capacity expansions of existing refineries, and by imports. No new refineries have been built in the U.S. in over twenty years and product imports have reached over 2 million barrels per day. Economic growth in other countries has reduced the availability of products to U.S. consumers and increased competition for imports. Recent petroleum product prices have reached and sustained record highs, driven by a growing shortfall in supply. There are a number of reasons that this shortfall is a major concern for the U.S., most of which have been documented in abundance recently in the press. It is perhaps sufficient to state that shortfalls create economic hardship and slow the economy. It is also a strategic issue for the U.S. to grow imports and increase the threat of shortages and embargos.

One of the major solutions to this growing shortfall is to provide additional domestic refining capacity.

The problems and impediments preventing the growth and investment for new refining capacity in the U.S. are significant. Despite this, a new refinery project, the Arizona Clean Fuels (ACF) project, has been proposed and will be completing engineering design consistent with the final Air Permit expected to be issued later this year. This project will be used below to highlight specific costs and permitting requirements.

New Refinery Construction Considerations

There are four general areas of consideration that drive the feasibility and timing of new refining projects:

1. Overall Project economics driven by product values, feedstock costs, and operating costs,
2. Technology choices driven by crude slate, target product mix, legislated and target product quality requirements (and projected changes) – a lengthy process of project development, engineering and construction,
3. Public Acceptance – significant reluctance in most areas of the U.S. to allow a new refinery "in my back yard". Public communication and hearings processes are lengthy and often confrontational,
4. Permitting processes for environmental permits, access permits, construction permits and zoning, etc. – driven by federal, state, and local legislation and zoning.

Refining Economics

Historical refining margins in the U.S. have, on average and in general, not been adequate to support new refinery construction. Returns on Capital Employed have been in the 5% to 7% range. Capacity expansions and modifications have been economic due to leverage on base infrastructure and facility investments.

Refinery sales transactions over the past ten years have, on average, been at about 25% of the cost of new-build facilities. Condition of the plants, local markets, and a company's perspective on future cash flows drive the valuation process. These facilities often require significant additional investment to ensure reliable operation and compliance with regulatory requirements.

Refineries are by their nature very costly facilities. The proposed ACF refinery which will produce about 150,000 barrels per day of gasoline, diesel, and jet fuel products, will cost over $2 billion with an additional $500 million required for crude oil and product pipelines. Rapidly growing demand for petroleum products in the southwestern U.S. makes this project economic.

Technology Choices

The refining industry is not traditionally viewed as "high tech". However, the need for high quality products and significant flexibility to process wide ranges of crude oils, and the need to implement state-of-the-art environmental controls, has led to the development of very sophisticated processes. There are several process licensors and choices for each type of facility that a refiner needs. Also, due to the high cost of each process facility, extensive studies and comparisons are required to match a refiner's products and processing objectives.

One area where the industry has led in major technology developments is in the "Best Available Control Technology" for emissions as defined in and required by the Clean Air Act. Every refinery modification and new process unit has required the development and application of specific control technology.

The development of the Arizona Clean Fuels project included an extensive analysis of emission sources and inclusion of the Best Available Control Technology. This will be the first refinery where all sources will be addressed at the same time in this manner.

Public Acceptance

A major hurdle to the construction of a new oil refinery is to overcome the historic public perceptions of oil refineries and to obtain public acceptance. Generally, the public has a "not in my back yard" attitude to oil refineries. Certainly, refineries of the past have, to some extent, earned this reaction from the public. Modern facilities have overcome the shortcomings of these previous refineries. The refining industry has developed and implemented emissions controls, operating practices, and outreach programs to address the concerns of both government agencies and the public. Certainly these programs and projects have increased costs, but have been viewed by the industry as necessary.

Refineries have significant benefit to the public by generation of both direct and indirect jobs and economic activity. Local communities can benefit significantly from the operation of a refinery.

A new refinery, such as the Arizona Clean Fuels project, with the control and monitoring required by current regulations will have minimal impact on the surrounding environment.

The proposed locations in Yuma County, Arizona, are remote from population concentrations. The project has gained support from local politicians and business leaders.

Permitting Processes

Certainly the most-often noted issue in new refinery construction is that of the extensive permitting that is required. Generally, permits are required from multiple agencies at the federal, state and local levels. Also permits are required not only for the refinery but also for pipeline and utility services to and from the site. The permitting processes are lengthy and costly. Project developers are also not in control of the pace and timing of permit review and issue and this uncertainty can lead to project delays and cost escalation.

The most extensive and important permit is often the "Air Permit" that is usually issued by the relevant state agency and outlines all requirements for compliance to the Clean Air Act and New Source Performance Standards with emission levels, reporting and Best Available Control Technology requirements. The extensive scope of this permit requires detailed air modeling, technical review of all facilities, and agreement on the Best Available Control Technology. For example, the Arizona Clean Fuels permit application was submitted to the Arizona Department of Environmental Quality on December 22, 1999, and a Draft Permit issued on October 10, 2003 – a time period of almost four years. In response to the declaration of large portions of Maricopa County as a "Non-Attainment Zone" for federal Ozone standards in the summer of 2003, the proposed refinery was moved to a site in Yuma County and a revision to this Draft Permit is still pending. Following its proposal, reviews, public hearings, and final permit drafting will take several months.

Fortunately, some other federal and state agencies review and comment on the permit and project coincident with the preparation of the Air Permit. For example the EPA, the U.S. Forest Service and the National Park Service will be consulted by ADEQ. However, all of these agencies have seen increased demands on their time and reviews don't always meet the expected timeframes thereby extending the permitting schedule. In the western_United States, for example, EPA Region IX encompasses the most dramatic growth seen anywhere in the country. However, large projects that would support and provide jobs for that growing population can be held up for years by the air permitting process alone. This Regional EPA office has a limited number of technical staff members who must review and approve the air permits for every project in California, Nevada, Arizona, Hawaii, and Guam. Similarly, the National Park Service, Bureau of Land Management, and U.S. Forest Service must compete for the services of only a few federal staff members who have the technical expertise and responsibility to review all proposed major source air permits for projects across the entire western half of the country. This coupled with the lack of regulated or recommended timing requirements for permit issue leads to significant delays. Finally, although industry recognizes the statutory requirement for these agencies to ensure compliance with all regulations, there often appears to be more attention paid to the concerns of a small minority of constituents rather than a balanced review.

Although the Air Permit is one of the most important permits for any project, there are many other rigorous permits that must be obtained for both refinery and pipeline projects from a multitude of agencies. For example:

- NEPA Compliance from a controlling agency such as the Bureau of Land Management
- Land Use Permits from controlling agencies and jurisdictions
- National Historic Preservation Act Compliance
- Access permits from Bureau of Land Management, U.S. Army Corps of Engineers, and State Land Commissions as well as private land owners.
- Military Agency approvals if military facilities involved.

A listing of permits required by the Arizona Clean Fuels refinery and pipeline projects shows about thirty permits required excluding local zoning, access and construction permits. The majority of these permits are not initiated until the Air Permit is issued, since it finalizes the basis for the project. The timing of these can be extensive and is estimated to be about eighteen to twenty-four months. Although design engineering can be done in parallel to these permitting activities, no significant construction can begin until they are in place. Construction of a large refinery such as ACF proposes takes about three years. This sequential process results in long lead times for project development and completion.

Indisputably, the refining industry in the U.S. has not constructed a new grass roots refinery for over twenty years. Refining economics have generally not supported new refinery costs and the industry has focused on expansions of existing refineries. Major investments in Clean Fuels production and regulatory programs have also absorbed much of the industry capital. The total capital cost of an economically-sized facility of about 150,000 barrels per day is approaching $3 billion.

The complexity of the refining processes and technology choices results in lengthy project development times which can be one to two years. Following this project definition, corporate strategic decisions, public reviews, local government discussions, and multi-level permitting process typically take four to five years before a final "go-decision" can be made. Engineering and construction on a significant project is a major undertaking and takes three to four years. Total project time from inception to startup is in the order of ten years.

The massive investments required for development of a new refinery project coupled with uncertainty on timing and final approval of permits, issues of public acceptance and market uncertainty in the future, have deterred the refining industry from new projects.

Some efficiencies may be possible in the overall development timing. Internal corporate engineering and construction efficiencies may reduce overall project timing. Reducing the number of agencies involved in major project permitting through the "lead agency" approach and ensuring internal accountability for permit issue timing could reduce time and workload on all agencies involved.

FINAL ARIZONA PERMIT

Governor Janet Napolitano
State of Arizona

Stephen A. Owens, Director
Arizona Department of Environmental Quality

1110 W. Washington St.
Phoenix, AZ 85007

(602) 771-2338 Voice
(602) 771-2366 Fax

ARIZONA DEPARTMENT OF ENVIRONMENTAL QUALITY
AIR QUALITY CLASS I PERMIT

COMPANY NAME:	Arizona Clean Fuels Yuma, LLC
FACILITY NAME:	Arizona Clean Fuels
PERMIT NUMBER:	1001205
DATE ISSUED:	**April 14, 2005**
EXPIRATION DATE:	**April 14, 2010**

SUMMARY

This operating permit is issued to Arizona Clean Fuels Yuma, LLC, the Permittee, for operation of the Arizona Clean Fuels petroleum refinery. The permitted facility will be located on an approximately 1,450-acre site approximately 40 miles east of Yuma, near the town of Tacna, in Yuma County.

The permitted facility will have the capacity to refine approximately 150,000 barrels per day (4.6 million gallons per day) of crude oil and approximately 1.8 million gallons per day of other petroleum-based materials. The primary products of the refinery will be gasoline, jet fuel, propane, and diesel fuel.

The major process units at the proposed refinery will include a crude distillation unit, a delayed coking unit, a hydrocracker, a distillate hydrotreater, a naphtha hydrotreater, a naphtha catalytic reforming unit, a butane conversion unit, a benzene reduction unit, and an isomerization unit. Supporting process units will include a gas concentration unit, a hydrogen generation unit, an amine regeneration unit, two sulfur recovery units, a sour water stripper, two steam boilers, a wastewater treatment plant, a tank farm, product loading facilities, a mechanical-draft wet cooling tower, and three internal-combustion engines used to drive emergency equipment.

Emission units, emitting activities, and pollution control equipment at the permitted refinery will include the following:

- Fifty-one storage tanks for petroleum liquids. Five of these tanks will be equipped with vapor control systems vented to compressors and forty-six will be equipped with internal floating roofs and vapor control systems vented to a thermal oxidizer;
- Two steam boilers fired with natural gas. These boilers will be equipped with low-NO_X burners and flue gas recirculation;
- Eighteen process heaters fired with refinery fuel gas. Each of these heaters will be equipped with low-NO_X burners; nine will also be equipped with selective catalytic reduction;
- Two sulfur recovery units, equipped with a common tail gas treatment unit and thermal oxidizer;

- Catalyst regenerators at the catalytic reforming unit and the butane conversion unit, each equipped with a caustic wet scrubber;
- Equipment leaks, emissions from which will be minimized through implementation of a leak detection and repair program;
- Coker pit;
- Coke storage silo, equipped with a fabric filter baghouse;
- Coke railcar loading system;
- Two emergency flares;
- Truck loading rack, equipped with a vapor control system and thermal oxidizer;
- Wastewater treatment plant, equipped with a thermal oxidizer;
- Wastewater treatment plant solids dryer, equipped with a fabric filter baghouse;
- Three diesel-fired reciprocating internal combustion engines used to drive two emergency fire water pumps and an emergency electric generator;
- A mechanical-draft wet cooling tower, equipped with a high-efficiency drift eliminator; and
- Vehicle traffic on paved and unpaved roads.

All definitions, terms, and conditions used in this permit conform to those in the Arizona Administrative Code (A.A.C.) R18-2-101 and Title 40 of the Code of Federal Regulations (CFR), except as otherwise defined in this permit. Unless noted otherwise, references cited in the permit conditions refer to the A.A.C. All material permit conditions have been identified within the permit by a double underline. All terms and conditions in this permit are enforceable by the Administrator of the U.S. Environmental Protection Agency, except for those terms and conditions that have been designated as "State Requirements."

The Arizona Clean Fuels Yuma, LLC petroleum refinery will be a major source because the potential emission rates of the following pollutants are greater than 100 tons per year: particulate matter (PM), PM_{10}; nitrogen oxides (NO_X), carbon monoxide (CO), and volatile organic compounds (VOC). In addition, the proposed refinery has potential emission rates of hazardous air pollutants in excess of 25 tons per year in total.

This permit is issued in accordance with Titles I and V of the Clean Air Act, and Title 49, Chapter 3 of the Arizona Revised Statutes (ARS).

ARIZONA DEPARTMENT OF ENVIRONMENTAL QUALITY

1110 West Washington Street· Phoenix, Arizona 85007
(602) 771-2300 · www.adeq.state.az.us

Janet Napolitano
Governor

Stephen A. Owens
Director

VIA FEDERAL EXPRESS

AQD:CTS:94383

July 29, 2004

Jeff Donofrio
Committee on Energy and Commerce Democratic Staff
U.S. House of Representatives
2322 RHOB
Washington, D.C. 20515

Subject: Chronology and History of the Proposed Arizona Clean Fuels Refinery

Dear Mr. Donofrio:

As was promised in a July 21, 2004, conference call with you and your colleagues, I have enclosed with this letter a chronological summary of the documents that relate to the Arizona Department of Environmental Quality's efforts to process air quality applications for a proposed new refinery to be constructed and operated by Arizona Clean Fuels. In addition to the chronological summary, I have also enclosed copies of the documents used to develop this summary.

I hope that this information provides you with the history associated with this proposed project, and should you need any additional information, please contact me at (602) 771-2308.

Sincerely,

Nancy C. Wrona, Director
Air Quality Division

NCW:ecm

Enclosures

Chronology of Documents for
Arizona Clean Fuels (a.k.a. Maricopa Refining Company)

Document Title	Issuance Date
Air Quality Installation Permit Number 1228 Synopsis: Permit issued to Maricopa Refining Co. (a.k.a. Arizona Clean Fuels) allowing construction and installation of equipment.	**January 16, 1992**
Class I Permit Application Cover Letter Synopsis: Cover letter from Dames and Moore (now URS Corporation, a.k.a. URS), Arizona Clean Fuels' contractor, applying for a new air quality installation and operating permit.	**December 23, 1999**
Permit Application Incompleteness Letter Synopsis: Letter from Arizona Department of Environmental Quality (ADEQ) to Arizona Clean Fuels requesting additional information in support of the December 23, 1999, permit application.	**January 31, 2000**
Memorandum Regarding Preliminary BACT Review Synopsis: Comments from RTP Environmental Associates (RTP), ADEQ's contractor, to ADEQ, Arizona Clean Fuels and URS regarding the Best Available Control Technology (BACT) review performed in the December 23, 1999, permit application.	**March 17, 2000**
Revised Sections of Permit Application Cover Letter Synopsis: Letter from URS to Arizona Clean Fuels and ADEQ responding to some of the comments in RTP's March 17, 2000, memorandum.	**June 29, 2001**
Memorandum Regarding Preliminary BACT Review Synopsis: Additional comments from RTP to ADEQ, Arizona Clean Fuels, and URS responding to URS's June 29, 2001, submittal.	**August 2, 2001**
Permit Application Addendum Cover Letter Synopsis: Cover letter for a new application addendum submitted by URS, containing some responses to RTP's August 2, 2001, comments, as well as some information requested in ADEQ's January 31, 2000, incompleteness letter.	**October 31, 2001**
Permit Application Addendum Cover Letter Synopsis: Cover letter for a new application addendum submitted by URS, containing additional responses to RTP's August 2, 2001, request for information.	**November 16, 2001**

Chronology of Documents for
Arizona Clean Fuels (a.k.a. Maricopa Refining Company)

Document Title	Issuance Date
Permit Application Addendum Cover Letter Synopsis: Cover letter for a new application addendum submitted by URS, containing additional responses to RTP's comments, as well as some information requested in ADEQ's January 31, 2000, incompleteness letter.	**March 14, 2002**
Response to Comments Letter Synopsis: Letter from URS to RTP supplementing the October 2001, November 2001 and March 2002 permit application addendums.	**April 24, 2002**
Permit Application Completeness Letter Synopsis: Letter from ADEQ to Arizona Clean Fuels, indicating that, based on all the information received on or before August 23, 2002, the application was deemed complete.	**September 4, 2002**
Letter Regarding Predicted Impacts on Nearby Community Synopsis: Letter from Gallagher and Kennedy, Arizona Clean Fuels' attorneys, explaining the company's willingness to relocate a local school and community center in order to minimize predicted impacts on the nearby community.	**September 5, 2003**
Letter Regarding Relocation of Proposed Refinery Synopsis: Letter from Gallagher and Kennedy to ADEQ explaining Arizona Clean Fuels intent to relocate the proposed project to Yuma, Arizona, and that a new, site-specific permit application would be resubmitted in the future.	**October 30, 2003**
Letters Regarding Licensing Time Frames Synopsis: Letters between ADEQ, Arizona Clean Fuels, Office of the Attorney General, and Gallagher and Kennedy, agreeing to restart the permitting timeframes upon receipt of a new permit application.	**April 5-6, 2004**
New Application Cover Letter Synopsis: Cover letter from URS Corporation on behalf of Arizona Clean Fuels, submitting a new application for an air quality installation and operating permit.	**July 14, 2004**